How the Cold War Transformed Philosophy of Science
To the Icy Slopes of Logic

This intriguing and ground-breaking book is the first in-depth study of the development of philosophy of science in the United States during the Cold War. It documents the political vitality of logical empiricism and Otto Neurath's Unity of Science movement when these projects emigrated to the United States in the 1930s and follows their depoliticization by a convergence of intellectual, cultural, and political forces in the 1950s. Students of logical empiricism and the Vienna Circle often treat these as strictly intellectual nonpolitical projects. In fact, the refugee philosophers of science were highly active politically and debated questions about values inside and outside science, as a result of which their philosophy of science was scrutinized politically from both within and without the profession, by such institutions as J. Edgar Hoover's FBI.

Based on extensive archival research, this book constitutes a major chapter in American intellectual history during the Cold War. It reveals how an unlikely combination of intellectual and political forces taking root in Cold War anticommunism shaped both the curricula of colleges and even the research undertaken by leading philosophers.

It will prove absorbing reading to philosophers and historians of science, intellectual historians, and scholars of Cold War studies.

George A. Reisch is an independent scholar.

Every action, in the middle of the twentieth century, presupposes and involves the adoption of an attitude with regard to the Soviet enterprise.

Raymond Aron, *The Opium of the Intellectuals*, 55

How the Cold War Transformed Philosophy of Science

To the Icy Slopes of Logic

GEORGE A. REISCH

CAMBRIDGE
UNIVERSITY PRESS

CAMBRIDGE UNIVERSITY PRESS
Cambridge, New York, Melbourne, Madrid, Cape Town, Singapore,
São Paulo, Delhi, Dubai, Tokyo, Mexico City

Cambridge University Press
32 Avenue of the Americas, New York, NY 10013-2473, USA

www.cambridge.org
Information on this title: www.cambridge.org/9780521546898

First published 2005

A catalog record for this publication is available from the British Library

Library of Congress Cataloging in Publication data

Reisch, George A., 1962–
How the Cold War transformed philosophy of science : to the icy slopes of logic /
George A. Reisch.
p. cm.
Includes bibliographical references and index.
ISBN 0-521-83797-9 (hardback : alk. paper) – ISBN 0-521-54689-3 (pbk. : alk. paper)
1. Science – Philosophy – History – 20th century. 2. Cold War – Influence.
3. United States – Intellectual life – 20th century. 4. Logical positivism. I. Title.

Q174.8.R45 2005
501 – dc22 2004052545

ISBN 978-0-521-83797-2 Hardback
ISBN 978-0-521-54689-8 Paperback

Contents

Preface and Acknowledgments

A few days after finalizing the chapters of this book, I happened to watch a television documentary about string theory, one of the latest approaches by which physicists are pursuing a unified theory of nature. By conceiving subatomic particles as loops or pieces of string, instead of dimensionless points or spherically symmetric fields of force, the program explained, physicists have found new possibilities for mathematically connecting nature's forces. Some think the long sought-for unification of general relativity and quantum mechanics may soon come into view.

For one who had just written a book about the Unity of Science movement of the 1930s and '40s, this documentary brimmed with significance. Were they alive today and sitting in front of my television with me, I realized, the philosophers who led this movement – Otto Neurath, Rudolf Carnap, Philipp Frank, and Charles Morris – would have been fascinated. The science would have impressed them, but so would the efforts of public television to popularize contemporary physics and its unificationist impulse. Their Unity of Science movement was, in part, an effort to do just that.

On the other hand, these philosophers might well be disappointed were they to come back to life. For unlike public television, the discipline of philosophy of science they helped to cultivate in North America no longer holds the unity of science among its core issues and concerns. Especially during the postmodern 1980s and '90s, after all, one of the more celebrated concepts in the humanities was *disunity*. Unity came to mean, among other things, exclusion of subaltern cultures and ideas and conservative, elitist disdain for the particularities and vitalities of different cultures. In tune with the times, some philosophers of science marshaled observations from ecology, biology, and even high-energy physics to de-pict scientific communities as something like a patchwork of urban ethnic neighborhoods with different languages, practices, and goals – contigu-ous but not continuous and hardly a collective quest for a general, unified understanding of nature.

For the resurrected logical empiricists sitting in my living room, the point is not simply that this disunified picture rejects their ideal of unity. In their day, like ours, the sciences were not well unified and they did not claim that some complete, unified theory of everything lay just around the corner. Instead, they would be more disappointed by the contemporary disunity between *science* and *philosophy* suggested by this interest in par-ticularism and disconnection. As this documentary about string theory reminded, the impulse to create simple, unified understandings of nature is as much a mainspring of science today as it was for Copernicus, Newton, Darwin, and other heroes of science's history. But many in contemporary science studies believe otherwise, their dissent enabled by an insular, spe-cialized academic culture. To understand science, many scholars in the humanities believe, one needs only the right metatheory of knowledge (of usually French, German, or Italian provenance). One does not need, in particular, to cross boundaries and the quadrangle to learn how science's practitioners understand what they do, unmediated by metatheoretical reinterpretations.

One logical empiricist featured in this book tried to address these sev-eral disconnections in the late 1940s. Then teaching both physics and philosophy at Harvard, Philipp Frank observed that science professors and, in turn, their students were beginning to perceive philosophers as impractical and uninformed about science. Philosophers fed this percep-tion, Frank suggested, as they carved out special problem areas of their own concerning language and formal logic. They felt little need to keep abreast of science, Frank reported, especially because its pace seemed always to accelerate and its discoveries revealed strikingly counterintu-itive puzzles. Even worse, neither party seemed willing to join forces and educate the public about the complexities of scientific methods, theo-ries, and their interpretations. Believing that historic opportunities were being missed, Frank spent the last two decades of his life promoting log-ical empiricism as a tool to help to unify the "two cultures" of scientists and humanists and to equip students with a critical understanding of science. In an age of atomic weapons and Cold War ideology, Frank be-lieved, such an understanding was necessary for a healthy, productive democracy.

Besides these cultural considerations, Frank and his fellow logical empiricists – even their philosophical rival Karl Popper – would have been impressed by some technical considerations raised in this presentation of string theory. When the program came to ongoing debate over whether string theory (or parts of it) can be empirically tested, they would have felt at home. One physicist took this debate quite seriously as he told the camera, "If you can't test your theory, it's not science." Popper would have emphatically agreed, while Carnap, knowing that things were never quite so simple, would have objected, perhaps, that we must distinguish testability from confirmability. The large, loud, and famously combative Neurath would have been so startled to find himself agreeing with Popper in this instance that he might have spilled coffee on his sweater even without barking his usual objections – "Metaphysics!" or "Absolutism!" – at his colleagues. Indeed, these philosophers argued often with each other, sometimes with great emotion and hurt feelings. But this is because they shared the conviction that philosophy of science mattered beyond the confines of the academy. In a world given to superstition, wars, social reaction, and persecution, they wished to introduce a new kind of philosophy with cultural as well as practical, scientific strengths – one of which was the power to help to clarify issues in scientific practice. They would have been gratified, therefore, to see that twenty-first-century scientists still need the philosophical tools they designed (such as criteria of meaning, testability, or confirmability) to help to evaluate knowledge claims and to avoid the often deceptive traps of metaphysics and pseudoscience.

Yet this sense of familiarity also would have contained a surprise, if not a disappointment. "If you can't test your theory," this physicist actually said, "it's not science. It's *philosophy*." Philosophy *itself* in his outlook represented a backwater of untestable, irrelevant claims of the sort that logical empiricists spent much of their careers urging scientists and philosophers to avoid. Finding themselves variously forgotten, unheard, or ignored in both science and philosophy of science, these philosophers could only conclude that, for all their cultural and scientific ambitions, something had gone wrong.

Politics, in part, is what went wrong. This book does not pretend to offer a complete account of all the events and circumstances of intellectual, social, economic, and other kinds of history that bear on postwar trends in philosophy of science. But it does propose that any convincing account must include the politics of anticommunism that, as the following chapters show, wind through and, in a sense, unify some of the experiences and circumstances marking the rise and fall of the Unity of Science movement during the Cold War. To those who suppose that philosophy evolves according to its own intellectual rules, untouched by the irrationalities of politics, commerce, and fashion outside the ivory tower, this claim may seem dubious at the outset. Philosophers of science, and especially those who appreciate the historical contributions of logical empiricism, tend to be intellectually precise and conceptually scrupulous. Were political goals and values to infect their profession, they would be identified and discarded faster than one can say "*das Nichts selbst nichtet.*"

The claim, however, is not that logical empiricists failed to uphold their well-known strictures separating philosophy from politics and thus became susceptible to political influence. Rather, the claim is that the profession's adoption of those strictures was, in one sense, a response to anticommunist forces that were extremely powerful and are now largely forgotten. One aim of this book, therefore, is to survey the pathways by which Cold War anticommunism and its instantiation known as McCarthyism worked their way through intellectual and academic life in the decades immediately following World War II.

As historian Ellen Schrecker has documented, administrators and academics across the disciplines participated in the Cold War "hysteria" over the "red menace." What sociologist C. Wright Mills perhaps more aptly called the "new American celebration" was fueled by patriotism, fear of nuclear warfare, and confident declarations from Washington and the conservative press that the United States was indeed at war with a powerful communist nation actively seeking world domination. The weapons being used were not guns and bombs but rather strategies for geopolitical control, technological competition, and propaganda. Since both Moscow and Washington were adept at propaganda and covert operations, fears that communist operatives could infiltrate American institutions (such as higher education) and bring down Western capitalism without firing a shot did not necessarily seem exaggerated. During these same years, CIA operatives, sometimes without military assistance, orchestrated coups and installed governments in nations such as Iran and Guatemala.

These fears thrived in popular culture surrounding academia. It was nearly universally believed that Moscow sponsored spies, financed (and thus controlled) many American civic and cultural organizations, and deployed its advanced scientific technologies in a quest for global and – with the satellite *Sputnik* in 1957 – extraterrestrial domination. The Soviets were also believed to be experts in techniques of psychological manipulation, more popularly known as "brainwashing" and "mind control."

Thus anyone – friends, neighbors, university professors – could succumb to this secret conspiracy to topple American democracy from within. Even those who strived to be "neutral" about the Cold War's epic ideological confrontations – between open and closed societies, between democracy and totalitarianism, between free markets and economic planning – chose a risky path. For by failing to condemn communism and failing to help steel one's compatriots against its pernicious influence, neutralists often seemed to be on the side of the conspirators. In general, only public and professional affirmations of anticommunism could protect one from being suspect as "pink" or "red."

For those who are more familiar with the Vietnam War or the destruction of the World Trade Center than with Sputnik and the Cuban missile crisis, the remarkable power of anticommunism to minimize dissent and cultivate anti-intellectualism and political conformity in 1950s America may nonetheless seem somewhat familiar. In the public eye, the social and political risks of appearing "soft on communism" during the Cold War were not unlike contemporary risks of appearing sympathetic to terrorism. In the wake of World War II, as in the wake of September 11, 2001, national leaders defined events in stark, moral terms: Invaders who lacked the freedom and social and religious values embodied by the United States actively sought to destroy it. Those who objected that geopolitical tensions and causes of terrorism were more complicated and that understanding and managing them required historical, sociological, and economic knowledge of nations and peoples were often viewed with suspicion.

In the eyes of some, logical empiricism and its Unity of Science movement seemed suspicious, too. Logical empiricism was originally a project that self-consciously sought engagement not only with science but with progressive social and cultural developments (both in Europe of the 1920s and in North America of the 1930s and '40s). In the space of about ten years, however, from roughly 1949 to 1959, it became the scrupulously nonpolitical project in applied logic and semantics that most philosophers today associate with the name "logical empiricism" or "logical positivism." Since several logical empiricists' careers crossed paths with anticommunist politics on campus, in major philanthropic organizations, and in J. Edgar Hoover's FBI, there is evidence that anticommunism was a force behind this transformation. It affected the kind and range of problems that philosophers of science pursued, the methods and tools employed, and the relations between philosophy of science and science itself.

A word about this "transformation" will help to introduce further the claims being made. Philosophers of biology distinguish between evolutionary change underpinned by *transformation* and *selection* within a population. Here, "transformation" is used nontechnically to refer to a process of professional and disciplinary change that was, for the most part, selectionist. The population in question contained the American and European philosophers of science who together cultivated logical empiricism as it thrived in late 1930s North America. Some, such as Otto Neurath, Philipp Frank, and Charles Morris, shared the belief that logical empiricism, or philosophy of science more broadly, should embrace not only formal, abstract studies of scientific theory and scientific language, but also socially and politically relevant topics (such as the study of values in science, the sociology of science, and the logical structure and evidentiary content of ideologies and ideological claims). These and other topics, and the task of popularizing them within other disciplines and the general public, belonged to the Unity of Science movement that they promoted beginning in the mid-1930s. While nearly all logical empiricists were happy to be involved one way or another in this movement, one subpopulation (including, in different ways, Carnap, Reichenbach, Feigl, and Richard Rudner) favored a narrower discipline, confined to topics such as induction, explanation, and technical semantics, which they took to be ill-suited, if not categorically inappropriate, for treating matters of ideology and social life. The transformation in question largely consists of a loss of influence and leadership of the first group and the rise and success of the latter. Thus, these leaders of the profession did not, to use the popular expression, simply "cave in" to political pressure and transform their beliefs and research in that sense.

This study is based on historical sources, usually archival and unpublished. As professional intellectuals in all fields know, there is often a difference in tone as well as in content between what scholars tell each other in formal lectures and publications and what they say in private conversation or correspondence. Under cover of the noble practice of historiography, this book is largely a protocol of reading philosophers' mail. This invasion of privacy brings us objectively closer than the published record to the history of logical empiricism in North America. But it also comes with subjective liabilities. This book is selective. Some figures in the history of American philosophy of science, such as Edgar Zilsel, Victor Kraft, Egon Brunswik, and Carl Hempel are barely (or only just now) mentioned. Nor is close attention paid to Hans Reichenbach's counterpart to the Vienna Circle, the Berlin Society for Empirical Philosophy.

British philosophical leftism is also only sampled as it intersects with the Unity of Science movement in North America.

This book is also somewhat sympathetic, perhaps inevitably, to those who struggled to sustain their projects in politically and intellectually hostile climates. One reader found it excessively sympathetic to Otto Neurath and the antiformalistic harangues he sometimes sent to his more talented and articulate colleagues, especially Carnap. In formal logic and semantics, it is true, Neurath was not as talented as many of his colleagues. Similar cases could be made for Frank and Morris. What these chapters show, however, is that Cold War intellectual life was no meritocracy guaranteed to promote the best over the also-rans. With major campuses conducting formal hearings and FBI agents interviewing faculty and department secretaries about suspicious professors, intellectual life in the 1950s mixed scholarship, fear, peer pressure, ostracism, and, sometimes, overt bullying by colleagues. Winners and losers over the long term were not always determined according to intellectual talent.

What sympathy there is for Neurath, Frank, and Morris in these chapters is only partly sentimental. Their interests in the historical and sociological aspects of scientific (and philosophical) thought are enormously suggestive and worthy of contemporary study. Especially when compared with the "received view" of logical empiricism that fully abstracted knowledge from its social and historical contexts, something about Neurath's and Frank's historicism and contextualism seems almost certainly correct if only because contextual understanding is required to make sense of how and why their insights and projects were eclipsed in the first place. For philosophers of science who wish their discipline enjoyed more public authority and credibility, and more productive and understanding engagement with practicing scientists, some such contextualism would seem to be invaluable. For once the profession's contemporary boundaries and values are historicized and contexualized, they can hardly be seen as necessary and immobile. They can be contested and adjusted as surely as they were once transformed in different social and cultural circumstances, in another time.

For support, conversations, and criticisms of the research that led eventually to this book, I would like to thank many persons. Robert Richards, Howard Stein, and Dan Garber advised the doctoral dissertation written at the University of Chicago out of which it grew. Most of the research that led to this book was supported by the National Science Foundation, grant number SES0000222. Many others encouraged, and sometimes corrected, my evolving views about the history of logical empiricism and the Unity of Science movement. These include Don Howard, Thomas Uebel, Michael Friedman, Alan Richardson, Gary Hardcastle, Richard Creath, André Carus, Nathan Hauser, David Stump, Seth Sharpless, Michael Stöltzner, Hans-Joachim Dahms, Veronika Hofer, Elliott Sober, Steve Fuller, Abraham Edel, Tom Ryckman, Ralph Gregory, John McCumber, George Mallen, Robert Cohen, Fred Beuttler, and David Hollinger. I additionally thank Friedrich Stadler and Elisabeth Nemeth at the Institut Wiener Kreis; Michael Davis, Warren Schmaus, Bob Ladenson, John Ongley, and Jack Snapper at the Illinois Institute of Technology; and two anonymous referees who proposed and encouraged worthwhile revisions. I also thank the staffs at the following archival facilities for permission to quote from documents in their collections. When not stated explicitly in the text, the collections involved are referenced according to the following abbreviations:

ASP RC: Rudolf Carnap Collection, Archive of Scientific Philosophy, Hillman Library, University of Pittsburgh, Pittsburgh, Pennsylvania.

CMP: Charles Morris Papers, owned by the Peirce Edition Project, Indiana University Purdue University Indianapolis, Indianapolis. (The Charles Morris Papers are presently unprocessed.)

HFP: Herbert Feigl Papers, University Archives, University of Minnesota, Twin Cities Campus, Minneapolis.

JRMC: Jacob Rader Marcus Center of the American Jewish Archives, Cincinnati Campus, Hebrew Union College, Jewish Institute of Religion, Cincinnati, Ohio.

ONN: Otto Neurath Nachlass (Wiener Kreis Archiv), Rijksarchief in Noord-Holland, Haarlem, The Netherlands.

RAC: Rockefeller Archive Center, Sleepy Hollow, New York.

USMP, UCPP, PP: Unity of Science movement papers, University of Chicago Press Papers, University of Chicago Presidents' Papers, 1925–45, Department of Special Collections, Regenstein Library, University of Chicago, Chicago, Illinois.

1

An Introduction to Logical Empiricism and the Unity of Science Movement in the Cold War

For those interested in the history of philosophy of science, logical empiricism holds a special attraction. Like old sepia-toned photographs of ancestors who made our lives possible by surviving wars, emigrations, and the vicissitudes of times gone by, logical empiricism holds the nostalgic allure of the smoky Viennese cafés where much of it took shape some eighty years ago. The setting and the story are irresistible. In the Vienna of Freud, Schoenberg, Wittgenstein, and other twentieth-century luminaries, the philosophers, mathematicians, and logicians making up the Vienna Circle were surrounded by intellectual creativity. They themselves were on the front lines of the century's exciting developments in physics and logic. The core members included Moritz Schlick, Rudolf Carnap, Kurt Gödel, Philipp Frank, and Otto Neurath, while their colleagues and devotees in Europe and America included Hans Reichenbach, Carl Hempel, Ernest Nagel, and W. V. O. Quine. Until the circle's dissolution and demise in the early 1930s, these present and future leaders in philosophy met regularly at the University of Vienna and at various cafés to debate their ideas about knowledge, science, logic, and language. As they sipped coffee and lit their pipes, they ignited nothing less than a revolution in philosophy and bequeathed to us the discipline we know today as philosophy of science.

Nostalgia, of course, carries little philosophical weight. Most contemporary philosophers, however much they may appreciate logical empiricism as their profession's founding movement, agree that in the 1950s and '60s logical empiricism was revealed to be a catalog of mistakes, misjudgments, and oversimplifications about science and epistemology. Much has changed in philosophy of science. Most visibly, the cafés of the 1920s have given way to styrofoam coffee cups and fluorescent lights of corporate hotels where philosophers of science, now representing a well-established academic field, convene to network, debate issues, and conduct their business in higher education.

Yet recent research has shown that the profession's journey from European cafés to corporate hotels involved more than growth in membership, a change of national venue, and improved, revised beliefs about science and epistemology. It also involved sweeping and substantive changes that are only now coming into focus. The more we learn about early logical empiricism – its basic values, goals, methods, and the sense of historical mission shared by some of its practitioners – the more foreign and distant it seems when compared with contemporary philosophy of science. Thus, two general questions continue to drive studies about the Vienna Circle and early logical empiricism: What, precisely, was logical empiricism originally all about? and, the main topic of this book, How did philosophy of science evolve into the very different form it takes today?

Compelling answers to the first question began appearing in the 1970s when historians and philosophers began to recover and interpret the rich history of logical empiricism.[1] With such a wide cast of characters, whose specialties lay in philosophy, logic, mathematics, and social science, it has become clear that most early logical empiricists, though not all, were as passionate about problems in culture and politics as they were about technical philosophy and epistemology. Neurath, Carnap, and Frank, in particular, actively sought to forge personal, intellectual, and institutional connections between logical empiricism and various cultural and political institutions and movements in Europe. These include Carnap's lifelong interest in artificial international languages and Neurath's work in museums, public education, and the ISOTYPE system of visual iconography, whose graphic descendants are now ubiquitous in airports, shopping malls, and other public spaces. Neurath, Carnap, Herbert Feigl, and Hans Reichenbach were invited to lecture at the Bauhaus, while Neurath additionally collaborated with the Belgiand International Congress for Modern Architecture (CIAM) (Faludi 1989; Galison 1990). There were also debates with Marxists (including Lenin) and Critical Theorists of the Frankfurt school (Lenin 1908; Horkheimer 1937; Dahms 1994) as well

---

[1] For a recent and useful compendium of biographical and philosophical information about the Vienna Circle and its associates, see Stadler 2001. For an overview of the scholarly "rediscovery" of logical empiricism, see Uebel 1991, and of its political aspects, see Heidelberger and Stadler 2003.

as Philipp Frank's attempts to befriend neo-Thomist critics of scientism and positivism at the annual Conferences on Science, Philosophy and Religion in New York City in the 1940s (Frank 1950). At least two logical empiricists, moreover, did not just debate matters of political theory or national and economic policy. Neurath had a tumultuous and nearly fatal role in the Bavarian socialist revolution of 1919 and was later hired by Moscow for his ISOTYPE talents. Hans Reichenbach's socialist student activism at the University of Berlin cost him his chances of later gaining employment there.[2]

The Vienna Circle specifically reached out to the wider public to promote their critique of traditional philosophy and to popularize their *Wissenschaftliche Weltauffassung*, or scientific world-conception, as a replacement. They did so in Vienna through the Ernst Mach Society and its public lectures, and they did so in Europe and America via Otto Neurath's Unity of Science movement. The movement promoted the task of unifying and coordinating the sciences so that they could be better used as tools for the deliberate shaping and planning of modern life. And it sought to cultivate epistemological and scientific sophistication among even ordinary citizens so that they might better evaluate obscurantist rhetoric from reactionary and antiscientific quarters and better contribute to planning a future unified science that would assist society's collective goals.

Together, logical empiricism and Neurath's Unity of Science movement were in the business of *Aufklärung* (Scott 1987; Uebel 1998). They sought nothing less than to specify and to help fulfill the promise of the eighteenth-century French Enlightenment while taking full advantage of twentieth-century developments in science, logic, social thought, and politics. This constructive, enlightenment agenda is the main subject of this book. For only by once again putting these ambitions of logical empiricism in plain view can we see both how much philosophy of science has changed in the last half of the twentieth century and, in turn, what kinds of conditions and forces were involved in its transformation.

The Conventional Wisdom about Logical Empiricism

Before introducing this book's main thesis, it is helpful to consider some of the conventional – and mostly misleading – wisdom about logical empiricism. Before this recent flowering of interest and research, its cultural

---

[2] On Neurath's politics, see Cartwright et al. 1996; on Reichenbach, see Traigar 1984. An informative, if spiteful, account of Neurath's work in the U.S.S.R. is Chislett 1992.

scope and scientific ambitions were obscured by several circumstances, especially in the English-speaking philosophical world. One factor remains logical empiricism's attacks on traditional and contemporary metaphysics and pseudoscience. These were vivid displays of analytical fireworks that helped to stamp the project with a negative, eliminative character. In addition, until Neurath's writings began to be translated and published in English in the 1970s, his constructive interests in unified science and politics and his finely tuned epistemological insights about language and science were obscured by his reputation as the "original neo-positivist caveman" (Uebel 1991, 5) who thumped his club on the ground and muttered Machisms such as "blue here now." Another factor was the influence of Rudolf Carnap's *Der Logische Aufbau der Welt* (Carnap 1969), which, however much it naturally and deservedly captures philosophical attention, is wrongly taken as a paradigm for logical empiricism as a whole. Taken together, these and other factors helped to create an impression that logical empiricism, even despite its subsequent liberalizations and changes, was an early, phenomenalistic moment in the history of Western epistemology, and little more.

Popular secondary writings also helped to obscure logical empiricism's cultural engagements. Karl Popper's influential *Logic of Scientific Discovery* (Popper 1935) and his widely read essay, "Science: Conjectures and Refutations" (Popper 1969), trumpeted his conceit that he alone diagnosed an inductivist fallacy at logical empiricism's core (thereby reinforcing the view that its project was essentially, if not also exclusively, epistemological). A. J. Ayer's even more widely read *Language, Truth and Logic* (Ayer 1936) presented logical empiricism as mainly Carnapian philosophy of science (up through logical syntax (Carnap 1937c)) viewed through the lens of Wittgensteinian ordinary language philosophy. Philosophy's point and purpose, and hence logical empiricism's, Ayer explained, was merely (but not unimportantly) to assist the progress of science, whenever called on, by providing clarifying analyses of scientific language (Ayer 1936, 152). While Ayer's account was faithful to the movement's iconoclasm – its rejection of metaphysics, its flirtations with verificationism and foundationalism, and its rejection of a synthetic a priori – it does not mention logical empiricism's constructive ambitions. Save for two footnotes, Neurath's voice is missing from *Language, Truth and Logic* because Ayer sought "to emphasize not so much the unity of science" – the topic and goal dearest to Neurath – "as the unity of philosophy with science" (ibid., p. 151). That logical empiricism was conceived by its founders in part to assist the coordination and coordinated use of scientific knowledge, to help modernize and improve life, education, and social and economic organization, is a fact no reader of Ayer's book will surmise.

If Ayer collapsed logical empiricism's broad agenda into a narrow but active scientific project, by the 1970s logical empiricism was reduced further. No longer a participant in science, it was remembered as a school of commentary *about* science. Suppe's compendium *The Structure of Scientific Theories* (Suppe 1977), which sits near *Language, Truth and Logic* on every philosopher of science's bookshelf, presented logical empiricism as a set of propositions about science and its methods. Much as some members of the Vienna Circle feared, as we see below, logical empiricism became, and was remembered as, a sect whose doctrines were verificationism, inductivism, and phenomenalism. Suppe wrote that this narrow, strictly epistemological agenda exhausted logical empiricism's legacy:

For over thirty years logical positivism . . . provided the basic framework for posing problems about the nature of scientific knowledge and also imposed constraints on what would count as appropriate solutions to these problems: Singular knowledge of directly observable phenomena was nonproblematic, whereas the remaining knowledge science purported to provide was problematic at best. (Suppe 1977, 617)

By the late 1960s when Suppe wrote this, logical empiricism was widely considered defunct and this characterization of the program provided a convenient way to understand its demise. What the program offered for analyzing "the remaining knowledge science purported to provide" were models of explanation, reduction, induction, and confirmation that were themselves found wanting. Two influential works, Quine's "Two Dogmas of Empiricism" (Quine 1951) and Kuhn's famous *Structure of Scientific Revolutions* (Kuhn 1962), were by then helping to solidify consensus. Among other problems, logical empiricism was internally crippled, according to Quine, by the unspecifiability (without moving in a circle) of an analytic-synthetic distinction. According to Kuhn, it was unable to elucidate science's conceptual holism and the alleged theoretical and linguistic discontinuities that punctuate science's history and, many presumed, its essential nature. Logical empiricism was in sad shape. It had lost its connections to scientific practice, could hardly stand up under its own conceptual weight, and the science it aimed to interpret was shown by historical research to be merely an idealized fiction existing only in philosophers' imaginations.

A New Explanation for the Demise of Logical Empiricism

Knowing as we do now that logical empiricism was originally a philosophical project with cultural and social ambitions, the time is ripe to inquire how the discipline was transformed and how these cultural and social ambitions were lost. The answer defended here is that it was transformed during the 1950s at least partly, if not mainly, by political pressures that were common throughout civic as well as intellectual life during the Cold War following World War II. In large part, these pressures led logical empiricism to shed its cultural and social engagements by shedding Neurath's Unity of Science movement. The movement was not merely a public, scientific front for an otherwise independent philosophical program. It helped to determine which kinds of questions and research topics were pursued, and how they were pursued, at the heart of philosophy of science.

This is not to say that, were it not for the Cold War, contemporary philosophy of science would now be some kind of nonacademic public servant. The claim, rather, is that logical empiricism originally aspired *both* to technical, philosophical sophistication as well as to engagement with scientists and modern social and economic trends. The Cold War, this book argues, made that agenda impossible and effectively forced the discipline to take the apolitical, highly abstract form remembered in Suppe's *Structure of Scientific Theories*. The chasm that yawns between that book and the Vienna Circle's combative manifesto, *Wissenschaftliche Weltauffassung*, in other words, was created by the Cold War. Nor does this interpretation dismiss the perspicuity of Quine's, Kuhn's, and other criticisms of logical empiricism. It does claim, however, that the power of these political forces must be acknowledged and that we begin to assemble, as sketched below, a more complicated and more accurate story of philosophy of science in the twentieth century.

One historiographic (and ultimately metaphysical) aside may help to dismantle a prejudice that this thesis is likely to meet. It comes, appropriately, from Neurath, who, as we see below, fought many battles with other philosophers whose influence and reputation came to overshadow his own. One guiding element in these debates was Neurath's multifaceted pluralism and, especially, his criticisms of what he called "absolutism." For example, Neurath criticized Carnap's and Tarski's semantic theory of truth (which held, for example, that the statement "the snow is white" is true if and only if the snow is white) on the ground that it erected a dual order in which language speaks first about itself, and then the world, in order to allow a comparison of these reports and a determination of whether truth-conditions obtain.

Neurath objected because, he insisted, a healthy empiricism cannot ever – even in philosophical abstraction – ignore the practical conditions in which language and science operate. Thus, in his famously cumbersome model of protocol statements –

Otto's protocol at 3:17 o'clock: [At 3:16 o'clock Otto said to himself: (at 3:15 o'clock there was a table in the room perceived by Otto)] (Neurath 1932/33, 93) –

the statement's outermost report is always about a specific person and what they believe they see and know about the snow or table before them. For Neurath, there is no legitimate dualism of language and the world that a theory of truth may invoke. Knowledge, speech, language, and behavior remain always, as Nancy Cartwright and Thomas Uebel have emphasized for our understanding of Neurath, in the same "earthly plane."[3]

Here lay, for example, one of Neurath's antipathies to Popperianism. In Popper's metaphysics of first, second, and third worlds, the last is populated like Plato's heaven by objective concepts or objects studied by generations of philosophers and scientists. Pythagoras and contemporary students in seventh grade, Popper reasoned, can know and understand Pythagoras' theorem as *the same thing* because it enjoys ontological status as an enduring, timeless object. Neurath would have none of this, and neither will any philosopher who is sympathetic with this book's political thesis.[4] For if philosophy of science is devoted to the study of anything like such an ontological domain of metaphysical objects or conditions – truth, explanation, confirmation, meaningfulness, analyticity, and so on – the claim that political forces controlled its career in America will always be trumped by the reply that political forces could cause, at most, a temporary diversion in philosophy's historical development. Politics could never fundamentally change the discipline precisely because political forces cannot (and therefore did not) connect with the otherworldly objects that, as philosophers investigate them, guide philosophical practice.

This multiplication of worlds is fatuous metaphysics, Neurath would say, much as he barked "Metaphysics!" "Metaphysics!" (and later, to save

---

[3] Cartwright and Uebel 1996. The phrase comes from Neurath, who praised Marxist naturalistic methodology in which "everything lies in the same earthly plane" (Neurath 1928, 295).

[4] For more on Neurath's debate with Popper, see Neurath 1935 and Cat 1995.

his voice, just "M!") at the Thursday evening meetings of the Vienna Circle.[5] It is metaphysical for Neurath because it has no place within an honest, empirical, and scientific depiction of philosophy of science as something that (some) human beings *do* in our earthly plane. Philosophy of science must be conceived as a set of practices, values, goals, and jargons that are chosen, utilized, and (hopefully) improved by individuals for their intellectual pursuits. These practices are taught to others and modified by debate as well as by often undetected historical or sociological pressures. All of these processes and the agents that sustain them exist in the same earthly plane, right alongside culture, society, and politics.

As this book shows, many choices made by first-generation logical empiricists and their students were made alongside intellectual, institutional, and personal pressures arising directly out of the Cold War and McCarthyism. This will explain both how philosophy of science was radically changed and depoliticized by these pressures and how this thesis ought to seem no more implausible than the better known fact that Hollywood movie making was also transformed by McCarthyism. There is neither a heavenly Idea of entertainment that controls the history of cinema nor a timeless, objective domain of intellectual pursuits and values lording over the philosophy of science.

The Unity of Science Movement and the
International Encyclopedia of Unified Science

Logical empiricism came to America during the 1930s. With the exception of Herbert Feigl, who emigrated in 1930, the main wave began mid-decade and included Rudolf Carnap in 1935, Karl Menger in 1936, Carl Hempel in 1937, Hans Reichenbach, Felix Kaufman, Gustav Bergmann, and Philipp Frank in 1938, and Kurt Gödel and Edgar Zilsel in 1939 (Stadler 2001). Most came to America as participants in Neurath's Unity of Science movement. Though Neurath himself never emigrated to America (despite the advice and wishes of his colleagues), he promoted and organized the movement from Holland, and later England,

---

[5] Neurath's version of this famous anecdote bears repeating. During the "Wittgenstein period," Neurath recalled in 1944, "again and again" he objected, "this is metaphysics," during the group discussions of the *Tractatus*. "It became dull and Hahn suggested I should speak of M. only to shorten the sounds and since I too often said M,' he suggested I should only remark when I am satisfied by saying, NM' [for not metaphysics']" (Neurath to Carnap and Morris, November 18, 1944, ASP RC 102-55-06).

while taking several trips to America as this wave of emigration came ashore. The movement thus became a kind of institutional home-away-from-home for the émigré philosophers, one that helped them maintain the contact, dialogue, and intellectual focus they had provided for each other in Vienna, Berlin, and Prague.[6] As we see below, it also facilitated connections between the émigrés and American philosophers who, in some cases, were already in pursuit of a socially and politically engaged program in philosophy of science.

The Unity of Science movement was also the public, pedagogic, and scientific voice of logical empiricism. It consisted of a series of International Congresses for the Unity of Science (held in Prague, 1934; Paris, 1935; Copenhagen, 1936; Paris, 1937; Cambridge, England, 1938; Cambridge, Massachusetts, 1939; and Chicago, 1941); publications such as the *International Encyclopedia of Unified Science* and a short-lived English incarnation of *Erkenntnis* titled the *Journal of Unified Science*; regular announcements and items appearing in journals such as *Philosophy of Science* and *Synthese*; and some coverage in the popular media (such as *Time* and the *New York Times*). The logical empiricists were received in America both as representatives of a new social and cultural movement and as intellectuals, philosophers, and logicians.

To contemporary philosophers, the familiar item in this list is the *International Encyclopedia of Unified Science*, which for decades was mentioned on or near the title page of Kuhn's famous *Structure of Scientific Revolutions*. Though from its beginning it had an influential life of its own, Kuhn's book was originally commissioned as a monograph for the *Encyclopedia* after the task of writing a historical monograph had been passed from the Italian philosopher and historian Federigo Enriques, to George Sarton (who declined), to I. B. Cohen, and, finally, to Kuhn. Though no one has produced a detailed, historical account of how Kuhn's monograph and its ideas were influenced by the Unity of Science movement, enough has been learned to dismiss one persistent commonplace, namely, that Kuhn refuted logical empiricism, Trojan-horse style.[7] Kuhn's book was written in the last years of the 1950s and was published in 1962 when the encyclopedia project was moribund, roughly a decade after its last burst of vitality in the early 1950s (Reisch 1995). Something else, therefore, already killed the *Encyclopedia* and the larger Unity of Science movement. The culprit is plainly suggested by the dates of the International Congresses listed

[6] This point was impressed on me by Abraham Edel, personal correspondence.
[7] See, e.g., Friedman 1991; Reisch 1991; Irzik and Grünberg 1995.

above: The second world war nearly halted the movement, and despite earnest efforts by some of its leaders, pressures of the Cold War prevented it from ever recovering momentum.

Over the course of its history, these leaders were Otto Neurath, Rudolf Carnap, Philipp Frank, and the American pragmatist philosopher Charles Morris. Morris, at the University of Chicago, was extremely helpful in assisting the emigration of logical empiricism. He lectured and wrote often in the 1930s about the new Unity of Science movement, its cultural and political importance, and its proper place alongside American pragmatism as part of a comprehensive theory of signs. Following Charles Sanders Peirce, Morris called this theory "semiotic" and tirelessly promoted it as the future of philosophy (Morris 1937). At the same time, Morris assisted his colleagues' emigrations. He first met them in Prague in 1934, at the Eighth International Congress of Philosophy where Neurath held his first meeting on behalf of the Unity of Science movement and the new encyclopedia project. Morris advised those planning to come to America that they should promptly publish an article or book in English before seeking a position in an American college or university. Several took his advice and his offers to help. Morris soon found himself arranging translations, putting authors in contact with publishers, and writing letters to friends and colleagues in the United States who might possibly hire a scientific philosopher.[8] With Morris's help, Reichenbach found a position at UCLA, Frank took a nontenured position at Harvard, and, in 1936, Carnap arrived at Morris's University of Chicago (after a one-year position at Harvard). Besides all this activity, as well as his own writing and teaching, Morris enticed the University of Chicago Press to publish Neurath's new *International Encyclopedia of Unified Science*.

With good reason, Morris hoped that his university would become the center of the Unity of Science movement in America. From Chicago, he and Carnap edited the *Encyclopedia* as its monographs began appearing in 1938, while Neurath, its editor-in-chief, lived in Holland. Morris also assumed most of the negotiations with the University of Chicago Press, negotiations that were often complicated and strained, especially those concerning the movement's plan to rescue *Erkenntnis* – logical empiricism's

---

[8] Morris describes editing a manuscript of Reichenbach's *Experience and Prediction* in Morris to Reichenbach, June 8, 1937, CMP. "I thank you very much for your continuing efforts to find a position for me in the USA" (Reichenbach to Morris, July 5, 1937, CMP). Morris also helped Philipp Frank to translate and to publish essays prior to his emigration.

original European voice – by purchasing it from the German publisher Felix Meiner (Reisch 1995). Still, despite these and other difficulties, the *Encyclopedia* was initially a great success. Wary about committing to a long-term project that might not carry its financial weight, the Press agreed to publish the *Encyclopedia* on the condition that they receive at least 250 advanced subscriptions. That hurdle was easily cleared. Some 500 had been received for the first, introductory unit, titled *Foundations of the Unity of Science*, which contains the twenty monographs of the *Encyclopedia* that exist today. Individual monographs were also selling briskly in book-stores.[9] Although they fell behind their original schedule to publish one monograph per month – in part because of Neurath's distance[10] – the editors were pleased and the Press never doubted its decision to accept the project.

The *Encyclopedia* and the movement were celebrated in New York City, arguably the midcentury heart of the nation's intellectual life. John Dewey was the senior philosopher among several, including Ernest Nagel, Sidney Hook, Horace Kallen, and Meyer Schapiro, who helped the famous group of New York Intellectuals define the trends and values of the nation's then highly politicized intellectual life. Some leftist intellectuals and philosophers, to be sure, criticized Neurath and logical empiricists, usually on the grounds that they were not leftist, radical, or dialectical materialist enough. But in the mainstream of New York philosophy, defined by students of Dewey and Morris Cohen, the new philosophical émigrés and their projects were applauded and utilized. Dewey, Hook, and Nagel, for example, variously enlisted logical empiricists and logical empiricism in their battles against neo-Thomism, the popular movement promoted by Mortimer Adler and University of Chicago President Robert Maynard Hutchins (whose own series of monographs, *Great Books of the Western World*, can be seen as competing with Neurath's new *Encyclopedia*). For those who did not socialize or correspond personally with Neurath, Carnap, and the others, they were introduced to logical empiricism and

---

[9] As of March 31, 1939, some 547 subscriptions and over 1,000 copies of all monographs published had been sold (To William B. Harrell from Bean, March 6, 1939, UCPP, box 346, folder 1). By 1945, roughly 1,800 of each published monograph had been sold (To MDA from JS, January 19, 1945, UCPP, box 346, folder 4).

[10] For Neurath to review and edit each monograph, the proofs would have to be mailed to Holland and returned. In addition, Neurath was extraordinarily busy. In 1939, his ISOTYPE history of modern life appeared (Neurath 1939). Morris, as well as the Press, were sometimes frustrated by these delays. (See, e.g., Bean to Neurath, April 21, 1938, UCPP, box 348, folder 3.)

Neurath's movement through articles in the *Partisan Review* or by Neurath's cousin, science writer Waldemar Kaempffert, who praised Neurath and the new *Encyclopedia* in the *New York Times.*

By 1939, the *Encyclopedia* began to take shape. Morris, Neurath, and Carnap persuaded the Press to announce the first nonintroductory unit: six volumes titled *Methods of Science.* In a draft of a prospectus, Morris explained that these volumes would be devoted to specific sciences and problems within them related to the unity of science. As a whole, the volumes would be

concerned with the development of a unified scientific language, with the presentation of the results of logical analysis in various sciences, with problems relevant to the foundations of the sciences, with the analysis and interrelation of central scientific concepts, with questions of scientific procedure, and with the sense in which science forms a unified whole.[11]

Neurath's plans at the time show how broad and influential he hoped the *Encyclopedia* would be. In the third unit, Morris later recalled, the new encyclopedists would take stock of the "*actual state of systematization* within the special sciences, and the connections which obtained between them." Unit four would consist of ten volumes treating education, engineering, medicine, and law. All of these professions, Neurath hoped, would find a home in the Unity of Science movement.[12]

Morris, Neurath, and Carnap also hoped that specific collaborative methods could be built in to the *Encyclopedia* as it grew and gained momentum. Though the early monographs were mainly read and edited by the three of them, these new monographs, their press release explained, would be circulated more widely prior to publication:

In order to avoid simple misunderstandings, the authors will have an opportunity to discuss each other's contribution before publication, so that there remains only the kernel of what seem to be genuine differences. In this way the crucial solved and unsolved problems in current methods of science will be made to stand out in various fields and in science as a whole.[13]

Like scientists, the new encyclopedists would strive to minimize spurious misunderstandings and maximize their collective intellectual power and efficiency.

[11] Morris, prospectus draft, UCPP, box 346, folder 1. This prospectus was never distributed.
[12] Morris 1960, 519, 520.
[13] Morris, prospectus draft, UCPP, box 346, folder 1.

The International Congresses

The *Encyclopedia* and its collaborative dialogues would also be supported by the International Congresses for the Unity of Science. The first in 1935, held at the Sorbonne in Paris, drew some 170 participants.[14] Besides the leading logical empiricists from Vienna, Prague, and Berlin and their American supporters, the conference drew leading philosophical lights from France, England, Italy, Poland, Scandinavia, and Holland. Session titles included Scientific Philosophy and Logical Empiricism, the Unity of Science (and the new encyclopedia), Language and Pseudo-problems, Induction and Probability, Logic and Experience, Philosophy of Mathematics, Logic, and History of Logic and of Scientific Philosophy. The conference mapped out the wide array of topics that the movement would address for roughly the next five years.

Subsequent congresses sometimes had a narrower focus. The second, in Copenhagen in 1936, was dedicated to philosophy of physics and biology and, in particular, the Copenhagen Interpretation of quantum mechanics. Niels Bohr, the Nobel Prize–winning author of the Copenhagen Interpretation, easily attended since the congress was held at his spacious home (see Figure 1). Though fewer Americans were present (many had spent precious Depression dollars a year before to attend in Paris), the cast of characters remained wide and international. The third congress, held again in Paris in 1937, was dedicated to the planning and conception of the *Encyclopedia* and core issues in logical empiricism. Large sessions were held on Unity of Science and Logic and Mathematics, while smaller sessions covered topics in physics, biology, and psychology.

The congresses were increasingly affected by the instabilities and uncertainties that preceded the war. News of Moritz Schlick's murder by a disturbed student reached his colleagues while they were at the Copenhagen congress, while the *Anschluss* of Austria to Nazi Germany occurred a few months before the Fourth International Congress. That congress was organized by L. Susan Stebbing and held at Cambridge University in England. It was dedicated (appropriately, given Wittgenstein's influence in British philosophy) to the topic of scientific language. This was the last congress held outside the United States.

Charles Morris organized the fifth congress at Harvard in 1939. It drew around 200 participants, many from California, Chicago, Harvard, Yale,

---

[14] This and other information about the International Congresses is usefully presented in Stadler 2001.

FIGURE 1. A photograph of participants at the Second International Congress for the Unity of Science, held in Copenhagen, 1936. Joergen Joergensen is standing in front, while Philipp Frank and Niels Bohr are seated front, right. Others pictured include Otto Neurath (third from left in the fourth row); Karl Popper (seen just to the left of Joergensen); and mathematician Harald Bohr (two rows behind his brother Niels). (Reproduced courtesy of the University of Chicago, Department of Special Collections, USMP.)

and the New York universities. Again, the conference focused on the unity of science thesis and methods for unifying the sciences as well as issues in logic and formal philosophy of science. Morris used the opportunity to broaden the movement and include topics in social science – "socio-humanistic sciences," he called them – including the scientific study of values pioneered by American pragmatists and urgently emphasized by John Dewey. As the organizer, he published an article prior to the congress in which he detailed his liberalizing agenda (Morris 1938b). But the issues he raised were soon overwhelmed by worldwide political tensions. On the eve of the conference, the participants learned that war in Europe was all but guaranteed. The next day, Horace Kallen of the New School for Social Research, a philosopher who had befriended both Neurath and Morris, presented his surprising thesis that the Unity of Science movement itself amounted to a kind of authoritarian totalitarianism that was dangerously allied to fascist ideologies in Italy, Spain, and Nazi Germany.

For most, however, the politics of the movement were not totalitarian but rather humanitarian, progressive, and pacifist. In 1941, after war had broken out, Carnap's student Milton Singer and Reichenbach's student Abraham Kaplan wrote about the Harvard congress in an article titled "Unifying Science in a Disunified World" (Singer and Kaplan 1941). They detailed the movement's importance for science and education, and they clearly admired its internationalistic and humanitarian values. Morris conveyed the same attitude in the promotional flyer he wrote for the movement's sixth and final congress, held at the University of Chicago in 1941: "The Organizing Committee feels that the present world condition enhances rather than restricts the need for the vigorous continuation of the unity of science movement."[15] Given Morris's broad, humanitarian ambitions for the movement, this congress appropriately featured sessions such as Science and Valuation, Science and Ethics, Historical Topics, and one talk addressing Science and Democracy.[16]

The War and the Demise of the Movement

The war hampered the encyclopedia project and the activities of the movement in several ways. European authors usually had more important problems to worry about than completing the monographs they had promised to Neurath, and the slowness and unreliability of mail drastically slowed communication among the authors, editors, and the Press. The journal *Synthese*, which carried a regular "Unity of Science Forum," like many other European journals ceased publishing until after the war. An even larger hurdle appeared, however, in May 1940 when Neurath narrowly escaped from occupied Holland. Having misjudged how much time he would have to relocate his home and his ISOTYPE workshop in advance of the encroaching Nazis, Neurath and his assistant (and future wife) Marie Reidemeister escaped in a small, overcrowded fishing boat they chanced on in Rotterdam just before it shoved off. They drifted until picked up by an English naval ship. Because of their Austrian nationality, they were treated as prisoners of war and spent several months interned in England. Their savior was L. Susan Stebbing, who found them a lawyer

[15] Promotional flyer, UCPP, box 346, folder 3.
[16] Announcement, "Final Notice: Sixth International Congress for the Unity of Science," UCPP, box 346, folder 2.

who appealed to authorities for their release and arranged for their marriage. Several months later, with financial and emotional support from Stebbing and other friends and colleagues, the Neuraths settled to live and work in Oxford, England.

Though Neurath was able to resume his editorial duties by the summer of 1941, the project soon floundered again in 1943 when the University of Chicago Press decided to suspend it. With only nine of the twenty monographs published and with Morris and Neurath giving the Press nothing more than promises about monographs in the pipeline, they decided that the project was getting too expensive (paper, for example, was in short supply) and that subscribers were being shortchanged. Monographs were appearing too far behind the announced schedule of one per month, and the Press believed that the substitute authors whom the editors had enlisted were less than first-rate (Reisch 1995).

After receiving the news, Neurath became furious and deftly persuaded the press to change its mind. He made it clear that, if necessary, he could take the *Encyclopedia* to another publisher. Holland would soon be liberated, he surmised, and he could perhaps take the *Encyclopedia* to his "very faithful Dutch Publisher," Van Stockum & Zoon, who had published the first issue of the movement's *Journal of Unified Science*. He pressed this notion of being "faithful" to the movement by regarding the *Encyclopedia* as a war effort. He praised what had been the project's reassuring, "business as usual" spirit, and he reminded them that this "real international enterprise" was sustained "partly by refugees" who would be discouraged and demoralized: "[T]he war is going on very well and victory comes nearer every day. It would be like *defeatism* to suspend now anything." Neurath did not choose these most effective words merely for the occasion. He often wrote privately to Morris about the movement in similar terms: "During wartime science and logical analysis cannot rest.... We have to prepare future peace life, particularly in Europe." "This nazified Germany and Europe will need some good dishes, we shall present them."[17]

Obstacles appeared in front of the movement roughly every two years. After the outbreak of war in 1939, Neurath's internment in 1941, and the Press's suspension of the *Encyclopedia* in 1943, disaster stuck again with news of Neurath's sudden death in late 1945, days after his sixty-third birthday. Besides the shock and loss to his friends, Neurath died at a critical, unstable time immediately after the war when the profession

---

[17] Neurath to Morris, January 7, 1942, USMP, box 2, folder 7; Neurath to Morris, December 28, 1942, CMP.

was poised to move and grow in different possible directions. Neurath was in the midst of two separate disputes that, in retrospect, arguably helped to figure the historical outcome. With Carnap, he was mired in an increasingly personal argument that had begun in 1942 with the publication of Carnap's *Introduction to Semantics* (Carnap 1942). Neurath first complained that the book was filled with unacceptable metaphysics, and the ensuing dispute erupted further in 1943 when Neurath learned that his encyclopedia monograph *Foundations of the Social Sciences* (Neurath 1944) had been printed with a disclaimer, requested by Carnap, saying that Carnap had not edited the monograph. (To help mollify the University of Chicago Press, Morris rushed the monograph into print without Carnap's having had a chance to edit it.) Neurath took this gesture as a personal insult, the meaning of which was that Neurath and his ideas were second-rate and not worthy of association with Carnap's highly respected name and work. For Neurath, at least, more than personal feelings were involved, for he was concerned about the formal, abstract direction that philosophy of science was taking, due in no small part, he believed, to the influence and leadership of Carnap. Neurath's own distinctive interests in empiricism and the unification of the sciences, he worried, were being eclipsed by Carnap's more formal, "scholastic" style of work.

During the last months of his life, Neurath was also mired in an equally frustrating exchange with Horace Kallen over his charges that logical empiricism and the Unity of Science movement were "totalitarian." Resuming the debate Kallen had begun in 1939, Neurath could still make little sense of Kallen's view that logical empiricism was ready-at-hand to assist the march of fascism and totalitarianism. He was even more agitated because Kallen had read some of Neurath's writings and reported on them as if Neurath quite simply wished to *legislate* rules and terminological reforms for all of science. Neurath's project was both more subtle and essentially democratic in its method, though Kallen could not see, and possibly chose not to see, that this was so. From at least two sides, therefore, Neurath felt alienated and increasingly powerless to guide the movement of which he was leader. In the midst of these two stressful disputes, he died suddenly of a stroke in December 1945.

The Unity of Science Movement in the Cold War

Given the fatigue of war, the shock and sadness of Neurath's death, and some ensuing surprises (such as the fact that Neurath had not secured official contracts with his encyclopedia authors), it was not until 1947 that the movement and the *Encyclopedia* began to stir. This time Philipp Frank, Neurath's close friend and the philosopher of science whose views and style most matched his own, joined the leadership by helping to re-establish in America the Institute for the Unity of Science that Neurath had maintained in Holland and England. While Frank taught physics and philosophy of science at Harvard, he and Morris circulated plans among their colleagues to re-establish the Institute in Boston. With Frank initially at the helm, the new Institute would decentralize and distribute the move-ment's leadership among a changing or cycling roster of officers.[18] This, it was hoped, would help avoid catastrophic breakdowns in leadership in the future and help bring new, younger talent into the movement.

At the time, Morris was traveling and writing as a Rockefeller fellow. He therefore had access to grant officers whom he helped to persuade to support the movement and its new Institute. The Institute would sponsor the *Encyclopedia*, organize future congresses for the unity of science, and pursue some new projects. Frank, in particular, was eager to promote research in sociology of science and to produce a dictionary of scientific terms. He also organized essay contests to help popularize the Institute and bring students into the fold.

Yet the Institute did not thrive. There were problems with Frank's leadership, and, more importantly, the very idea of the Institute and its Neurathian mission appears to have lost popularity among important philosophers (including Feigl and Reichenbach), who sought a more technical, less public profile for philosophy of science. As Frank strug-gled to balance the Institute's more popular agenda with the more pro-fessional agenda of his colleagues, most of the Institute's projects fell by the wayside. The essay contests were an embarrassing failure, no progress was made toward advancing the sociology of science, and the Institute's Rockefeller funding lasted only through 1955. Nor did the Institute ac-celerate or promote the *Encyclopedia*, which limped along, shouldered by Morris and Carnap, until the last of its twenty monographs appeared in 1970.

One central reason why the Institute and the movement failed to thrive in the early 1950s is that a repressive McCarthyite "climate of fear" swept

---

[18] University of Chicago Press memorandum, September 13, 1946, UCPP, box 346, folder 4. Reporting conversations between Morris and the Press, the memo notes that in the wake of Neurath's death, the movement's "plans are to decentralize the organization" by creating the Institute, which "will probably be headed by Carnap." Philipp Frank, however, led the effort to establish the Institute and became its president.

through American political, popular, and intellectual landscapes. The climate was so inhospitable and professionally perilous that leaders of the movement, with the exception of Frank, as we see below, effectively chose not to invest their energy and careers in the task of revitalizing the Unity of Science movement. In hindsight, seeds of this change could be detected in the early 1930s as some of the American intellectuals who exalted Marxism and traveled to the Soviet Union to see at first hand the fruits of the revolution began to qualify their beliefs and hopes. In the mid-1930s, the dissenters were still few and the Unity of Science movement was nonetheless admired by the nation's leftist intellectuals and philosophers. Still, doubts and worries continued to accumulate through the decade. With the much-admired Trotsky in exile, rampant rumors of collectivization disasters, and Stalin brazenly persecuting his rivals in show trials, the stage was set for a dramatic shift in the intellectual left's perceptions of Russia. For many it occurred in 1939 with news of the Hitler-Stalin nonaggression pact. The great and glorious revolution, many concluded, had been hijacked by Stalin and a band of thugs aiming to subject the world to their dictatorship. Months later, Hitler invaded Poland and Horace Kallen would denounce his friend Neurath's Unity of Science movement as "totalitarian."

Not all leftists so converted to anti-Stalinism and anticommunism. Those who did, however, were often angry with and aggressive toward those who, they believed, out of blindness, stupidity, or lack of patriotism, remained in league with the Soviets. The ardor with which Sidney Hook attacked "fellow travelers" and with which Kallen attacked "totalitarian" unified science was soon matched by the hardball anticommunism that J. Edgar Hoover, Senator Joseph McCarthy, and other professional anti-communists played in the public domain. Beginning in the late 1940s, they attacked intellectuals, politicians, and scientists whom they believed were engaged one way or another in Soviet espionage. There was no mere parallel between the anticommunist crusades of McCarthy and the FBI, on the one hand, and the "antitotalitarian" agenda of Hook, Kallen, and other intellectuals. State and federal policies and laws designed to com-bat communism affected nearly all major research universities and made it practically impossible, without genuine risk to one's professional and social standing, to be sympathetic to Marxism or socialism either inside or outside the classroom.

In these and other ways, an intellectual and political culture that first warmly received the Unity of Science movement in the 1930s turned against it and thus helped to guarantee that, despite the efforts of Frank, it would never regain momentum in the postwar world. Several factors and pressures came together to achieve this result. One was the fact that unified science was a popular goal. It was not exclusively logical empiricism's. Some version of the unity of science thesis was shared by Marxists of all stripes, with the result that the topic and its practical goal was more "pink" during the Cold War than the decades before. Another factor concerns professionalization and the goal of cultivating core problems and methods that would define, legitimize, and preserve a place for philosophy of science in Cold War academic culture. A third is widespread rejection of "collectivism" by intellectuals across the disciplines and the celebration of "individualism" and liberty in politics and social theory. The values and methods of the Unity of Science movement were simply out of step with the medley of anticommunist, anticollectivist, and antiscientistic themes that dominated Cold War America. At the many universities that required faculty to sign patriotic loyalty oaths, anticommunism was not merely a mood or attitude, but rather an official feature of institutional life and work.

One reason the effects of campus anticommunism on philosophy of science and other disciplines have remained obscure is that the social and institutional mechanisms in play are hardly noble or admirable. It is easy to defend the personal integrity of many of the philosophers treated here, but it is not easy to defend the behavior of the philosophical and academic professions as a whole during the McCarthy years. The AAUP and APA were anemic in their efforts to defend philosophers attacked by anticommunists and dismissed from their jobs (McCumber 1996; 2001). As a whole, the academy and higher education engaged in something like an orgy of patriotic conformism that will offend even casual supporters of late twentieth-century political correctness:

Professors and administrators ignored the stated ideals of their calling and overrode the civil liberties of their colleagues and employees in the service of such supposedly higher values as institutional loyalty and national security. In retrospect, it is easy to accuse these people of hypocrisy...but most of the men and women who participated in or condoned the firing of their controversial colleagues did so because they sincerely believed that what they were doing was in the nation's interest. Patriotism, not expedience, sustained the academic community's willingness to collaborate with McCarthyism.... When, by the late fifties, the hearings and dismissals [at colleges and universities] tapered off, it was not because they encountered resistance but because they were no longer necessary. All was quiet on the academic front. (Schrecker 1986, 340–41)

The few academics who remain living to recall these upheavals do so not often, nor with relish. Judging from secondary accounts and memoirs (such as Sidney Hook's (1987)), many wounds never healed and scores were still being settled in the 1990s. Those in the profession who had conversations in the early 1950s with FBI agents about the patriotism of certain philosophers – including Carnap, Frank, William Malisoff, and Albert Blumberg – probably hoped that these conversations would remain unknown.[19]

Finally, one of the most remarkable recent discoveries about Cold War intellectual life is that not all of the pressures of anticommunism were negative, repressive, and prohibitionary. To complement Schrecker's pioneering study of academic anticommunism, Frances Stonor Saunders has explored the positive rewards of anti-Stalinism for intellectuals and artists who participated in the long-running Congress for Cultural Freedom. This institution of the "cultural Cold War" was waged jointly by a handful of influential American and European scholars (including, for a while, Sidney Hook) and U.S. government experts in military intelligence. Combining the brains of Sidney Hook, Daniel Burnham, and other anti-Stalinist intellectuals with the financial brawn of the CIA and major philanthropic organizations, organizers of this congress generously sponsored anticommunist liberalism throughout Europe and Asia in the form of publications, conferences, and exhibitions.

To the Icy Slopes of Logic

The following chapters together examine how, in light of these various pressures and circumstances, logical empiricism took the apolitical, technical, and professional form it had taken by the end of the 1950s. The main event in this transformation is the death of the Unity of Science movement. What survived the Cold War was logical empiricism *without* Neurath's Unity of Science movement, a logical empiricism stripped of the points of contact it had begun to cultivate in the United States with scientists, the public, and with other progressive, liberal movements. By the late 1950s, we see below, leading philosophers of science typically distanced philosophy of science proper from normative concerns of ethics and politics using arguments and suppositions that would not have gone unchallenged by Neurath, Frank, Morris, Dewey, and others in the 1930s.

---

[19] FBI files on these investigations were requested via the Freedom of Information Act.

In the 1960s and after, however, these philosophers were either dead or lacked influence or students willing to carry their torches into the profession's future.

It is at this point in the story of logical empiricism that celebrated arguments of Kuhn and Quine need to be reconsidered and recontextualized. This book does not undertake this task, but it suggests some general parameters. Briefly, it suggests that these critiques became possible and trenchant only because logical empiricism had taken the recent course it had. Kuhn complained that the logical empiricist "image of science by which we are now possessed" (Kuhn 1962, 1) was an idealized caricature that did not acknowledge science's vital connections to laboratory practice and to psychological and sociological dynamics within scientific communities. But Kuhn overlooked or was perhaps unaware of the fact that the program he critiqued in the late 1950s had only recently downplayed the interests of Morris, Neurath, and Frank in science's connections to social, historical, and economic life and their hopes that these topics would thrive among the discipline's core issues. As one recent analysis suggests, Kuhn's celebrated influence is not due to some discovery of connections among science, society, and history to which logical empiricists were simply blind. Rather, the success of *The Structure of Scientific Revolutions* was arguably due to the kind of relationship it posited between science and society, one that comported well with Cold War innovations in federal funding of science and military research (Fuller 2000).

Quine was correct that a distinction between analytic and synthetic statements was crucial for maintaining the mature, logical empiricist conception of a theory (articulated in Carnap 1939 and 1956, for example) as a formal structure tethered to experience via synthetic propositions. Without this, the structure collapses, as Quine put it, into a metaphorical web whose threads are all, more or less, analytic and synthetic (Quine 1951). But Quine's criticism that this distinction is corrupt because it cannot be formally specified without moving in a circle assumed that some foundational, noncircular specification is the only adequate kind of specification. An alternative, it would appear, lay in pragmatic approaches of the sort championed, again, by Dewey, Morris, Neurath, and, especially at the time of Quine's critique, Frank, who was then promoting pragmatism (specifically, Bridgman's operationism) as a lingua franca for philosophy of science. As Howard Stein commented when recalling an exchange he observed between Carnap and Quine, philosophy of science under Quine became more concerned with the critique of doctrine and less with creating tools (designing languages, in Carnap's case) whose value is to be judged at least partly, if not mainly, by their pragmatic utility.[20] However ineffectual or unheard of, the efforts of Frank and Morris, in their own ways, promoted a synthesis of pragmatism and analytic philosophy long before recent attempts.

The Plan of the Book

The chapters that follow are arranged roughly chronologically to create a narrative arc depicting the rise and fall of the Unity of Science movement in North America. To establish background for the claim that political forces helped to drive its downfall, the first chapters document the political and ideological vitality of the movement in Europe and, mainly, in the United States. Chapter 2 introduces the main proponents and organizers of the Unity of Science movement – Neurath, Carnap, Frank, and Morris – along with some political aspects of their work and careers. Chapter 3 surveys the leftist philosophical scene in the 1930s that is explored in subsequent chapters. It also describes the warm reception and healthy collaboration between Neurath's movement and the influential philosophers and intellectuals working in New York City in the mid-1930s. On the basis of correspondence among Carnap, Morris, Neurath, Nagel, Dewey, and others in the late 1930s, chapters 3 and 4 suggest that the first years of the new *International Encyclopedia* were a short but nonetheless golden age for the movement, a time when American pragmatism and logical empiricism collaborated and together sought to promote liberal, progressive goals for Western culture. As Dewey put it in his first contribution to Neurath's *Encyclopedia*, the unity of science was a kind of "social problem" that both groups were dedicated to solving.

Chapters 5, 6, and 7 continue to examine the leftist philosophical scene in North America in the 1930s and '40s. Chapter 5 examines some radical philosophers, mainly Albert Blumberg and William Malisoff, whose careers intertwine with that of logical empiricism, while chapters 6 and 7 explore the leftward regions of that scene, ranging from the radical intellectuals who wrote for *Science & Society* to openly communist philosophers who far outflanked all these philosophers in their commitments to dialectical materialism and the Communist Party. While these three chapters may be skimmed or skipped by readers who are more interested in the central story of the Unity of Science movement, they document

---

[20] Stein 1992. For an account of the Cold War's effects on academic philosophy that traces these effects largely to Quine and his influence, see McCumber 2001.

the proximate, even collegial relations, in some cases, of the movement to radical Marxist intellectuals who, however much they criticized logical empiricism's methods, shared a devotion to unity among the sciences. In some cases, such as those of Malisoff and British philosopher Maurice Cornforth, these figures reappear in later chapters to be variously rejected and criticized by professional philosophers of science in the 1950s.

Chapter 8 introduces the general intellectual climate of the Cold War and the roots of that climate in the anti-Stalinist conversions of Marxist socialists who once adored the Soviet Union. It also examines emerging anti-intellectualism in Cold War culture, based in part on growing rejection of pragmatism and logical empiricism's shared scientific approach to understanding values. This values debate, which recurs throughout the central narrative, divides intellectuals according to whether they believe that science and its methods can or cannot answer (or help answer) questions about ethical, social, and political values.

Chapter 9 returns to the interior of the Unity of Science movement to show how one former leftist directed much of his anger and bitterness toward Moscow directly at Neurath and his Unity of Science movement. Horace Kallen's denunciation of the project as "totalitarian" is explored to help explain both the "communistic" reputation the movement eventually received and Neurath's eventual marginalization as a philosopher given to authoritarianism in both habit and doctrine. Chapters 10 and 11 then chart a postwar schism between Neurath and Philipp Frank, on the one side, and the majority of their logical empiricist colleagues, on the other. The first explores Neurath's critique of Carnap's (and Tarski's) semantical conception of truth and traces that critique to Neurath's hopes that philosophy of science and the Unity of Science movement would contribute to cultural and educational reforms involved in postwar reconstruction of Europe. The second documents Neurath's alliance with Frank, their shared critique of excessively formal, "scholastic" philosophy of science, and Frank's lifelong effort to promote a philosophy of science in North America as a bridge in higher education between science and the humanities.

The last chapters follow Morris, Carnap, and especially Frank in their various efforts to revive the Unity of Science movement after the war and explore several of the ways that anticommunist pressures opposed them. These pressures can be grouped into three kinds or levels that are described in chapters 12 and 13. The first is anticollectivism in social and economic theory (illustrated here by the immensely popular writings of Friedrich Hayek); the second is anticommunism in popular culture and on American campuses; and the third consists of personal campaigns directed specifically at these philosophers. Chapter 13 is devoted to examining how Morris, Carnap, and Frank differently experienced these pressures in the form of loyalty oaths, anticommunist investigations undertaken by the FBI, and accusations and complaints from colleagues.

Against this backdrop of anticommunist pressures and dangers, chapter 14 depicts a struggle for dominance among three factions vying to shape the content and style of postwar philosophy of science. These include Frank, with his new Institute for the Unity of Science; Reichenbach, Feigl, and others, who together tended to oppose Frank's plans and projects in favor of more technical topics and professional protocols; and C. West Churchman, who succeeded Malisoff as editor at *Philosophy of Science*. As shown in chapter 15, Frank eventually lost this contest. His efforts to lead his new Institute for the Unity of Science were encumbered by conflicts with his colleagues, loss of funding, and the decline of his own reputation as a philosopher. Chapter 16 then examines a parallel loss of influence on the part of Charles Morris and his movement away from technical philosophy of science toward social science and the study of values.

With Frank, Morris, and Neurath largely out of the picture, chapter 17 surveys the developments and circumstances marking the final death of the Unity of Science movement and its goals and ambitions within professional philosophy of science. These include the official dissolution of Frank's Institute, the rechartering of the Philosophy of Science Association, and the connections forged between logical empiricism and government-funded military research epitomized by the RAND Corporation. While some logical empiricists availed themselves of these research opportunities, the chapter shows that a more or less official consensus emerged among the profession's leaders: Matters of ethics, society, and politics are officially outside the boundaries of professional philosophy of science. Despite that demarcation, however, the chapter suggests that logical empiricism's mature, axiomatic view of knowledge (or theories) shows a sympathy with Cold War dichotomies – as understood by Sidney Hook, for example – between irreconcilable, "absolute" values and ideologies. Cold War Logical empiricism did not take sides in these political battles, but it agreed (in a way) that there were irreconcilable sides to be taken.

Chapter 18 concludes by examining several issues involved in the transformation of philosophy of science described here and deserving of further study or scrutiny. Contextual issues include the rise of the postwar university and the concomitant decline of unaffiliated "public intellectuals." More technical issues include contemporary interest in the "disunity of science," the conventional division between analytic and continental philosophy, and the manner in which the goals and values of the Unity of Science movement should be seen as directly opposed to the "absolutism" that guided Cold War politics and the professionalization of philosophy of science. Had history taken a different path, it is argued, and had the Unity of Science movement and its supporters not been marginalized as they were, the arguments for this general depoliticization would have become at least less representative of the discipline as a whole, if not less convincing.

2

Otto Neurath, Charles Morris, Rudolf Carnap, and Philipp Frank

Political Philosophers of Science

Logical empiricists themselves indicated that logical empiricism and the Unity of Science movement had not only intellectual (or narrowly episte-mological) ambitions, but also social, cultural, and – broadly construed – political ambitions. In his autobiography, Carnap wrote, "All of us in the [Vienna] Circle were strongly interested in social and political progress. Most of us, myself included, were socialists" (1963a, 23). For most, more-over, their socialist politics or outlooks were in various ways connected to their philosophical projects. Had the Vienna Circle's manifesto, *Wis-senschaftliche Weltauffassung*, written by Carnap, Neurath, and Hans Hahn in 1929, been sooner translated and published in America, the progres-sive, socialist outlook of logical empiricism might have been better known to American philosophers of science. The manifesto sketched a broad, modernist aesthetic that connected the tasks of eliminating metaphysics, reforming philosophy, and unifying the sciences:

The endeavor is to link and harmonise the achievements of individual investi-gators in their various fields of science. From this aim follows the emphasis on *collective efforts*, and also the emphasis on what can be grasped intersubjectively; from this springs the search for a neutral system of formulae, for a symbolism freed from the slag of historical languages; and also the search for a total system of con-cepts. Neatness and clarity are striven for, and dark distances and unfathomable depths rejected. In science there are no "depths"; there is surface everywhere: all experience forms a complex network, which cannot always be surveyed and can often be grasped only in parts. Everything is accessible to man; and man is the measure of all things. (Neurath, Carnap, and Hahn 1929, 306)

Prometheus and Protagoras were similarly united in other cultural move-ments that Carnap, Neurath, and Hahn believed were sympathetic with logical empiricism. Like these other projects, the scientific world outlook "fulfills a demand of the day":

We have to fashion intellectual tools for everyday life, for the daily life of the scholar but also for the daily life of all those who in some way join in working at the conscious re-shaping of life. The vitality that shows itself in the efforts for a rational transformation of the social and economic order, permeates the movement for a scientific world-conception too. (ibid., p. 305)

There were more than mere parallels between this goal of unified science and movements for economic socialization, education reform, and the internationalization and "unification of mankind." They shared an "inner link" and intersected in the idea that the sciences, free of metaphysics and strengthened theoretically and practically through unifying connections among them, would be a most effective tool in this conscious, rational transformation of the world (ibid.).

That is why when Neurath, Carnap, Frank, and others brought logical empiricism and the Unity of Science movement to North America, they consciously brought with them more than a narrow, specialized project in philosophy. Their philosophical projects differed, and so too did the political values each attached to those projects or the political implications they derived from them. But a survey of Neurath's, Frank's, Carnap's, and Morris's different conceptions of how philosophy of science – done right and given the place it deserves in modern life – could help make the world better reveals nonetheless a general unity to their hopes and ambitions.

Otto Neurath (1882–1945): Philosophy and Politics

Otto Neurath's philosophical reputation suffered after the 1930s because of several circumstances. Besides his knack for expressing ideas with less than crystal clarity, and a strong, aggressive personality that was either adored or reviled by colleagues, Neurath's conceptions of science and knowledge were in some ways more at home in the postmodern 1990s than in the 1950s and early '60s when logical empiricism – as a formal, specialized, and nonpolitical program – dominated philosophy of science. For Neurath was suspicious and critical of formalism and his program openly reached for engagement with both scientific practice and social and political life. In these respects, Neurath's voice and his ambitions were in tension with Carnap's – a tension that was sometimes productive and at other times (as discussed in chapter 10) frustrating and disappointing for both of them. Over a decade after his friend Neurath died, Carnap recalled their many debates and Neurath's regular "emphasis on the connection between our philosophical activity and the great historical processes going on in the world" and his "most lively and stimulating talks" about social problems and their connections to logical empiricism (1963a, 23).

As a philosopher critical of overly formal, systematic idealizations of scientific theory, Neurath's views are best introduced as a collection of themes, more or less interconnected, that informed his lifelong crusade against metaphysics and the various disputes and debates he took up.

Meaning Holism

Neurath upheld a holistic theory of meaning according to which words and sentences gain meaning in virtue of connections to other words and sentences. Language must not be regarded as reaching out to and imbibing meaning from some extralinguistic real world. As he put it during the Vienna Circle's protocol sentence debates,

> Statements are always compared with statements, certainly not with some "reality," nor with "things." If a statement is made, it is to be confronted with the totality of existing statements. If it agrees with them, it is joined to them; if it does not agree, it is called "untrue" and rejected; or the existing complex of statements of science is modified so that the new statement can be incorporated; the latter decision is mostly taken with hesitation. There can be no other concept of "truth" for science. (Neurath 1931b, 53)

Whether Neurath took semantic agreement to be the substance of scientific truth, only a marker of scientific truth, or, as Thomas Uebel has defended Neurath, as an altogether separate approach to a practical theory of sentence *acceptance* (Uebel 2004), it is certain that Neurath strived to avoid commitment to, and even hypothetical discourse about, a realistic metaphysics of extralinguistic circumstances or entities that language somehow grasped.

Pluralism versus Absolutism

Especially later in his career, Neurath emphasized pluralism as an antidote to "absolutism." Usually because they mistakenly believed that scientific theories reveal to us determinate metaphysical realities, absolutists supposed that concepts, statements, and scientific theories have singular, exhaustively specifiable meanings. Pluralists properly abstained from a realism promising knowledge about some single, absolutely specifiable world and – in light of meaning holism – understood that a word or statement had plural and ambiguous meanings, depending on who was using the language and which meanings and semantic threads within that language they had in mind.

"Empiricism" versus "Systematization"

Empiricists, for Neurath, understood that, in light of pluralism, scientific language generally resists being neatly fitted into a logical system or model. Neurath's position was subtle, however, for he admitted regularly (as any observer of scientific theory must) that parts of scientific language can be productively and legitimately systematized. His pluralism, however, warned that such systematizations were not self-legitimating or exclusive. The same areas of science could be systematized in substantively different ways that, for example, connected to or incorporated concepts from other parts of science and had, as a result, different practical strengths. In addition, such systematizations were never comprehensive across the sciences. Neurath urged philosophers (such as Carnap) whose talents lay in analyzing or constructing systematizations that in the world of science they remain only "islands of systematization"[1] unconnected to others and surrounded by fluid ambiguity of the bulk of our ordinary and even specialized scientific languages.

Encyclopedism

Neurath's suspicions of "systematization" directly introduce his "encyclopedic" model of unified science – a model he opposed to "pyramidism," which takes a future unified science to be a comprehensive and systematic ordering or hierarchy of sciences. Obviously wedded in Neurath's mind to the project of producing an actual encyclopedia of the sciences, Neurath's pluralism eventuates in such a project:

If we reject the rationalistic anticipation of the system of the sciences, if we reject the notion of a philosophical system which is to legislate for the sciences, what is the maximum coordination of the sciences which remains possible? The only answer that can be given for the time being is: An Encyclopedia of the Sciences. (Neurath 1937, 176–77)

[1] Neurath to Carnap, September 25, 1943, ASP RC 102-55-03.

Such an encyclopedia would not rule out future systematizations and organizations of different sciences, but its historicity and capacity to evolve and change (in future editions and added volumes) usefully avoided the supposition that the goal of science in the foreseeable future is some final, definitive, and true science, much less one that corresponds, as a metaphysical realist and absolutist might suppose, to some one, true world.

Decisionism versus "Pseudorationalism"

All these themes help to make sense of Neurath's critique of "pseudorationalism." In what remains one of the most telling of Neurath's writings, an early essay of 1913 – "The Lost Wanderers of Descartes and the Auxiliary Motive" (Neurath 1913) – Neurath argued that progress in science requires free decisions, the necessity of which are typically obscured by absolutism and illusory faiths in epistemic foundations. Where Descartes believed that only in special cases – such as being lost in a forest – must we act exclusively on provisional and uncertain (or unclear and indistinct) knowledge, Neurath urged that this is our regular, epistemic condition. It was only a philosophical conceit that led Descartes and his followers to suppose that reason remains in principle available to analyze any problematic situation, determine and rank outcomes, and so relieve us of our decision-making responsibilities. Instead, Neurath insisted, we must live with pluralistic ambiguity of our language and, for the Unity of Science movement, ongoing and unending choices about where and how to connect and unify the sciences.

It is perhaps this decisionistic aspect to Neurath's conception that best makes sense of his project as an existential, historical project. The future of science depends on those who choose to participate in it and utilize its fruits, not only because there are others who oppose science and would eliminate it if they could, but also because there exist no extra-human forces or principles to guarantee its future and coax it to unfold as some ideal mirror of reality. Scientific knowledge is a tool that humans make ultimately for their own use, and as such is properly conceived as part of a collective, social enterprise – a "movement" – that renders it as politically and socially engaged as other institutions created and actively sustained by human beings.

Neurath's Political Activities and Reputation

Neurath's entrée into politics was not primarily by way of his philosophy or his Unity of Science movement. For the most part, he lived and worked outside academia and was directly involved in social or economic projects. Trained as a political economist in Berlin under Eduard Meyer, Neurath taught for only a few years before embarking on what would become a series of political and administrative projects. He worked for the office of war economy in the Austrian government and, in 1919, as president of the Central Economic Office in the short-lived socialist government of Bavaria, attempting to implement a currency-less economy-in-kind. After being convicted of treason by the subsequent government, Neurath served a short sentence and returned to Vienna to become active in Viennese communal politics.[2]

By 1923, Neurath's lifelong interest in visual iconography led him to found his Social and Economic Museum in Vienna, where he and his assistants (including Marie Reidemeister, Josef Frank – the architect and brother of Philipp Frank – and the graphic designer Gerd Arntz) developed the ISOTYPE system. This language – International System of Typographic Pictorial Education – was designed primarily for public exhibits and to convey a maximum of demographic and economic information (to adults, foreigners, and even illiterates) using a minimum of text (see Figure 2). Neurath pursued these populist and socialist projects while he participated in the early years of the Vienna Circle when it was led by Moritz Schlick (Stadler 2001, 195–98).

Because of Neurath's wide range of skills and interests, we find him at several different locations and sources of twentieth century modernism. He lectured at the Dessau Bauhaus (Galison 1990; Dahms 2004), worked with Cornelius Van Eesteren's Belgian Congress of International Modern Architecture (Faludi 1989), and was enlisted by Moscow to help prepare propaganda about the first Five Year Plan (Chislett 1992). Neurath was in Moscow in February 1934 when Dollfuss's regime took power in Vienna and promptly consolidated its power by attacking its internal enemies. The Verein Ernst Mach, the free-thinker organization that the Circle adopted as a kind of public, educational forum, was summarily shut down (despite Schlick's formal protests) on the grounds that it was allied with the Social Democratic Party (Stadler 2001, 347). Soon, authorities were rifling through files in Neurath's offices most likely, as Marie Neurath later explained, because Otto had been denounced as a Communist. She then cabled Neurath not to return to Vienna and they began the next phase of their lives in The Hague, where (half expecting

---

[2] For biographical information about Neurath, see Cartwright et al. 1996, part 1; Neurath 1973, chap. 1; and Stadler 2001.

In War Seasonal Fluctuations Disappear
Quarterly Coal-Production in the United States

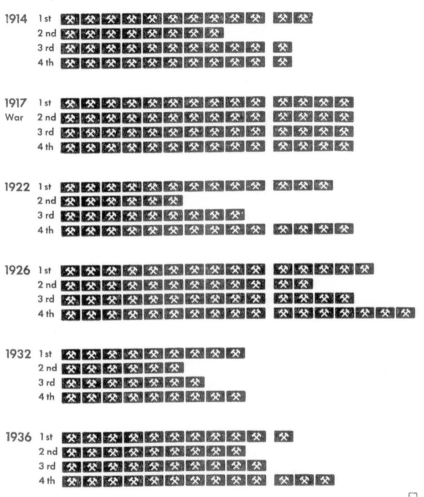

Each symbol represents 10 million short tons of coal, produced quarterly

FIGURE 2. An example of Neurath's ISOTYPE diagrams visually illustrating a historical economic trend, in this case the production of coal (Neurath 1939, 87).

such an outcome) Neurath had one year before established his International Foundation for Visual Education. Within this Foundation, Neurath set up the Institute for the Unity of Science, from which he guided the nascent Unity of Science movement and continued his career as a museum and exhibit expert (Neurath 1973, 62–63). He supported himself with a range projects such as contract work for the Mexican government, for Compton's Picture Encyclopedia, and for the National Tuberculosis Association.

Though Neurath was selective about the elements of Marxist doctrine that he did and did not accept, one need not look far in his writings to see his overall leftist orientation. He regularly contributed writings to *Der Kampf*, the mouthpiece of the Austrian Social Democrats, and wrote essays such as "Personal Life and Class Struggle," in which the goal of unifying the sciences has both scientific and revolutionary value. "Marxism," he wrote, is "the carrier of the scientific attitude," and "the proletariat is the bearer of science without metaphysics" (Neurath 1928, 297). A unified science that excluded metaphysics would thus disarm obscuring religious and pseudoscientific beliefs about the world and about society and empower the proletariat: "[T]he better the proletariat grasps the social engineering relations of our order and surveys its own chances, the more successfully it can fight."[3]

Neurath's interest in Marxism was no secret to observers of the Unity of Science movement, either. In 1937, for example, Morris reported to Neurath that some members of Morris's Logic of Science Discussion Group at the University of Chicago had objected to the word "international" in the title of the new encyclopedia that Morris was probably advertising to his colleagues. There were several reasons for the objection, Morris noted, one of which concerned the word's "political implications which some persons do not like."[4] Indeed, the word "international" abounded in leftist discourse. The evolution of communism was (and remains) periodized into "Internationals" (such as Stalin's Third and Trotsky's hoped-for "Fourth International") that bonded national Communist Parties into collaborating communities; the Communist International (or "Comintern") directed party activities from Moscow; while the official publishing firm of the Communist Party was "International Publishers." The title of the famous socialist anthem is *L'Internationale*. On the campus of the University of Chicago, which had two years before experienced formal and highly publicized anticommunist investigations, such "political implications" of the word "international" would have easily come to mind.[5]

---

[3] Neurath 1928, 296. For more analysis of Neurath's Marxism, see Cartwright et al. 1996.

[4] Morris to Neurath, April 17, 1937, USMP, box 1, folder 16; Morris to Neurath, June 20 and 22, 1937, USMP, box 1, folder 16.

[5] These investigations at the University of Chicago are described in chapter 13.

Neurath dismissed Morris's worry by arguing that the word had no political meaning in this case. It simply characterized the authorship of the new encyclopedia. Officially, the project had no political orientation. Its promotional literature, written usually by Morris and edited and approved by Neurath, stated that its goals and the science-unifying tools of logical and linguistic analysis were ideology-free. One promotional flyer seems to echo Morris and Neurath's conversation:

Collaborators of various nationalities have been invited: only their personal competence has been considered or the benefits to be obtained from a variety of cultural viewpoints – their political views or the political ideologies of the countries they come from have not entered into consideration, since the *Encyclopedia* is a scientific and not a political enterprise. Each collaborator will, of course, be responsible only for the ideas which he himself expresses.[6]

Whether this nod to impartiality and nonpartisanship was intended to mollify critics such as those Morris encountered or possibly to satisfy Carnap's probable insistence that the encyclopedia officially demarcate philosophy and politics, we should not conclude that Morris, Neurath, and Carnap genuinely believed that their encyclopedia project had no connection at all to their shared and their individual socialist views.

First, this policy describes a neutralism about the selection of "collaborators" and states that they were recruited without reference to their politics. Nothing in this would prevent collaborators from having strong political views, however, in which case (as the next sentence states) those views should not be taken to represent those of the others. That is the primary sense in which the encyclopedia was a "scientific and not a political enterprise." Second, we should not read into Morris's distinction our own, contemporary understanding of apolitical technical work in philosophy – according to which each essay and contribution would make no reference to, and have no bearing on, social or political or cultural problems or topics – if only because that postwar mode of philosophy had not yet been born.

The kind of partisanship that Morris and, most likely, the University of Chicago Press wished to avoid was rather the kind that Herbert Feigl wished to avoid when in 1936 he invited Neurath to take a break from his encyclopedia work in Chicago and come to Feigl's University of Iowa to give a talk and participate in a roundtable discussion. Indeed, Feigl

---

[6] Promotional flyer, UCPP, box 347, folder 2.

urged Neurath to be "as politically neutral as possible." But even in the heartland of America, Feigl did not urge Neurath to pretend he did not have Marxist leanings, but rather to make himself and his project acceptable to an audience that, it would appear, could be easily intimidated by revolutionary politics. Marxism should enter the presentation, Feigl insisted, "only in sweetened form."[7] That is the form Marxism takes in the encyclopedia's promotional literature, as well. It was not about revolutionary politics in any direct sense, but it was directly about goals and concerns – modernity, education, and the place of science in society – that were unmistakably linked to liberal and socialistic progressivism. The encyclopedia's goals, the flyer continued, were to help science to "perform adequately its educational role in the modern world," to buttress "the general educational implications of the Unity of Science Movement," and "to reach those persons upon whom the future of science depends."[8]

To whatever extent Morris, Neurath, and Feigl hoped to minimize the political reputation of the Unity of Science movement and its forthcoming encyclopedia, we see in subsequent chapters that there was little doubt about the project's progressive, socialistic credentials. Socialists were sympathetic to the goal of unifying the sciences precisely because science was widely seen as a tool to plan, to create, and to manage desired social and economic structures of modern life. Many quickly reached out to the movement as kindred political spirits. And were the goals of Morris's promotional flyer to present the new encyclopedia to the public as a forum for apolitical intellectuals, those were quickly foiled. When *Time* magazine reported on the publication of the encyclopedia's first two monographs in August 1938, it pointed out without fanfare or emphasis that the "hulking, booming Otto Neurath" who led the new encyclopedia project had "strong socialist leanings in politics."[9]

Even in the early 1940s, the darkest days of the movement and the encyclopedia, Neurath never lost his faith that through science and social and economic planning the world was ripe for transformation. He bolstered his reputation as an advocate of large-scale planning with his

[7] Feigl to Neurath, September 19, 1936, ONN. Feigl invited Neurath to lecture and then participate in roundtable discussion: "Für ersteren Schlage ich 'Empirical Sociology' (oder dgl.e) vor – (soll politisch moglichst 'neutral' sein!) für letztere entweder 'Encyclopedia' oder 'Pict. Statistics.'" Below in the same letter, Feigl urged Neurath to present Marxism "nur in verzucherter Form (Sie kapieren doch?)."
[8] Press release, UCPP, box 347, folder 2.
[9] *Time*, August 1, 1938, page unknown.

essay "International Planning for Freedom" in 1942 by arguing that lasting peace would be secured only by setting up overlapping economic commonwealths throughout Europe (Neurath 1942). He also joined the editorial board of the *New Commonwealth Quarterly*, a journal dedicated to planned, postwar reorganization of Europe. In the last year of his life, he wrote articles in the *Journal of Education* arguing that curricula in Germany (as well as all of Europe) needed to be restructured (in scientific and antimetaphysical ways) to help prevent any future for fascism. With J. A. Lauwerys, reader in education at the University of London, he also published a series of articles titled "Nazi Text-books and the Future" and "Plato's Republic and German Education" that specifically attacked Platonic idealism as both a philosophical and psychological source of Nazism. Not surprisingly, ensuing debate in the journal with classicists and teachers was lively.[10]

Neurath called Platonism "totalitarian" both because of the regimented style of life in Plato's ideal Republic and for reasons directly growing out of Neurath's philosophy of science. For Neurath, a properly scientific world conception would eschew ontological realism and thus discourage individuals from sacrificing their own and others' happiness and humanity for the sake of allegedly "higher" or ontologically superior agencies or causes. Neurath worked during the last years of his life on a manuscript detailing the history of persecution. He drew on his experiences as a refugee, the carnage he observed during the two world wars, and this notion that metaphysical misconceptions at least assist, if not lie at the root of, such tragedies.[11]

In Neurath's mind, these political projects dovetailed with his work on behalf of the Unity of Science movement. Especially when the University of Chicago Press was growing restless because new monographs were not appearing, Morris wondered whether Neurath's focus was drifting. It may have drifted away from the encyclopedia but not from the goals of the movement, which was well poised to provide reconstructed Europe with a scientific world-view. "As you know," he reassured Morris in 1942, "I do not cease to manage things and I shall think that we have a lot to do for Europe after this war."[12] This helps to explain Neurath's decision to remain in England and also his decision (strongly resented by Morris

---

[10] Neurath 1945a; Neurath and Lauwerys 1944; 1945.

[11] See Reisch 2003b. Neurath's thoughts on metaphysics and persecution are discussed in chapter 10.

[12] Neurath to Morris, July 17, 1942, ASP RC 102-56-04.

and Carnap) to publish the *Journal of Unified Science* with Basil-Blackwell, not the University of Chicago Press, as Morris and Carnap desired. And, as discussed in chapter 10, Neurath's vision of a politically and culturally engaged movement also supported his complaints about Carnap's work in semantics.

In some respects, Neurath wanted the encyclopedia to play a leading symbolic and methodological role in promoting a postwar *Wissenschaftliche Weltauffassung*. During the war, he once suggested to Morris and Carnap that future volumes should be restructured in a less didactic, more informal way. They should include not only monographic treatments by individuals, but also dialogues among scientists and philosophers about important issues. These dialogues would make the methodological point that Neurath increasingly felt that his colleagues were missing (or at least failing to emphasize sufficiently in their writings): In a world without epistemic foundations, scientific decisions and the actions based on them must emerge from public discussion and debate, from argumentation that leads to agreement and coordinated action. Paths of action are to be collectively, cooperative *made* and not (as a naïve realism would have it) *discovered* as if there were one singular or optimal solution to any problem. In this regard, he saw the encyclopedia as potentially nothing less than an educational microcosm for the management and use of scientific knowledge in modern life. "You see," Neurath explained to Morris and Carnap, "I think we should present ourselves as able to organize orchestration when asking people to support orchestration in social life."[13] Unfortunately, Neurath's hopes to position the unity of science movement on the postwar world stage – as both an actor and an exemplar – were cut off by his sudden death in late 1945.

Charles Morris (1901–1979): Signs, Symbols, and Diplomacy

Morris's intellectual and political personality in the years before the Cold War was dominated by his Deweyan faith in science as a powerful and effective tool for shaping modern life. Though he would later develop a broad, international humanism based on his interests in Buddhism, in William Sheldon's research on body types, and his own aspirations as an

[13] Neurath to Carnap and Morris, November 18, 1944, CMP. Neurath used the word "orchestration" after he adopted it from Kallen's arguments against the Unity of Science movement, treated in chapter 9.

empirical researcher with the "science of man," Morris's agenda in the 1930s consisted mainly of two tasks: the integration of pragmatism and logical empiricism within a general theory of signs (or "semiotic") and the defense of science and scientific philosophy against its enemies both within and without the American intellectual scene (Morris 1937; Petrilli 1992; Reisch 1995, 2001b).

Morris had always wanted to promote philosophy as a leader in the growth and direction of modern culture. After receiving his Ph.D. under George Herbert Mead at the University of Chicago, and after a six-year tenure at Rice University at Houston, he returned to Chicago's department in 1931. Morris was justifiably nervous about returning to his alma mater, for he accepted the department's offer in the wake of resignations tendered by nearly the entire senior faculty in philosophy.[14] The resignations capped a series of disputes between the department and the university president, Robert Maynard Hutchins. Hutchins was eager to make the university the finest in the world. But he was no fan of Chicago's philosophy department. Like his good friend Mortimer Adler, whom he met when he was dean of law at Yale and Adler was completing his Ph.D. in psychology at Columbia, Hutchins disdained pragmatism and all things scientific in philosophy. Persuaded by Adler's many and persuasive sermons that keys to educational and cultural reform lay plainly in the pages of St. Thomas's *Summa*, Hutchins gave Adler a job at Chicago. The department, however, was suspicious of Adler's qualifications, angry about his (and Hutchins's) agenda for refashioning the department along Thomistic lines, and furious after learning that Adler's salary was roughly three times what they expected it to be. Chairman J. H. Tufts resigned in 1930 and he was soon followed by Mead, E. A. Burtt, and Arthur Murphy.[15]

Morris never persuaded Hutchins to admire science, pragmatism, and the Unity of Science movement. But he was persistent. Shortly after arriving back in Chicago, he proposed to Hutchins a venture designed to re-establish philosophy in its proper "directive role in human life." He had in mind an institute – "something similar to the British Institute of Philosophy but centered at the University of Chicago."[16] His sales pitch

---

[14] Morris was nervous about being perceived as opportunistic. He explained "the grounds for the decision" to accept Chicago's offer in a letter he sent to friends and colleagues (Morris to various, March 30, 1931, CMP).

[15] Ashmore 1989, 85–87.

[16] Morris, generic fund-raising letter, March 7, 1932, CMP.

to a prospective patron in 1932 shows his optimism, his allegiance to Deweyan pragmatism, and some institute envy:

We have here a great Oriental Institute, backed by millions. I would like to see the problems of the present and the future taken as seriously as we take the problems of the past. I do not know how much can be done to give intelligence a directive role in human life, but we shall never know until we try. . . . We need a new mind for the new time; what we need now are engineers of ideas as drastic as our practical ones.[17]

In his proposal to Hutchins, he specified that the institute would require $50,000 each year and would support research by faculty and visiting scholars in philosophy – "Logic, ethics, categorian analysis, cosmology, philosophy of the sciences, axiology, esthetics, social philosophy, [and] methodology" – as well as public lectures on topics such as "The Future of Democracy, Philosophic Ways of Life, the Ethics of the Various Professions, the Division of Professional Labor, the Physical World, the World of Society."[18]

The general task of the institute was to respond to the world situation, one that Morris had viewed since his doctoral dissertation (at least) through Spenglerian lenses. According to Spengler and his influential *Decline of the West*, cultures undergo cyclical patterns of growth and decay. The West's present decline, Morris believed, was caused by its lack of a useful and comprehensive framework with which to organize the wealth of new scientific information produced during this period of "pervasive and drastic [cultural] change." Anticipating the perils of overspecialization that he would later mention in his prospectus for the encyclopedia, he explained to Hutchins that "the elaboration of knowledge is necessary for the wise consideration of [cultural and social] goals, but the increase of knowledge alone . . . is not sufficient." The institute would do what was necessary: It would organize and encourage specialists to explore possibilities for "vast intellectual synthesis" and thus help to prepare "a new cultural epoch in the West." Were this opportunity neglected and philosophy not given a chance to provide coherent cultural "vision" and "a way of life," the West would remain in "a trough in the cultural wave."[19]

Hutchins remained unimpressed. Morris's vision of an institute of philosophy remained just a vision. But in roughly two years, when Morris made contact with his future colleagues in the Unity of Science

[17] Morris to Richard Riggs Day, April 25, 1932, CMP.
[18] Morris, "Institute of Philosophy," proposal, March 1, 1934, PP, box 106, folder 14.
[19] Morris, "Institute of Philosophy," proposal, March 1, 1934, PP, box 106, folder 14.

movement, he almost certainly had to have seen their project as a partial realization of his political and intellectual plans. Inspired most likely by Carnap's work in logical syntax, Morris turned from his ongoing work in theories of mind (Morris 1932) and his work on his teacher George Herbert Mead (Morris 1934a) to articulate a technical theory of signs and symbols. Morris intended to generalize Carnap's syntactic program and address not only formal (syntactical) meaning, but also empirical (or semantic) and practical and ethical (pragmatic) meanings of signs. Ever seeking to build bridges between and reconcile opposing philosophical (and cultural) points of view – "synthesis" is perhaps the most frequent substantive term in Morris's philosophical writings – Morris's program embraced syntax, semantics, and pragmatics under the rubric "semiotic" and formed an intellectual umbrella under which American pragmatism, European logical empiricism and the Unity of Science movement could collaborate. Morris promoted semiotic in several papers in the mid-1930s, some of which he read at Neurath's international congresses, and in his second contribution to the encyclopedia, "Foundations of the Theory of Signs," in 1938 (Morris 1937; Morris 1938a).

In his quest to cultivate the movement in America, Neurath could not have wished for a more reliable, energetic, and yet deferential colleague. Not only did Morris actively assist his logical empiricist colleagues' emigration, but he articulated an overarching philosophical project that welcomed all varieties of scientific philosophy as important complements to American philosophical currents. Though Neurath was at times suspicious about Morris's projects, Neurath always appreciated Morris as an able collaborator who also understood the cultural and historical potential of the Unity of Science movement. Perhaps not coincidentally, when Morris began his friendship and collaboration with Neurath and Carnap, he himself began to write more aggressively about the political and cultural importance he saw in philosophy and he confidently pursued contact with like-minded philosophers and intellectuals around the world.

One was in his backyard. By 1937, Morris had begun to collaborate with the artist Laslo Maholy-Nagy, who had re-established the German Bauhaus in Chicago's South Side, not far from Morris's University of Chicago. When Maholy-Nagy asked Morris to help him develop a scientific curriculum at the New Bauhaus, he jumped at the chance and helped Maholy-Nagy to recruit Carnap, the physicist Carl Eckhart, and biologist Ralph Gerard variously to lecture and to develop the curriculum. Morris then wrote to the university's vice president to explain these developments and say that, if necessary to avoid conflicts with his university contract, he would teach at the New Bauhaus without pay. "The universities must in the future align themselves with the creative arts as they have done with the sciences. Even where connections cannot be made officially they can be explored personally."[20] Morris pursued this rapprochement between science and art throughout his life: through his postwar research on values, his own creative writings, and collaboration with the artist and designer Gyorgy Kepes.[21]

Other contacts Morris pursued were in the Soviet Union. After first meeting Neurath, Carnap, and other logical empiricists in Prague in 1934, Morris traveled to Leningrad and the Institute of Philosophy of the Communist Academy. Ever the diplomat, Morris reported to Neurath his "distinct impression that scientific philosophy will continually grow in Russia, and that relations between Russia and scientific philosophy in other countries will become better." The problem had been Lenin's direct and well-known criticism of logical empiricism as a variant of Machian idealism, one that later became a party-line criticism of logical empiricism from dialectical materialists (Lenin 1908). As far as Morris was concerned, such disputes should not prevent future collaboration. He asked Neurath to send Arnost Kolman, the Czech-born head of the institute, an invitation to the upcoming first International Congress in 1935. Though Kolman would later lock horns with Sidney Hook and other anticommunist philosophers during the Cold War, Morris happily reported that this leader of the Soviet philosophical establishment was "much interested" in the Unity of Science movement.[22]

That same year, Morris published the short pamphlet *Pragmatism and the Crisis of Democracy* in which he diagnosed the social, economic, and political problems of Western culture and recommended a version of pragmatism as a cure. Again following Spengler, Morris explained that the West struggled under two conflicting sets of values: "Two foreign cultures [Roman and "Graeco-Arabic"] fought on the soil of the west" and stability had yet to be won (Morris 1934b, 21). The resulting turmoil had yielded three schools of thought – Thomism, fascism, and Marxist communism – none of which alone yielded a satisfactory program. Invoking an anatomical metaphor, Morris described Thomism as a "philosophy of the isolated head" (ibid., p. 4) that rejects science and instead embraces

[20] Morris to Woodward, October 13, 1937, CMP.
[21] See, e.g., Morris 1956; 1966; Morris and Sciadini 1956; Galison 1990, 746–49.
[22] Morris to Neurath, September 24, 1934, ONN. Hook describes his encounters with Kolman in Hook 1987, 407, 414–16. There is no evidence that Kolman came to the First International Congress in Paris, 1935.

nonscientific philosophical tools for solving social and cultural problems. (Understandably, he did not mention Hutchins or Adler in his account.) The second school was a "philosophy of the blood." Morris had in mind German and Italian fascisms that, instead of looking back in time to a medieval or ancient golden age, as do the neo-Thomists, believe that the great days of the West "are gone, as well as behind." Fascists await the arrival of a "new Caesar" who will stave off "internal disintegration" and create a new national, tribal greatness (pp. 5–6). Finally, Morris described "the philosophy of brawn" that looks to the day when "the exploited of the world will wrest power" from their capitalist oppressors and enjoy a Marxist, communist paradise (p. 7).

The answer to these three, inadequate approaches, Morris explained, was a synthetic and pragmatic "philosophy of the heart." Just as the heart maintains the brain, blood, and muscles together, this outlook accepts portions of truth from each philosophy and weaves them into an optimistic and plainly Deweyan pragmatism:

Where its opponents see the West as old and decadent, [this philosophy of the heart] sees youth and untapped powers. It feels great constructive movement in art, philosophy, science, religion, and social organization at work. It sets its vision upon the attainment in the West of a new and distinctive cultural synthesis.... It sees in intelligence a formative principle of reconstruction. *It is essentially the marriage of the scientific habit of mind with the moral ideal of democracy.* (p. 8)

Morris wrote this essay most likely in 1933 or 1934 while he was first corresponding with Rudolf Carnap and making plans to meet him and other logical empiricists in the upcoming months. He was no doubt familiar with the Vienna Circle's manifesto, *Wissenschaftliche Weltauffassung*, in which Carnap, Neurath, and Hans Hahn made similar (and similarly optimistic and ambitious) proclamations. Their scientific world-conception would transform, enlighten, and invigorate culture in much the same way as Morris's philosophy of the heart:

We witness the spirit of the scientific world-conception penetrating in growing measure the forms of personal and public life, in education, upbringing, architecture, and the shaping of economic and social life according to rational principles. *The Scientific world-conception serves life, and life receives it.* (Neurath et al. 1929, 317–18)

Morris fully agreed that the future of philosophy lay in the direction of science and that the future of society lay in the direction of a scientifically informed democratic socialism. In his *Pragmatism and the Crisis of Democracy*, for example, he criticized "unrestricted capitalism." It is overly individualistic, and not "socially and morally conceived and oriented" (Morris 1934b, 20) as an economic system should be.

Morris's "The Cultural Significance of Science"

Despite its political subject matter, Morris's pamphlet was relatively tame and reserved. So, too, were his initial writings for the Unity of Science movement. In an article he published in *Erkenntnis*, for example, Morris deliberately avoided discussing what he saw as "the social significance of the union of scientific forces" in contemporary life because Philipp Frank had warned him that the journal was being scrutinized by Nazi authorities and was perhaps – Frank correctly predicted – going to be suppressed.[23] Perhaps for that and similar reasons, shortly after he returned from the First International Congress for the Unity of Science in Paris, Morris's enthusiasm about the "cultural significance of science" boiled over in a talk he delivered at the University of Minnesota (Reisch 2001b).

Morris began by connecting the worsening political crises in Europe to growing uncertainties in philosophy, art, politics, ethics, and social science. Amid this intellectual turmoil, he explained, science was under attack: in Europe, by fascists and totalitarians, and in America where, in the words of Eric Temple Bell, there was "resentment against all things rational and scientific." Morris thus set out to defend science against these onslaughts. In short, the lecture is Morris's manifesto for scientism.

One reason that Morris never published "The Cultural Significance of Science" is that it presents a more unflinching account of the conflicts and politics driving academic philosophy than his ever-polite and nonconfrontational published writings. For instance, the smooth lines of "pragmatism" that he presented in his pamphlet *Pragmatism and the Crisis of Democracy* give way to what he calls a "family quarrel" between Dewey, on the one hand, and Morris's colleagues at Chicago, Charner Perry and Frank Knight, on the other. The dispute involves fundamental disagreement over the nature of science and the philosophical respectability of scientism. He also more pointedly connects antiscientific reaction throughout the world to the intellectual politics at the University of Chicago. Morris hinted that Hutchins and Adler, who were among

---

[23] Morris recounts this conversation with Frank in (Morris to Neurath, November 15, 1934, ONN). The article in question was Morris 1935. *Erkenntnis* was published from 1930 to 1939, though its life in Nazi Germany with publisher Felix Meiner was finished after 1938. The last volume was published in Holland with Van Stockum & Zoon.

the leading representatives of the neo-Thomist movement of the 1930s and '40s,[24] exhibited the same philosophical and emotional tendencies in their veneration of Aristotle and Thomas that drove the followers of Mussolini and Hitler.

Morris's central claim was that the cultural significance of science (and, a fortiori, scientism in philosophy, art, and ethics) was science's power to create civil and social values. That is why science had to be defended.

The major question before science today is whether it is to be made simply one instrument of and servant of the new nationalistic tendencies, or whether science and the scientific habit of mind is to play the more creative role of itself providing the mentality of the future. (Reisch 2001b)

By embracing science and defending it against fascist totalitarianism, the Unity of Science movement would thus help support democracy in the future. At the same time, the movement would pursue the task he assigned to "philosophy" in his earlier proposals for an institute of philosophy at the University of Chicago. By assembling and interpreting scientific information from different specialities in a more unified whole, society could achieve a new cultural and intellectual synthesis to replace the turbulence and disunity of the present intellectual epoch. From this point of view, Morris's career in the mid-1930s had a remarkable coherence among its personal, philosophical, and political components: During the same years that Morris helped his new colleagues to emigrate from nations in which they were no longer welcome or safe, he believed that he was facilitating their participation in international intellectual cooperation that could, in the end, achieve a kind of salvation for all of Western culture.

Morris's Dissatisfaction with the Unity of Science Movement

Though Morris always pushed his colleagues to make the Unity of Science movement broad and ecumenical, he remained a loyal collaborator who stewarded the encyclopedia to its final days in the 1960s long after other components of the movement had become dormant. Still, ever the diplomat who understood that tensions and disagreements could exist side by side with unified collaboration, Morris began to encourage his colleagues to take a more Deweyan, cultural emphasis on social and

---

[24] Besides Adler and Hutchins, other prominent members of the neo-Thomist movement were Jacques Maritain, Etienne Gilson, Karl Barth, and, arguably, Henri Bergson and Pierre Duhem.

cultural values in their work on behalf of the movement. In the midst of organizing the International Congress to be held at Harvard in 1939, Morris argued in *Synthese's* Unity of Science Forum that the movement was losing its potential for transforming and improving culture. By remaining too narrowly wedded to the natural sciences and failing to pursue American pragmatism's quest to integrate "socio-humanistic studies with the wider corpus of scientific knowledge and procedure," the movement was failing to seize historic opportunities. After having extolled the potential of science for creating cultural values in his lecture in Minnesota, Morris was evidently dismayed to see his colleagues dragging their heels and neglecting issues concerning social science and values. Without leadership that the movement could provide, Morris believed, "creative cultural forces" could never gain strength and prestige from an alliance with natural science. Instead, these forces would remain suppressed in a narrowly scientistic age that mistakenly regarded studies of "mind, value, art, and moral behavior" as necessarily unscientific (Morris 1938a, 28). That was certainly not the case, Morris insisted. He reminded his colleagues that his tripartite semiotic theory of signs provided access to the narrowly technical as well as the broadly cultural meanings in scientific language.

Perhaps trying to lead by example, Morris himself explored those areas that he urged the movement to champion. Besides his work in art and design, he also marched bravely into the philosophy of religion where he attempted to bridge empirical science and religious thought. He did so by appealing to William Sheldon's constitution theory that posited, and claimed to verify empirically, correlations between an individual's body type (or "somatotype," as Sheldon called it) and his or her general personality or temperament. In his book *Paths of Life: Preface to a World Religion* (1942), Morris connected Sheldon's taxonomy of temperaments to traditional religious outlooks. Just as Morris's "philosophy of the heart" (described in his *Pragmatism and the Crisis of Democracy*) was a selective synthesis of Marxism, Thomism, and fascism, Morris recommended in *Paths of Life* that human beings consciously cultivate a personality type or temperament that synthesized and balanced the main features of all dominant temperaments. Thus Morris outlined nothing less than a new world religion – a cosmological orientation within the world and society – that Morris named after Maitreya, the future prophet of Buddhism.

Not surprisingly, Morris's idiosyncratic foray into religion and personality theory was not very popular among his logical empiricist colleagues.

Neurath was shocked over the book and found nothing in it he could agree with.[25] Carnap remained supportive and sympathetic with this and all of Morris's subsequent research – the two were always good friends – but did not engage Morris's writings on religion or Sheldon's theories in any substantive way. It was not only Morris's explorations of religion and value theory that caused his drift and marginalization in the profession during the 1950s and '60s. Morris's technical semiotic opus *Signs, Language, and Behavior* (Morris 1946b) was poorly received, and by the early 1950s his work had become primarily empirical and descriptive. He joined the ranks of the many theologians, sociologists, psychologists, and humanists promoting a new intercultural "science of man" in the face of political and cultural tensions of the Cold War. With funding for several years provided by the Rockefeller Foundation, Morris pursued questionnaire-based studies of college students in China, India, Europe, and America. Morris's Sheldonian speculations about "paths of life" thus became empirical studies into what he now called "ways of life" and were intended as a contribution toward the deliberate, intelligent planning of future cultural life. Morris did not abandon Sheldon's views, however, for he always looked for correlations between respondents' stated values and preferences and their Sheldonian somatotypes.[26] Morris's postwar research therefore remained both as scientific and as cultural as it was in the 1930s. One reason to visit the "great cultures" of India, China, and Russia, one of his proposals explained, was to facilitate "the international exchange of ideas and ideals, and in that way to help create the ideas and ideals appropriate to the contemporary world."[27]

Rudolf Carnap (1891–1970): Logic and Humanism

Of the three original leaders of the Unity of Science movement, Carnap was the most emphatic about demarcating philosophy and politics:

Logic, including applied logic, and the theory of knowledge, the analysis of language and the methodology of science, are, like science itself, neutral with respect

[25] Neurath expressed his skepticism about Sheldon's typologies for bodies and personalities in (Neurath to Morris, December 28, 1942, CMP). Neurath was skeptical about not only Sheldon, but also Morris's appeal to Spengler, against whom Neurath had earlier published an emphatic critique (Neurath 1921).

[26] Morris's postwar research in values is exemplified in Morris 1948b; 1951; 1956.

[27] Morris to Walter S. Rogers, Institute of Current World Affairs, New York City, March 21, 1943, CMP. Morris did not revisit Russia.

to practical aims, whether they are moral aims for the individual or political aims for a society. (1963a, 23)

His view of philosophy as neutral with respect to politics became central to logical empiricism's postwar reputation as a strictly philosophical program. Carnap also supported that view by example. His large body of work – his early writings in philosophy of space and time, his constructivist project in the *Logische Aufbau der Welt*, the syntactic program of *The Logical Syntax of Language*, and his mature writings in semantics and inductive logic – mapped out much of what came to count as professional philosophy of science. Carnap also offered the most considered and specific accounts of what "unity of science" or "unified science" might mean for scientists and philosophers. Thus, Carnap not only helped Neurath and Morris to organize the Unity of Science movement, but also ensured that alongside the movement's programmatic announcements there remained specified models of how scientific languages might be (or become) linked to each other, or how scientific procedures (such as induction) might be formulated as generic tools for all sciences to utilize. Throughout these writings, however, there are very few elaborations of the political or social relevance of logical empiricism or the unity of science, and there are plenty of reminders that politics and philosophy are strictly different enterprises.

Still, as we will see in chapter 13, Carnap was a committed and often active humanitarian who used his intellectual reputation to advance political and social causes or assist individuals whom he believed could benefit from his efforts. Several factors help to elucidate Carnap's unlikely neutralist activism. One is his ongoing debate with Neurath about the political relevance of logical empiricism and Neurath's adamant view that the two cannot be legitimately separated. Though Carnap noted that in "private conversations" he came to "closer contact with Neurath's ideas" about Marxism and relations between logical empiricism and the world's "great historical processes" (1963a, 23–24), their sometimes stormy intellectual friendship (as discussed in chapter 10) most likely led Carnap to defend important programmatic distinctions such as this one between politics and philosophy. If he did not, he may have reasoned, Neurath would casually dismiss them and sweep them aside.

Another factor concerns issues explored by David Hollinger, namely that the political neutrality of philosophy participates in "the ideal of a universal brotherhood of honest inquirers aloof from religious and ethnic sectarianism" (1996, 15). The ideal is a reaction to anti-Semitism, which played no small role in the decisions of several logical empiricists to emigrate to the United States. As we see below, Carnap's entanglements with the McCarthyite loyalty-oath controversies at the University of California find him expressing the similar view that only academic factors should be involved in matters of academic professionalism.

Carnap's neutralism can also be understood – and qualified – on the basis of his mature division of the study of language into syntax, semantics, and pragmatics. Formal, syntactic analysis, concerning only symbols and their internal rules of combination, can be conducted independently of semantic analysis, which further takes into consideration relations of meaning between symbols and the extra-symbolic world. Both syntax and semantics, then, can be undertaken in abstraction from "pragmatic" study further involving individuals who use language for various purposes, including ethical or social aims.[28] When Carnap introduced his Principle of Tolerance – his policy that "everyone is free to use the language most suited to his purpose" (1963a, 18) provided only that the language is internally coherent – he wrote, "in logic, there are no morals" (Carnap 1937c, 52).

Carnap's point was that there are no valid a priori prohibitions in the creation or analysis of languages. Anything goes. But that absence of prohibitions does not entail that technical philosophical analysis proceeds without grounding in ultimately pragmatic goals and values. This point becomes clearest in Carnap's late formulation of "language frameworks" according to which philosophical questions internal to the framework – its syntax and semantics – are distinct from external, pragmatic questions about the framework and its utility for achieving one's goals (Carnap 1950). If one is engaged in syntactic or semantic projects that are, in Carnap's sense, independent of morals or politics, those engagements are nonetheless not free of any pragmatic or moral commitment whatsoever, for one has chosen that framework and thus participates – as Neurath insisted to Carnap – in whatever "historic processes" made that framework available. Some choices and decisions, that is, are both constitutive of one's philosophical practice ("shall I study phenomenology or philosophy of biology?") and yet determined by extra-philosophical concerns, interests, beliefs, and so on. Carnap himself acknowledged this in the preface to his *Aufbau* where political and ideological concerns are (momentarily) pushed outside philosophy on the basis of a distinction

---

[28] Carnap 1939, 146. See also Carnap 1942, 9.

between justification of a thesis (within some framework) and matters of sociology, history, and psychology:

> It must be possible to give a rational foundation for each scientific thesis, but this does not mean that such a thesis must always be discovered rationally, that is, through an exercise of the understanding alone. After all, the basic orientation and the direction of interests are not the result of deliberation, but are determined by emotions, drives, dispositions, and general living conditions. This does not only hold for philosophy, but also the most rational of sciences, namely physics and mathematics. The decisive factor is, however, . . . the *justification* of a thesis. (Carnap 1969, xvii)

Carnap's formulation acknowledges that practicing philosophy of science may be connected in this way to matters of sociology, history, and psychology. Later in the *Aufbau*, Carnap spelled out some of these connections as he saw them in his own case. The new philosophy of science, he noted, involved an attitude and feeling shared by other movements and "walks of life." The collective spirit of the new philosophy, as Carnap described it, connected philosophy of science to other areas of modern life and also helped answer to the those "needs of the heart" (*Bedürfnisse des Gemüts*) that, since Lotze, were seen as the justification of extra-scientific metaphysics. In the new, modern project, Carnap suggested, these needs find some satisfaction in "the clarity of concepts, accuracy of methods, integrity of claims, achievement through cooperation, to which the individual subordinates himself."[29]

These remarks were neither obligatory nor perfunctory for Carnap. He had long thought about these real yet obscure relations between intellectual and emotional needs and capacities. Several years before, for example, when writing to Bertrand Russell, he complimented Russell for his outspoken pacifism and socialism by posing the question,

> Is it mere coincidence that the people who achieve the greatest clarity in the most abstract area of mathematical logic are also those who oppose, clearly and forcefully, the narrowing of the human spirit by means of affect and prejudice in the area of human relations?[30]

---

[29] Carnap 1969. The translation here is from Gabriel 2004, 11, where the phrase "cooperation, to which the individual subordinates himself" emphasizes collective ideals that Carnap imbibed, Gabriel suggests, from his boyhood experiences in the German youth movement.

[30] Carnap to Russell, July 29, 1922, ASP RC 102-68-31. For more on Carnap's early relationship to Russell, both personal and intellectual, see Reck 2004. For a recent look at Russell's political activism, see Russell 2002.

Clearly, Carnap did not think this coincidence was mere coincidence, but neither did he find its underlying cause transparent. The new approach in philosophy, as he put it in the *Aufbau*, recognized that intellect is embedded in larger, obscure emotional and social contexts. The approach is one that "demands clarity everywhere, but which realizes that the fabric of life can never be quite comprehended." Perhaps that is why Carnap's confidence that scientific philosophy would prosper, that this scientific "attitude will win the future," was a matter of "faith" (ibid., p. xviii).

Morris and Neurath were not always alone as they enthused about the political and social importance of science and their movement in the 1930s. In the aftermath of the First International Congress, while teaching at Harvard before going to the University of Chicago, Carnap was interviewed by Harvard geologist Kirtley Mather in a radio broadcast. Though Mather would later become one of the country's leading intellectual radicals in the 1950s, it was Carnap who steered their discussion about logic and clarity of thinking in an overtly political direction:

Mather: In my observation, a lot of muddy thinking is due to failure to see the inadequacy of the data on which far reaching conclusion are all too often based.

Carnap: Yes, there is no doubt that in daily life this important... condition for logical thinking is frequently neglected. Men expect a future which will satisfy their hopes and desires, even when such expectations are inadequately based on observed facts. In the same way, deceived by their desires, men count on just that behavior in others which would coincide with their own needs. It is in this way that we must explain the conduct of different nations, races, and social classes, since, unfortunately, their conduct is controlled more often by passions than by reflection upon the facts of psychology and the social sciences. Their expectations, inadequately founded, are usually followed by disappointments in the behavior of other parties: but the failures of their hopes, instead of leading to the correction of erroneous assumptions, frequently become the occasions for a childish reproval of opposing groups in the name of morality. (Carnap 1936a)

Of course, clarity of thought and understanding of human psychology do not implicate any particular political agenda. In this sense, again, Carnap's project was nonpolitical. But these remarks show that Carnap cared about the potential social and political effects of his philosophical work. He believed that were citizens and their leaders not so "deceived by their desires," better informed about the scientific foundations (or lack thereof) beneath their assumptions about other people, and better able to communicate with each other, peaceful cooperation among societies and nations might become easier to achieve. Though logic and philosophy of science may remain independent of politics in Carnap's project, politics is not independent of logic and philosophy of science.

In this and other cases, especially during the Cold War, as discussed in chapter 13, Carnap did not hide his leftist political orientation. In 1939, he agreed to be a signatory of the new Committee for Cultural Freedom organized by Sidney Hook. This committee (and a parallel group dominated by *Partisan Review* editors, called the League for Cultural Freedom and Socialism) attacked communist intellectuals for slavishly following the rapidly changing dictates of the Communist Party.[31] At this time, shortly after the Hitler-Stalin pact, popular distinctions on the left between fascism and Soviet socialism had begun to dissolve: The *Partisan Review* manifesto compared Stalinism and fascism ("To the deification of Hitler and Mussolini [these intellectuals] counterpose the deification of Stalin" (Cooney 1986, 143)), while Hook's manifesto more broadly targeted "totalitarianism" and the "totalitarian idea" in various guises:

Under varying labels and colors, but with an unvarying hatred for the free mind, the totalitarian idea is already enthroned in Germany, Italy, Russia, Japan, and Spain.... Through subsidized propaganda, through energetic agents, through political pressure, the totalitarian states succeed in infecting other countries with their false doctrines, in intimidating independent artists and scholars, and in spreading panic among intellectuals.... In fear or despair, they hasten to exalt one brand of intellectual servitude over another.... Instead of resisting and denouncing all attempts to straitjacket the human mind, they glorify, under deceptive slogans and names, the color or the cut of one straitjacket rather than another. (Hook 1987, 272)

From Carnap's point of view, the manifesto aligned with his own official disjunction between politics and intellectual activity. Lending support to

---

[31] Specifically, both groups targeted the writers and intellectuals in the League of American Writers, an organization born in 1935 out of the Communist Party–sponsored International Writers' Congresses that were designed to promote Stalinism to the American public as part of the party's wider "popular front" strategies. Most alarming to Hook was Moscow's control (and the intellectuals' acquiescence) over the very content of writings. In particular, radicalism had to be toned down; "the *specific* Marxist viewpoint must largely disappear" (Cooney 1986, 81). For the *Partisan Review* intellectuals, this development threatened their vision of a literary-socialist renaissance in America, driven by authentic American literature and genuine "proletarian" novels. Besides Carnap and Hook, others in philosophy or philosophy of science among the roughly 120 that signed Hook's manifesto included John Dewey, Arthur Bentley, Percy Bridgman, E. A. Burtt, G. Watts Cunningham, Irwin Edman, Jesse Holmes, Horace Kallen, Arthur Lovejoy, H. A. Overstreet, Ralph Barton Perry, George Sabine, and the sociologist Louis Wirth. For an insider's account of the League, see Folsom 1994.

Sidney Hook's increasingly vitriolic and anti-Soviet causes, however, was plainly a political gesture. As we see in chapter 13, this would not be the last time that Carnap was engaged in political matters by Sidney Hook. Yet it was the last time that Hook and Carnap would be on the same side of the political issues involved.

Philipp Frank (1884–1966): Unity of Philosophy and Science

Had circumstances been slightly different, Philipp Frank might enjoy today the reputation he deserves as a central figure in the history of twentieth-century philosophy of science. For at least two reasons, however, Frank was not for long in the profession's spotlight. First, he remained in Europe until the fall of 1938 and did not join Neurath, Carnap, and Morris as they launched the new encyclopedia and the movement in America. Frank was, nonetheless, a co-originator of logical empiricism. Before the more famous circle around Moritz Schlick began to form after Schlick arrived at the University of Vienna in 1922, Frank, Neurath, and Hahn met regularly for several years on Thursday evenings in Vienna coffeehouses to share their criticisms of philosophy and their visions of a reformist "rapprochement between philosophy and science"(Frank 1949b, 1). Frank, Neurath, and Hahn thus formed what historians now call the "first Vienna Circle" (Haller 1991). Another reason for Frank's relative obscurity is that he exerted himself more than others on behalf of the Unity of Science movement when it was dying. Though Frank is remembered as an inspiring teacher, and some of his writings and editions have continued to attract readers, it is fair to say – for reasons explored in later chapters – that Frank and his legacy went down with the ship.[32]

Frank was formally trained in mathematics and physics at the universities of Vienna and Göttingen and studied under the physicist-mathematicians Ludwig Boltzmann, Felix Klein, and David Hilbert. His devotion to mathematics and physics was matched by his interests in the writings of Mach, Poincaré, and Duhem – interests that fueled Frank's discussions with Neurath and Hahn until 1912. That year, Frank left Vienna to succeed the chair of his friend Albert Einstein at the University of Prague (Stadler 2001, 631). Frank then moved to Prague's Institute of Theoretical Physics where he worked until his emigration to the United States in the fall of 1938. Frank was not, however, out of the Vienna

[32] Physicist Guy Emery remembers Frank's classes fondly and pointed out to me that science writer Jeremy Bernstein was moved to dedicated his book (1967) to Frank.

Circle's loop during the years of Schlick's circle. He visited his colleagues in Vienna often and in 1929 organized the influential conference on the "Epistemology of the Exact Sciences." Held under the auspices of the German Physical Society in Prague, Frank used his role as local chairperson to push through a *Tagesordnung* jointly sponsored by the Ernst Mach Society and Reichenbach's Berlin-based Society for Empirical Philosophy (Frank 1949b, 40).

Frank's recollections of this session introduce one of the dominant goals of his entire career: the unity of science and philosophy. The physicists and mathematicians of the German Physical Society to whom Frank, Carnap, Feigl, Reichenbach, Neurath, and Hahn delivered their talks "did not particularly like the idea of combining this serious scientific meeting with such a foolish thing as philosophy" (1949b, 40). For them, Frank suggested, philosophy was a sprinkling of Kantian sentiments that had little to do with scientific theory or practice. Though these physicists sat through these talks in silence, Frank was encouraged later by receiving "a great many letters from scientists who expressed their great satisfaction that an attempt [had] been made toward a coherent world conception without contradictions between science and philosophy" (ibid., pp. 40, 41; Stadler 2001, 341).

Besides promoting an understanding of the intimate connections between science and philosophy, Frank also urged that science and philosophy together be understood as properly historical and social enterprises. With Duhem and Poincaré, Frank agreed that science hardly consisted in assembling experienced or observed facts – as held by the influential caricature of Machism that Frank sought to correct and counteract – because science necessarily required conceptual structures or "conventions" that served to organize and give sense to particular empirical facts (Frank 1949b, 7–16). Because different conventions could themselves underwrite very different kinds of scientific projects, and their adoption was never strictly dictated by empirical facts at hand, Frank saw the historical course of science – indeed, its very conceptual character at any one time – as being partly determined by choices scientists made.

For Frank as for Neurath, especially, the progress and conduct of science had to be understood as intertwined with not only "the great historical processes going on in the world," but also the values and habits and patterns of life that surround and support life. Alongside observed or experimentally created facts, a scientist's values and social context help to determine his or her intellectual choices and thus the larger direction of science in society. Thus Frank was naturally led to a number of issues and activities in the American intellectual scene that help to explain his enthusiasm for the Unity of Science movement. These include sociology of science, science education, the need to combat excessive specialization in the sciences, the gap between science and the humanities (including philosophy of science), and – the topic that arguably motivated Frank most during his career in North America – the need for humanists as well as scientists to better understand the role of values in both science and in social and political discourse. As we see in chapter 11, Frank was the lone member of the Unity of Science movement who regularly attended the annual New York conferences on "Science, Philosophy, Religion and Their Relations to the Democratic Way of Life." Beginning in 1940, these conferences, organized originally by Rabbi Louis Finkelstein and the Jewish Theological Seminary in New York, were a regular feature of the American postwar intellectual landscape (Beuttler 1997). For Frank, as a self-appointed ambassador of philosophy of science, they were an ideal forum in which he could forge intellectual connections with a broad array of intellectuals and tirelessly debunk the popular stereotypes of "positivism," value-neutral science, and "relativism" (popularly associated with Einstein's theories of relativity) that circulated among many intellectuals after the 1930s (and that still circulate today). Against the view that science properly stands apart from culture, history, statecraft, and all things humanistic, Frank urged attention to, and further study of, the mutual importance and connections among them.

The Socio-Political Agenda of the Unity of Science Movement

Based on these brief sketches of Neurath's, Morris's, Carnap's, and Frank's projects, one can see how their collaboration in the Unity of Science movement was facilitated by broad areas of sympathy among them, despite their differences and occasional arguments. Before they first met in 1934, Morris, on the one hand, and Neurath, Frank, and Carnap, on the other, shared a vision of a new, scientific philosophy that had the power to reform philosophy and – through unification and fighting overspecialization – science itself. Western society and culture, in turn, could benefit from the fruits of a scientific world conception that promised liberation from anachronistic and epistemologically corrupt beliefs and that would utilize science to achieve a world more stable, peaceful, and tolerable. Their Unity of Science movement as it thrived for the last half of the 1930s was an attempt to promote these possibilities to other intellectuals and educated, interested citizens and to help realize them by promoting cooperation and dialogue among specialists in different fields of science. If these sketches of its leaders' projects help to establish that their movement was as much political as intellectual and scientific, then there should be less doubt that this was so in light of the ways that the movement was received in the highly politicized intellectual milieu of New York City in the 1930s.

3

Leftist Philosophy of Science in America and the Reception of Logical Empiricism in New York City

A philosophy of science that aimed to be engaged in matters of politics and progressive social reform was not unique to the Vienna Circle and the Unity of Science movement in Europe. Prior to and simultaneous with the emigration of the movement to America in the 1930s, several different groups and camps of philosophers – both indigenous and exiled from Europe – pursued reformist and progressive programs that sought, as Marx famously wrote, not only to understand the world but to change it.

The purpose of this and the next three chapters is to survey the landscape of radical philosophy in 1930s America in order both to understand how logical empiricism and the Unity of Science movement were received in North America and to establish a context for understanding how left-leaning projects such as the Unity of Science movement came to be later rejected after the war by an academic establishment that had moved to the political right.[1] In particular, because this landscape of the 1930s was inhabited at one extreme by philosophers who proudly and aggressively proclaimed that the kind of philosophy they practiced was communist, and because all things communist came to be openly persecuted in American society in the 1950s, this survey will help to identify some of the political baggage that later accrued to the Unity of Science movement by association or proximity.

---

[1] Other useful surveys of the American intellectual left in the twentieth century include Aaron 1961 and Diggins 1992.

A Survey of the 1930s Philosophical Left

In contrast to the depression-era politics of the right, which was isolationist, free-market capitalist, antisocialist, and theologically conservative, the philosophical left of the 1930s was broadly united by values and beliefs that were internationalist and nonisolationist, liberal (if not socialist), scientistic and atheistic, modern, and, above all, antifascist. Not surprisingly, however, even those who endorsed the same concepts or slogans could contest the meanings proper to them. "Progressive reform" spoken of by a liberal democrat and a member of the Communist Party would most likely mean two different kinds of things.

To sort out some of these differences, leftist philosophers and the journals they championed (or that championed them) can be distinguished according to three axes or variables: first, their degree of political sympathy with Moscow and the official Communist Party of the U.S.A.; second, the extent to which they viewed philosophy as an integral part of their political outlook and goals; and, third, the depth of their commitment either to Marxism as an empirical science of society or to some version of dialectical materialism, Marxist-Leninism's metaphysical core. When these different considerations are superimposed, four more or less different camps appear to mark this philosophical and political terrain. Moving from the moderate to the extreme left, they are here called the liberal, pragmatic left; the socialist left; the radical academic left; and the radical communist left. As this and the next chapters show, logical empiricism and the Unity of Science movement were warmly received by many in the first two camps but criticized by the radical academic left and even more firmly rejected by the communist left.

The Liberal Pragmatic Left

First and closest to the political center lay the liberal, democratic, and socialist-friendly pragmatists in New York City, most of whom descended intellectually from Morris Cohen, John Dewey, or both. Owing to his extraordinary longevity and activity, Dewey himself belongs in this camp. Other influential figures are Dewey's student Sidney Hook, Cohen's student Ernest Nagel, and Horace Kallen, who was a student of William James. These philosophers were the leaders of the philosophical wing of the famous New York Intellectuals and, as discussed below in this chapter, they eagerly received Neurath, Carnap, and other logical empiricists when they emigrated to America or, in Neurath's case, visited New York City. In part, this warm reception reflected a shared commitment to scientism. Both groups embraced science as the epitome of knowledge and a tool – quite likely the only one available – to enlighten the public (and certain misguided intellectuals), improve modern life, and secure science and modernity against those who would destroy or eliminate it. Commitments to Marxism varied in this camp. Many who were enamored of it in the early '30s later came to be suspicious of Marxists, if not Marxism itself. Despite these differences, however, this camp shared commitments to philosophy as an independent, self-sustaining critical activity and to science and the importance of science for broader society. They contributed to philosophical journals, such as the *Journal of Philosophy*, the new *Philosophy of Science* founded by William Malisoff in 1934, and the Unity of Science movement's new *International Encyclopedia*, and sometimes contributed to more overly political journals, such as *Partisan Review* and *Science & Society*.

The Socialist Left

Second and farther to the left lay several philosophers who admired logical empiricism and the Unity of Science movement at least as much for their potential contributions to leftist radicalism as for their innovations in epistemology and philosophy. These philosophers were generally more committed to Marxism (in some form), and they took central themes of Marxism, such as class struggle and social and economic planning, to be crucial concerns for any adequate philosophy of science. Examples include William Gruen, who praised logical empiricism while introducing it to readers of *Partisan Review*, and John Somerville, who wrote for Malisoff's *Philosophy of Science*. Both enthusiastically promoted the movement as a tool for both improving Marxist theory and criticism and uniting (and thus empowering) intellectuals in their quest for a more socialist, internationalist world.

The Radical Academic Left

If *Partisan Review* and *Philosophy of Science* were platforms for philosophers who were (respectively) more or less engaged with Marxism, the journal *Science & Society* was a platform for Marxists engaged with philosophy and science. It was founded in New York in 1936 by an international group of intellectuals, many of them philosophers, who comprise a third camp, still farther to the left than these first two. Its founders and editors included Moritz Schlick's student Albert Blumberg, V. J. McGill, and Margaret Schlauch in America; J. D. Bernal, Lancelot Hogben, and Joseph Needham in Great Britain; and the American geneticist H. J. Muller, then working in the USSR. These scientists, philosophers, and historians were more convinced of the basic truth of Marxist theory than their less partisan colleagues on the left and they conceived their new journal as a tool to convert them. Those unsure about Marxism's social program, they wrote, "will find the answer to their doubts through their participation in this journal." They also hoped that *Science & Society* would reach beyond the academy and "solicit the active cooperation of all who are interested in its [Marxism's] progress."[2]

The philosophers around *Science & Society* agreed that science was a powerful, potential tool for social progress and that unifying the sciences was a worthy if not necessary undertaking. But they insisted that these goals be coupled to socialist politics and a metaphysics of dialectical materialism. Many, therefore, could not abide logical empiricism's rejection of metaphysics and its dismissal of one of the radical left's most celebrated metaphysical debates – Lenin's defense of materialism against Mach's (alleged) idealism – as an empty, time-wasting pseudo-question. In this regard, *Science & Society* was joined by other journals and monographs highly critical of logical empiricism and the Unity of Science movement. These include the volume *Philosophy for the Future*, edited by V. J. McGill, Marvin Farber, and Roy Wood Sellars, Max Eastman's journal *The New Masses*, and Max Horkheimer's *Zeitschrift für Sozialforschung*. These projects had different agendas for how Marxist theory and dialectical materialism were best to be understood, but they shared a dual commitment to Marxism and to philosophical analysis – a dual commitment that would result, they all hoped, in better, more compelling, convincing, and practically effective articulations of Marxist theory in the future. As the editors of *Science & Society* explained in their first editorial, they saw "Marxism as a dynamic, ever-developing movement of thought to which many minds must contribute."[3] In this camp, that is, Marxism was no sacred body of doctrine, but rather a promising theoretical program yet to be fully developed and applied to modern life.[4]

[2] Unsigned editorial, *Science & Society*, 1936, vol. 1, pp. i–ii.
[3] Unsigned editorial, *Science & Society*, 1936, vol. 1, pp. i.
[4] See Schrecker 1986 for a similar, if slightly more agnostic, assessment of *Science & Society*'s independence from party influence.

The Communist Left

This distinction introduces the most extreme group of philosophers on the Left – those who wrote and edited the journal *The Communist*. These philosophers issued similar criticisms of logical empiricism and the Unity of Science movement, but they distinguished themselves by insisting that Marxism and its social agenda required no creative interpretation on the part of intellectuals. Marxist enlightenment lay at hand in the writings and interpretations of Marx, Engels, and Lenin. In addition, the radical philosophical left's dual commitment to philosophical analysis and materialist metaphysics was collapsed in this camp, for it recognized no important distinction between philosophical practice and Communist party life. The party itself, these philosophers believed, was the living, breathing fusion of theory and practice wherein were understood and implemented Marxist-Leninist insights about class struggle and the evolution of society and nature.

Despite these differences, there are several instances of continuity or overlap among these camps. *Partisan Review* editors Philipp Rahv and William Philipps wrote an article for the second issue of *Science & Society*, while V. J. McGill straddled the second and third camps by both criticizing logical empiricism for political and metaphysical nonpartisanship and calling for closer cooperation between the Unity of Science movement and committed Marxist philosophers. Other figures remain best unassigned because their careers looped through (or near) these camps at different times. Albert Blumberg was an early philosophical champion of logical empiricism who helped to found *Science & Society*, dropped out of university life to work within the Communist Party as an administrator and teacher, and finally returned to academic philosophy as a professor at Rutgers. William Malisoff wrote philosophical books and articles, built up the profession by founding the journal *Philosophy of Science* and the early Philosophy of Science Association, and secretly met with KGB agents who were interested in his scientific expertise and his radical political contacts. Still, the cases of Blumberg and Malisoff support of this four-part taxonomy of the philosophical left. Partly for legal reasons, to be sure, but also because these camps were seen as robust and exclusive, neither Blumberg nor Malisoff *openly* combined their political and intellectual activities.

Partisan Review, Trotsky, and the Emergence of Anti-Stalinism

While this landscape provides a rough orientation and context for tracing the political career of logical empiricism, it necessarily fails to capture the dynamic, sometimes tumultuous, history of the intellectual left – a history marked by changing alliances among intellectuals, abrupt policy changes in Moscow, and shattering illusions about Stalin and Soviet projects. Some of these changes and events will be introduced as necessary in the subsequent chapters. One, however, is basic to the contours of the intellectual left: the exile and persecution of the Soviet revolutionary Leon Trotsky.

To most progressive, leftist intellectuals, Trotsky was important because the Soviet Union was important. In the wake of the calamity and tragedy of the First World War, the events of October 1917 were captivating and enigmatic. For the revolution that turned a mostly rural, agrarian society into a modern communist society seemed to defy the Marxist prediction that communism could arise only in the wake of capitalism's collapse. If history had taken a short cut, many supposed that the path was found by the Bolshevik intellectuals, of whom none was more impressive than Trotsky. Max Eastman described him in 1925 as a "genius" with "superior moral and intellectual revolutionary greatness" (quoted in Diggins 1992, 122). As John Patrick Diggins read the situation,

If Russian intellectuals could create a revolution in an agricultural country, what could stop American intellectuals from doing the same in an industrialized society? To America's Left, dejected and rendered powerless by the war, Lenin's stunning achievement was Marx's second coming. (ibid., p. 107)

American intellectuals were at the top of their game in the 1920s and '30s. The Bolshevik revolution showed how much knowledge, theory, ideas, and methods mattered crucially in the course of world history. They mattered as much as – in some ways possibly more than – statecraft, armies, and machine guns. As a result, intellectual life in New York in the '30s was so politicized that, according to film critic and intellectual Robert Warshow,

virtually all intellectual vitality was derived in one way or another from the Communist Party. If you were not somewhere within the party's wide orbit, then you were likely to be in the opposition, which mean that much of your thought and energy had to be diverted to maintaining yourself in opposition.[5]

The main figures in the scene Warshow recalled were the so-called New York Intellectuals, an influential group of philosophers (such as Dewey, Hook, Nagel, Kallen, Meyer Schapiro, and Max Eastman), literary critics, and writers (such as Lionel Trilling, Edmund White, Granville

[5] *Commentary*, December 1947. Quoted in Hook 1987, 136.

Hicks, James T. Farrell, and John Dos Passos) whose careers – not unlike Neurath's in Europe – were sometimes as much political as intellectual and academic. Writers reached for political engagement through organizations such as the John Reed Club and the Communist Party–sponsored League of American Writers. The New York Philosophers in some cases crossed fully over the line into leftist political activity and commentary. In the early 1930s, Sidney Hook, for example, helped to found the short-lived American Workers Party (whose mission was to cultivate an indigenous American counterpart to Bolshevism). Hook's teacher, Dewey, had earlier reported approvingly of revolutionary developments (especially in education) in a report of his travels to Russia (Dewey 1929). Dewey's approval surprised no one, given his long interest in progressive and radical causes. In the 1890s, before and during his chairmanship of the philosophy department at the University of Chicago, Dewey established his influential Laboratory School, lectured at Jane Addams's famous settlement house (Hull House), and later confessed that debates with Addams over how to interpret and respond to Chicago's violent Pullman strike (taking place not far from Dewey on Chicago's South Side) triggered fundamental revisions in his Hegelian understanding of social dialectics (Menand 2001, 306–14). Dewey's student Max Eastman took a similar path. He taught philosophy at Columbia University, edited the leftist magazine *Masses,* and became a party insider partly through his friendship with Trotsky (Diggins 1992, 121).

This continuity and mutual influence of intellectual and political life was reflected in *Partisan Review.* Run by young, mostly Jewish intellectuals, *Partisan Review* prized cosmopolitanism and internationalism, excoriated fascism in Germany, Spain, and Italy, and promoted a vision of modern life suffused with literary values. The journal also found an ally of sorts in the Unity of Science movement insofar as it also embraced a scientific, specifically Marxist-scientific, orientation in the world. *Partisan Review* was, its subtitle of the 1930s made plain, "A Quarterly of Literature and Marxism" that offered its readers, as one historian emphasized, a unified outlook that answered simultaneously to their literary, cultural, and political goals.[6]

The intellectual left and its journals paid close attention to Trotsky and his plight. After Lenin's death in 1924, struggles among the Bolsheviks were largely won by Josef Stalin, who consolidated his power

---

[6] Cooney 1986, 62. On the much contested history of *Partisan Review* and its official relationship(s) to the Communist Party, see Aaron 1961, esp. 297–303.

by persecuting his rivals. Trotsky was exiled in 1929 and spent the rest of his life moving around Europe and, finally, in Mexico as he and his many followers planned his "Fourth International."[7] Trotsky's second coming, he and his champions believed, would restore to the revolution both the integrity of Trotsky's intellectual talents as well as the socialistic and humanitarian ideals trampled by Stalin's dictatorship.

For the philosophical left, especially, Trotsky's case assumed center stage after the so-called Moscow Trials of 1936 and 1937 during which Stalin denounced and indicted Trotsky and nearly all of his former October revolutionaries for various kinds of treason. The trials led to exiles, executions, and purges throughout Soviet intellectual society and were impossible to ignore in New York City. But they did not elicit uniform responses. Some accepted the dictates of Moscow and saw Trotsky as a counterrevolutionary villain who deserved his condemnation and exile. Others remained agnostic and took refuge in the difficulty of separating fact from rumor during such volatile times. Others posed the question of whether noble socialistic ends might possibly justify Stalin's despicable means. Still others accepted no excuses or rationalizations for Stalin's behavior and concluded that he was simply a selfish, ruthless dictator who had betrayed the revolution. While this and other shocks to the left (such as the Nazi-Soviet nonaggression pact of 1939) led some intellectuals to withdraw from political engagements altogether, others saw these events as a call to arms and became angry and aggressive. One of their leaders was Sidney Hook.

Never one to back down from a challenge or a fight, Hook organized a publicity campaign to defend Trotsky against Stalin's charges. He helped to persuade Dewey, in turn, that only a figure of Dewey's stature could credibly and authoritatively investigate the charges and issue a public statement about them. Thus, in April 1937, Dewey and the Dewey Commission, as it was known, traveled to Coyoacan, Mexico, where the famous muralist Diego Rivera and his wife Frida Kahlo had offered refuge to Trotsky and his entourage. Dewey and the others interviewed Trotsky and found it obvious that Moscow's charges against him were empty. Trotsky, they explained in their official report, had been framed.

The affair and the assassination of Trotsky in 1940 helped to solidify the schisms and factions on the left. The magazines *The New International*

---

[7] A useful account of Trotsky's years in exile can be found in Feferman 1993, a biography of Trotsky's one-time assistant and bodyguard, philosopher Jean van Heijenoort.

and the *Fourth International* upheld Trotsky and his influence in organized labor, while *Partisan Review* and the journal *New Leader* were fora
in which Trotsky's admirers, or former admirers, articulated an anti-
Stalinist left. For those who opposed Trotsky, on the other hand, the
Dewey Commission and its supporters were seen as traitors to the cause
of socialism or dupes of bourgeois and capitalist propaganda. In the
eyes of the communist philosophical left, Dewey, Hook and the pragmatism they represented were alike denounced as incoherent and socially
reactionary.

The Reception of Otto Neurath in New York City

When logical empiricism and the Unity of Science movement came to
America beginning in the mid-1930s, they were received by an intellectual
culture that was almost routinely brimming with controversies such as the
those surrounding Trotsky. Late twentieth-century divisions between politics and philosophy are blurry and ineffective in this climate, a point made
clear by the correspondence between Ernest Nagel and Otto Neurath.
When Neurath first visited the United States in the autumn of 1936, he
was hosted by Nagel, who had met Neurath personally in Holland a year
before. Like Charles Morris, who had also traveled to Europe to introduce himself to the logical empiricists, Nagel was young and somewhat
starstruck by his association and friendship with the leaders of the new
scientific philosophy.[8] After a visit with Neurath in Holland, he profusely
thanked him when writing him goodbye from Paris. Nagel was thrilled
to have met Neurath and "the various men whose writings I have been
reading since I have come to intellectual maturity."[9]

Nagel made no secret about America's debt to these Europeans when
it came to philosophy of science. When Neurath wrote to ask Nagel for

---

[8] Morris's delight and humility about the new collegiality between American and European
philosophers is reflected in his "Opening Speech (for the American Delegates)" at the
First International Congress for the Unity of Science in Paris in 1935. He said, "[We]
appear before this congress not as beggars – though we do have much to learn and
overemphases to correct – but as representatives of those who have helped to sow the
seeds from which this congress itself is but one growth. The basic matter is not of course
one of giving or receiving: what is unique here is the development of a cooperative attitude
(as science is cooperative) to our common problems.... [All this] will prove in time to
be characteristic notes of a distinctive and important period in the history of science and
philosophy" (Morris 1936, 22).
[9] Nagel to Neurath, July 23, 1935, ONN.

an overview of the American philosophical scene, Nagel responded with
a list of some thirty names and a blanket qualification:

> In judging American tendencies I think it must be borne in mind that the techni-
> cal formulation of issues and the concern with the content of scientific method
> which characterize your group, are often lacking. But in spite of these and other
> differences, I think the list I am sending gives some idea of what is going on in
> directions which are of interest to you.[10]

When Neurath arrived in New York in the autumn of 1936, Nagel continued to guide him through the social and intellectual scene. After one
Saturday-evening gathering of philosophers in October – a reception for
Neurath, it would appear – Neurath asked Nagel to write up another
inventory of those he had just met. There was no need for Nagel to describe William Malisoff, for Neurath had already met him. So Nagel provided sketches of Hook, Meyer Schapiro, and Abraham Edel ("The above
three men are particularly good friends of mine"), as well as J. V. McGill,
Y. Krikorian, Daniel Bronstein, Albert Hoftstadter, William Gruen, Phillip
Wiener, Herbert Schneider, John Herman Randall, Jr., Horace Friess, and
John Allen Irving.[11]

As Nagel wrote his inventory, he knew Neurath was always on the look-
out for talent both intellectually and politically compatible with the Unity
of Science movement. This is most likely why his sketches included a brief
mention of each philosophers' politics. Some were "liberal with socialist
leanings," one had "materialist leanings," and one had "some sympathy
with some of the practical achievements of Italian fascism." Most, Nagel
wrote, "were Jewish and I think without exception have left sympathies
in politics."[12]

Neurath had little trouble winning them over. He persuaded Dewey,
who was put off by the noncognitivist view of ethics and values that he
and others associated with logical empiricism, to contribute twice to the
new encyclopedia. Neurath also became fast friends with Horace Kallen
and Sidney Hook, who were arguably the most politically outspoken
philosophers in New York and would later become the most outspoken
anticommunists. Hook offered to help translate and publish some of
Neurath's writings and attempted (with Nagel) to find Neurath a position at the New School for Social Research. Later in 1940, when Neurath
and his new wife Marie were interned as prisoners of war in England after

---

[10] Nagel to Neurath, October 14, 1934, ONN.
[11] Nagel to Neurath, October 13, 1936, ONN.
[12] Ibid.

fleeing Holland, most of these friends and colleagues sent them money to help reestablish a home and base for their ISOTYPE operations.[13] In one of his letters to Neurath, Hook excused himself for not writing at length and reassured Neurath that he and his colleagues thought often and well of Neurath.[14] The colleagues Hook had in mind probably extended beyond philosophers. That same year Neurath exchanged letters with novelist James T. Farrell, who had asked Neurath about the fashionable theory that Nietzsche had paved the way for the rise of Nazism.[15]

Neurath was not the only link between the Unity of Science movement and the New York philosophers. Morris was a devoted admirer of Dewey and maintained an intermittent correspondence with him. He also met often with Kallen, Hook, and Nagel during the years before the war.[16] Carnap and Morris joined Neurath in corresponding with Dewey about his two contributions to the new *International Encyclopedia* while Carnap and his wife Ina became frequent visitors and friends of Hook and Meyer Schapiro.[17]

The Unity of Science Movement and the Liberal Pragmatic Left

These personal and intellectual connections among the leaders of the Unity of Science movement and the New York Intellectuals were complemented by other personal and institutional connections between the two groups. Several of those Neurath asked to write for the new encyclopedia were contributors to *Partisan Review* and *Science & Society*. These included Nagel, Dewey, and Abraham Edel, who did write for Neurath, and Hook, Meyer Schapiro, and the British biologist Lancelot Hogben, who ultimately did not.[18] J. V. McGill of Hunter College, who was present

[13] Nagel mentions to Neurath that he and Waldemar Kaempffert requested Dewey's help to arrange an academic position for Neurath in the United States, and that Nagel planned to consult with the New School for Social Research (Nagel to Neurath, April 20, 1938, ONN). Neurath thanks Hook for his loan in Neurath to Hook, September 14, 1941, ONN.

[14] Hook to Neurath, June 27, 1939, ONN.

[15] Farrell to Neurath, April 16, 1939; Neurath to Farrell, July 5, 1939, ONN.

[16] Morris reports his contacts with these figures to Neurath in Morris to Neurath, March 3, January 24, 1940; January 31, 1943, ONN.

[17] Nagel mentioned this relationship in Nagel to Neurath, October 13, 1936, ONN.

[18] Nagel and Edel wrote book reviews in *Science & Society* (Nagel 1936; Nagel et al. 1937; Edel 1939). Schapiro never completed his monograph, which was to be about art and art criticism. In 1938, Neurath invited Hook to contribute a piece on "dialectics" in a future volume of the *Encyclopedia* (Neurath to Hook, February 18, 1938, ONN).

at the Saturday-night reception mentioned above, was surely interested in meeting Neurath because he had just written an article for the first issue of *Science & Society* that introduced this "most vigorous and influential" new movement on behalf of the unity of science to readers of the new Marxist journal (McGill 1936, 45).

Another young philosopher at the gathering, William Gruen, had just taken a seminar from Carnap at Harvard. Three years later, he published a substantial piece in *Partisan Review* titled, "What Is Logical Empiricism?" in which he praised logical empiricism and the Unity of Science movement for their intellectual, scientific, and political ambitions. The new encyclopedia, he wrote, would be "a cooperative work which promises to be one of the most important events in modern intellectual history," in part because "the philosophy of *unified science* has special bearing on social problems":

It widens the domain of scientific method to embrace all intellectual and practical enterprise. And in its anti-metaphysical methodology it constitutes a challenge not merely to traditional, speculative philosophy, but to every form of transcendentalism in the social sciences.

In Gruen's eyes, logical empiricism could do no wrong. It could clean Marxism's own stables of the "metaphysical conceptions" lurking within dialectical materialism and other programs that at best "obscure the practical issues between diverse philosophies of society" (Gruen 1939, 65). In much the way that Carnap saw logical syntax (and the Principle of Tolerance) as a means of reducing spurious arguing among philosophers, Gruen hoped that logical empiricism might reduce theoretical squabbling among intellectuals and critics and, in turn, enhance their collective, practical action. He even argued that logical empiricism was a valuable tool for literary and artistic criticism and outlined how it could help sharpen analyses of paintings and poems, even works of fiction (ibid., pp. 66, 68–69, 72). If Marxists brushed up on their physicalism, and art critics were better acquainted with logical reducibility of statements to observables, Gruen explained, then logical empiricism's "full advantages" could find "realization in the field of esthetics, ethics, and political thought" (p. 77).

Although he was the most exuberant, Gruen was not alone in seeing logical empiricism as having an important role to play in *Partisan Review*'s agenda. In a piece appearing a few issues later, editor Philipp Rahv again addressed questions about the intellectual integrity of dialectical materialism and invoked logical empiricist ideas as if he had not only edited but studied Gruen's introduction. Much like Carnap's or Neurath's early critiques of statements that were unverifiable and therefore metaphysical, Rahv criticized the metaphysical core of Marxist history for being "in no sense subject to experimental verification":

> In the writings of the classic Marxists it functions not as scientific method but as a source of metaphors relating to the ideas of change and transformation, and as a peculiar and often quite effective syntax of dramatic discourse. (Rahv 1940, 178–79)

Writing at a time when many disappointed and disillusioned intellectuals were re-evaluating their earlier commitments to Marxist theory, Rahv knew that many of his readers were considering abandoning it altogether. But that was not the right response, he explained, for Marxism's empirical components "have retained their vitality." There was, however, "diseased tissue" that "scientifically minded" criticism needed to cut away (ibid., pp. 177, 179).

While Rahv used logical empiricist ideas to separate healthy from diseased parts of Marxism, other authors in *Partisan Review* used them to distinguish worthwhile from worthless parts of the popular "general semantics movement."[19] Albert Wohlstetter and Morton White critically reviewed the writings of S. I. Hayakawa, Alfred Korzybski, and others to conclude that these "amateurs in semantics" used it "as a more or less get-rich-quick scheme for intellectual success in the social sciences." "Serious exponents of the study of meaning," on the other hand, "are concentrated for the most part in the Unity of Science Movement" (Wohlstetter and White 1939, 51, 52). Carnap, Tarski, Lukasiewicz, Philipp Frank, and Joseph Woodger toil genuinely in "the science of semantics," while the many pretenders "have not advanced social science one whit by their inept exploitation of the theory of meaning" (ibid., p. 57).

Given the intellectual and ideological sympathies among the New York philosophers and the logical empiricists in Neurath's movement, it was perhaps not difficult for Neurath to recruit several of the New Yorkers into the ranks of his new encyclopedists. Besides Dewey's two contributions (Dewey 1938; 1939), Nagel wrote a monograph on probability (Nagel 1939). In a few years, Meyer Schapiro would be on board to write about art and literary criticism and, in the late '50s, Abraham Edel was enlisted to write about science and ethics.

---

[19] This movement was so popular and enduring that Alfred Hitchcock scripted Tippi Hedron to remark (in *The Birds*) that she's taking "a course in general semantics at Berkeley."

Otto Neurath in the Popular Press

While *Partisan Review* promoted the Unity of Science movement in New York's politicized intellectual circles, popular publications covered it for a broader public. In this regard, Neurath was blessed with a cousin, Waldemar Kaempffert,[20] who was the *New York Time*'s science editor. Neurath appeared in at least two major articles – one in 1937 about Neurath and the new *Encyclopedia* and, a year later, in a review essay about the first two monographs. Both articles explained that the *Encyclopedia* was primarily a scientific project and that it would operate as a forum for scientists and philosophers to collaborate and collectively to build bridges connecting different parts of science. The goal was "to integrate the sciences so to unify them, so to dovetail them together, so that advances in the one will bring out advances in the other" (Kaempffert 1937b). This would happen, however, without invoking or creating any "super-science to legislate for all the known disciplines" (Kaempffert 1938). Thus Neurath's famous metaphor of science as a ship at sea appeared in Kaempffert's coverage:

In the past [scientists] have been independent navigators who paid little attention to whistles, flags, semaphores, wireless and other means of intercommunication. The encyclopedia is intended to tell each navigator what other ships are doing and what he can learn from their signals, their movements, their errands. A heterogeneous collection of vessels on a common ocean, but each going its own way, is to be converted into a homogenous fleet. [But] there will be no admiral to give orders. Only this encyclopedia is to serve as chart and compass.[21]

Where Neurath used the metaphor to emphasize that science, depicted as one ship, does and must proceed without metaphysical or epistemic foundations, Kaempffert's version (most likely written with input from Neurath) made a similar point by denying that there would be an "admiral to give orders." Understood as an attempt to plan the future of science, in other words, prospective plans would be created from within science and disseminated with the help of the *Encyclopedia*. They would not, emphatically, be imposed on science from without. Neurath made a

---

[20] Initially a writer for *Scientific American*, Kaempffert became the *New York Times*'s editor of science and engineering in 1927 and remained into the 1950s one of the most prominent science writers in the United States.

[21] Kaempffert 1938. On Neurath's metaphor of a science as a ship being continually rebuilt while at sea, see Cartwright et al. 1996, part 2.

similar point elsewhere: "Our program is the following: no system from above, but systematization from below."[22]

Kaempffert also emphasized that the project's scientific goals were socially and politically valuable:

Indeed this work must serve a higher purpose than that of amalgamating the sciences for the benefit of specialists. The men who are making this encyclopedia have no desire to enter the political arena but every desire to influence intellectual leaders. (Kaempffert 1938)

By influencing science and science education and producing a reference work accessible to "the educated public" (ibid.), the project would respect and nourish "a connection between science and everyday life" (Kaempffert 1937b). And it would do so on an international scale, as announced by its title. Nodding to growing worries about the rise of fascism in Europe and, in particular, the fascist view that science was properly a servant of national interests, Kaempffert saw the project as a servant of science's interests. It should "help insure [*sic*] freedom of scientific inquiry everywhere" (Kaempffert 1938).

In an article he wrote for *Survey Graphic*, a progressive magazine dedicated to "social interpretation" (as its subtitle put it) and pictorial documentation of social life around the world, Kaempffert explained that the movement's congresses and its *Encyclopedia* were indeed situated in "the political arena" – a Marxist arena concerned with "democratizing knowledge" and making both scientific research as well as the movement's own congresses available to "the common man":

In democratizing knowledge it is essential that science is won over. The laboratory workers constitute a modern priesthood which has thus far held itself aloof from the world. Here they are, little groups of specialists studying infinitesimal aspects of the cosmos – biologists experimenting with fruit flies. . . . They pay little attention to one another, these specialists. Yet they could make science far more powerful than it is, socially powerful, if they could be brought together in some comprehensive union.

The Unity of Science movement would help to remedy that situation by bringing together scientists and philosophers at the International Congresses in order for them to begin charting a more unified science. But there remained a further problem that the new encyclopedia would help to solve: "Congresses reach only the few who have the time and the money to travel. Hence the 'Encyclopedia for the Unification of Science' which the University of Chicago Press is now publishing" would reach out to all those who might be interested in the historic project. "It is the common man with whom he [Neurath] is concerned" (Kaempffert 1939, 540).

Kaempffert's was not the first celebration of Neurath and his work in *Survey Graphic*. Three years before, a two-page editorial titled "Social Showman" introduced the famously large-framed Neurath to readers as the "Big Man who created the little man" – the idealized human figure used throughout Neurath's ISOTYPE figures (see Fig. 3). Neurath's ISOTYPE work was naturally of interest to a magazine emphasizing visual information and communication, but there was more than aesthetic concerns driving this article's almost hyperbolic praise of Neurath and his projects. Neurath was a showman, it declared, who "inspired and formulated the most extraordinary designs ever used to give life to statistics, geography, natural resources and social forces." "His theatre is the world," it concluded, "and all of us are the actors. Few men of our time have laid their hand so close to the dramatic plot, elusive as a gypsy trail, that marks our destiny on this planet" ("Social Showman" 1936, 619).

Articles in *Time* magazine were less fawning, but still painted Neurath as an engaging socialistic champion of the common man. In a note reporting both about Neurath's ISOTYPE-based book *Modern Man in the Making* (Neurath 1939) and the Fifth International Congress at Harvard University, *Time* introduced Neurath as a "bald, booming, energy-oozing sociologist and scientific philosopher" on a quest to promote a unifying language for the different sciences ("Unity at Cambridge" 1939). This writer most likely plucked this imagery from an article a year before that explained that "hulking, booming Otto Neurath, who gives the impression of oozing vitality from every pore, is a social scientist of international distinction" as well as an activist with "strong socialist leanings in politics" ("Toward Unity" 1938).

Other Battles, Other Alliances: Hutchins, Adler, and Morris

The bonds and sense of shared mission between the New Yorkers and the Unity of Science movement were also nourished by their common enemies in the intellectual landscape of the mid-1930s. Although Morris had made the University of Chicago a center of the movement, it was also the home of Mortimer Adler and his friend Robert Maynard

Social Showman

all, a mathematician, disciplined in the natural sciences.

In bringing about the "renascence of hieroglyphics" Neurath wrested from the hierarchy of his own scientific colleagues their monopoly of learning. Our civilization, he feels, is still under the sway of a Middle Ages pattern in which a word language is the property of one class alone. In the Middle Ages, it was the monks with their Latin. Today it is the scientists whose polysyllabic books are over the heads of most of us. Yet if democratic cooperation toward the solution of complex problems is not to fail, we must all understand the great forces which affect our lives.

To Neurath the pie-chart, the bar-chart, the fluctuating lines of the graph-makers and the map makers are almost as inadequate as fancy words. Neurath is never without a huge pencil rooted to his hand to chart the ideas that transcend words. Not that he is inarticulate. A sociologist and the son of a sociologist, a logician and the husband of a logician, he knows and speaks the language of the scientists. In conversation, especially in German, his vocabulary carries all the nuances of ideas and images. In his English, which he has sharpened into the colloquial by reading dozens of detective stories on his voyages to America, he is precise, exact. But no matter how fluently one may describe something, he feels that to be fully understood, it must be visualized. Back of the automobile and the skyscraper, for example, lies r.p.m.s. and complex mathematical formulas, translated into action by the engineers. Back of Neurath's pictures and museum exhibits lie profound research, statistics transformed into ideas, ideas then designed into a picture narrative, a drama of social interpretation.

THE Big Man who created the Little Man is with us from overseas this fall. Though the two of them have not been heralded on Broadway, the Little Man is a sound actor—and Otto Neurath is his impresario. The significant Little Man, who represents a hundred thousand or maybe a million of us, made his first appearance in America in the pages of *Survey Graphic* in 1932. Row on row like the chorus of a review, or among the symbols of the things men buy and sell and eat and use, or perhaps at a halting place in man's long pilgrimage from savagery to peace and plenty, he has become the hero of act after act in which he may not even enter the stage at all. In business roles he signifies the man earning over $2500, or owning an automobile, or belonging to country clubs; or—in the field of social welfare—having a job, the toothache, tuberculosis, naturalization papers, or no job, or what not. But he is not really so elementary as all that. The Little Man was begot in a bed of statistics, trained by economists, and styled by imaginative designers. The true secret of his significance, and of the significance of all Neurath's methods, is that Neurath, the social showman, is also a philosopher. Authenticity in research, and precision in visualizing the finding of the experts, do not cramp the style of Neurath's large mind.

Like the man in the moon, whom he resembles when his great countenance is reflective, Neurath's orbit swings clear round the earth. He sets the imagination on fire as Van Loon, or H. G. Wells, sometimes does simply by giving us a surprising glimpse of ourselves in a moving social procession. But with this difference. Neurath is never carried away by his fancies, or his day dreams, into conjecture. He is, first of

NEURATH is a practical man. For years he has worked with research and educational organizations in half a dozen countries besides the United States. He created and directed the famous Gesellschafts-und Wirtschaftsmuseum of his native city of Vienna. It was in that museum that he blossomed out as a showman, the Barnum of man's fate and hope.

That was in 1924. Neurath was then, at forty-two, a mature social scientist of growing repute; his household was a center of practical as well as intellectual discussion. Wartime experience in the economic division in charge of civilian supplies in Poland had thrown him into first hand contact with the source and flow of commodities. His teaching experience, in Heidelberg and Vienna, had developed his gift for relating science to daily life. Then as general secretary of the Austrian Federation for Housing and Garden Cities, he expressed himself through posters and graphs and popular expositions that indicated his bent. When Vienna began its great program of rehabilitation and social welfare, Neurath started the Social and Economic Museum. There, with a group of first rate collaborators in graphic work, he directed

FIGURE 3. An article from the popular American magazine *Survey Graphic* celebrating Neurath's ISOTYPE techniques and his many visually oriented writings and museum exhibitions ("Social Showman" 1936, 618).

Hutchins, the celebrated wunderkind president of the university, who together championed philosophy as the savior of Western civilization just as emphatically as Dewey, Morris, and others in the Unity of Science movement. While the pragmatists and the logical empiricists wished to reform philosophy by making it scientific, Hutchins and Adler fiercely resisted all things scientific in philosophy. For them, science and scientific philosophy were value-free. Were they to ascend to cultural leadership as Dewey, Morris, and the Unity of Science movement believed they should, civilization would career into meaninglessness and, most likely, barbarism. They believed instead that intellectual and cultural life required a unifying framework that exalted values lying above and outside science. Adler convinced Hutchins that the future of philosophy lay in neo-Thomism, and Hutchins made their case in his book *The Higher Learning in America* (1936) where he defended this Thomistic program against other educational programs such as Dewey's.

Carnap arrived at the University of Chicago campus the same year that Hutchins's book appeared and found the neo-Thomist antics of Adler strange and surprising. At a department seminar, Carnap recalled, Adler

declared that he could demonstrate on the basis of purely metaphysical principles the impossibility of man's descent from "brute," i.e. subhuman forms of animals. I had of course no objection to someone's challenging a widely accepted scientific theory. What I found startling was rather the kind of arguments used. (Carnap 1963a, 42)

Carnap doubted that this kind of neo-Thomism "fitted well into the twentieth century" (ibid.) and quite likely saw no hope in trying to persuade Adler that epistemology and philosophy had in fact made progress in recent centuries.

About Hutchins, at least, Morris was more sanguine. Several times in the 1930s he tried to smooth relations between the movement and Hutchins, if only because Hutchins's fame, influence, and access to his university's cash would have been helpful to the *Encyclopedia* and the International Congresses. But the intellectual divide could not be bridged. Morris reported that Hutchins "did not think well of positivism, pragmatism, or any scientific philosophy" and would "have none of me" when Morris himself was proposed for chairman of the philosophy department.[23]

[23] Morris to Carnap, April 6, 1935, CMP.

In 1937, Morris wrote to Hutchins about the movement, knowing very well that Hutchins regarded its ideals and aims as being too narrow, technical, and "scientific." Morris had his own reservations about the narrow scope of the movement, but nonetheless defended it as a consequence of rigor and precision:

The desire for precision has led for the moment to a restriction of the field covered, and in this sense the movement does not at present deal with certain significant humanistic and philosophic problems. Hence it is not in my opinion the whole story, but it does show a desire to take seriously the problem of the integration of our intellectual life.... My pamphlet [for the *Encyclopedia* (Morris 1938a)] attempts to indicate a wider field than the Congresses have yet considered. I think that even in its present form the movement has some important educational implications.[24]

Morris focused on these "educational implications" in a paper he sent Hutchins two years later, "General Education and the Unity of Science Movement" (Morris 1939). The essay was Morris's attempt to reconcile the disputes between Dewey and Hutchins on the aims and content of education. But Hutchins stood his ground and more or less dismissed Morris's effort. He replied to Morris that he could not even get past Morris's and Dewey's very definition of science – it seemed to include everything:

My difficulty with it is in your and Mr. Dewey's definition of science and the scientific attitude. [It] seems to me to appropriate for science and to define as scientific an attitude which should be characteristic of all study and which therefore is not properly to be called scientific at all. Philosophy and history, for example, should be free from control by routine, prejudice, dogma, unexamined tradition and sheer self-interest. They are also dedicated to inquiring, to examining, to discriminating, and to drawing conclusions only on the basis of evidence after taking pains to gather all available evidence.... The attitude described as unscientific in this quotation is also unphilosophical and unhistorical.

As for Morris's general theory of semiotic, which Morris characterized as a "science of science," Hutchins found that he could not even conceive of such a thing. For if science is to be defined by "its method ... then it becomes apparent that there can be no such thing as a science of science." "One cannot develop about science a body of knowledge obtained by those laboratory methods which constitute, in my view, the distinguishing characteristic of science." Scientific study, Hutchins seemed to reason, is essentially conducted in a laboratory. Since science itself as human activity

[24] Morris to Hutchins, May 24, 1937, CMP.

takes place in the wider world, any "science of science" is practically or, he may have reasoned, logically impossible. With a flourish of Socratic maneuvering (of the sort for which Adler was even more well known), Hutchins lopped off the head of the semiotic project that Morris hoped could both unify pragmatism and logical empiricism and save Western culture.

What Hutchins failed to see (and Morris failed to emphasize) in their debate was that the movement promoted a scientific *orientation* and *outlook* for modern life. Neither Neurath nor Dewey supposed that the world could literally be examined in a test tube or subject to controlled experiments. Nor did Morris ask Hutchins precisely why science and scientific methods were, as Hutchins presumed, necessarily inapplicable to science itself. Still, the exchange illustrates how Hutchins took science to operate entirely on one side of a strict dichotomy between facts and values. Thus he closed his letter by dismissing Morris's claim that the Unity of Science movement could help advance thinking about "problems connected with morality and valuation." Not so, Hutchins shot back: "[T]he basic problem here is not how to reach a selected goal, but what goal to select."[25] Only an expansive, rationalistic philosophy such as Thomism, Hutchins believed, could handle problems of values. If scientistic, "technological," or "positivistic" programs were given culture's reins, only disaster could result.

According to Adler, one need only look to Europe for proof. Roughly one year later, in his infamous talk, "God and the Professors," Adler earnestly laid the blame for all of civilization's current ills at the feet of the college and university professors who had polluted the nation's educational system with "positivism." At the first Conference on Science, Philosophy and Religion and Their Relation to the Democratic Way of Life, Adler argued that the academy ought to find its grounding in Aristotelian percepts (for example, that "man is a rational animal"(Adler 1941, 124)) and Hutchins's proposed educational reforms (Hutchins 1936). But these reforms could never take root, Adler complained, given in the academy's self-perpetuating army of culture-destroying positivists:

The disorder of modern culture is a disorder in their minds, a disorder which manifests itself in the universities they have built, in the educational system they have devised, in the teaching they do, and which, through that teaching, perpetuates

[25] Hutchins to Morris, December 18, 1939, UCPP, box 106, folder 15.

itself and spreads out in ever widening circles from generation to generation. (Adler 1941, 123)

As Hutchins had told Morris, all this was clear once one distinguished between philosophical and cultural ends and mere scientific means:

> Theoretic philosophy delves more deeply into the nature of things than all the empirical sciences. . . . [P]ractical philosophy, dealing with ethical and political problems, is superior to applied science, because the latter at best gives us control over the physical means to be used, whereas practical philosophy determines the ends to be sought. . . . [F]or man the highest knowledge, and the most indispensable to his well-being, is the knowledge of God. (ibid., p. 124)

Frank heard Adler's talk, and it is likely that Einstein and Morris were in the audience as well, for they read papers at the conference after Adler. If they hoped for some kind of collaboration with Adler, they could not have been hopeful after hearing him speak. He rejected almost every feature of logical empiricism and the Unity of Science movement: their dismissal of theology and metaphysics, their empiricism, their rejection of a priori knowledge, and their view of science as a paramount source of information about the world. On the other hand, however, Adler was eager to promote a unity of knowledge, albeit one quite different from the movement's unity of science. While the movement rejected the imposition of any extra-scientific unifying scheme on the existing sciences, and especially any with a priorist epistemological ambitions, Adler and Hutchins wished to see the sciences take subordinate places in an Aristotelean schema of knowledge: "a hierarchy of studies, ordered educationally according to their intrinsic merits" (p. 124).

The controversy was heated not only because Adler and his scientistic opponents competed for the same ground of unity and cultural leadership, but because Adler seemed eager to attack positivists themselves along with their mischievous philosophical positivism. The terms Adler chose, moreover, were childish and, it would appear, deliberately incendiary. "Until the professors and their culture are liquidated," he wrote, "there can be no hope for modern culture (p. 134). As Europe descended into fascist occupation, Adler claimed that Nazism was a comparatively minor threat:

> The most serious threat to democracy is the positivism of its professors, which dominates every aspect of modern education and is the central corruption of modern culture. Democracy has much more to fear from the mentality of its teachers than from the nihilism of Hitler. (p. 128)

Neo-Thomism and the New York Philosophers

The New York philosophers were infuriated by Adler's attack. Hook published a rebuttal of his own with the title "The New Medievalism" (1940) and then rallied his colleagues to a more organized, collective counterattack in *Partisan Review*. Structured as a symposium with the title, "The New Failure of Nerve," the counterattack appeared in 1943 and featured articles by Hook, Dewey, and Nagel. Hook led the charge by sending Adler's claim back at him. It was Adler, Hutchins, and the neo-Thomists – not the pragmatists or logical positivists – who were abdicating their cultural and social responsibilities as philosophers. Borrowing Gilbert Murray's thesis that the ancients living prior to the Christian era had failed to embrace the intellectual and cultural responsibilities bequeathed to them by Hellenism, Hook argued that Adler and a cadre of theologically inclined intellectuals, including Jacques Maritain and Reinhold Niebuhr, were running scared from the responsibilities of science and critical thinking:

The new failure of nerve in contemporary culture is compounded of unwarranted hopes and unfounded beliefs. It is a desperate quest for a quick and all-inclusive faith that will save us from the trouble of thinking about difficult problems.

Nodding to the debates within logical empiricism (and Popperianism) about criteria for demarcating science from pseudoscience, and seconding Dewey's call for "intelligence" in modern life, Hook wrote that "these hopes, beliefs and faiths pretend to a knowledge which is not knowledge and to a superior insight not responsible to the checks of intelligence." The damage could be profound, Hook explained, if Adler succeeds in controlling philosophy's bid for cultural leadership. The very "possibility of intelligent human effort" could be doomed:

The more fervently [these hopes, beliefs and faiths] are held the more complete will be their failure. Out of them will grow a disillusion in the possibility of intelligent human effort so profound that even if Hitler is defeated, the blight of Hitlerism may rot the culture of his enemies. (Hook 1943, 23)

Dewey's article, "Anti-Naturalism in Extremis," supported Hook's and further criticized the neo-Thomists and all antinaturalists for losing faith in "human capacities" and peddling "escapism and humanistic defeatism"(Dewey 1943, 33, 39). In his defense of naturalism, Dewey especially considered the psychological appeal of antinaturalist a priorism, as he called it, and teased apart its supernatural and nontheological varieties. Both believed that above the inquiring, patient, ever-learning and tentative method of science there exists some organ of faculty which reveals ultimate and immutable truths, and that apart from the truths thus obtained there is no sure foundation for morals and for a humane order of society.

Supernaturalists, however, understood what the nontheologicals did not: There is no "complete agreement on the part of all absolutists as to standards, rules and ideals with respect to the specific content of ultimate truths." That is why they appeal to allegedly higher, authoritative and supernatural sources to clarify all ambiguities, to remove doubts, and, unfortunately for some, to sanction crusades to "wipe out heresies" (Dewey 1943, 36).

Nagel took up this psychological line of analysis as well:

In the midst of actual and impending disaster, men are inclined to listen to any voice speaking with sufficient authority; and during periods of social crisis, when rational methods of inquiry supply no immediate solutions for pressing problems, spokesmen for institutional and philosophic theologies find a ready audience for systematic disparagement of the achievements of empirical science. (Nagel 1943, 41)

As a result, "good sense has been sacrificed to ... malice." In his article "Malicious Philosophies of Science" (1943), Nagel defended science against the theologians and intellectuals who would place specific limits on its authority and application. They did so to self-aggrandize and make it appear that only their own intellectual and metaphysical systems – rather than public, empirical, and scientific knowledge – were uniquely able to solve social and cultural problems.

Nagel appealed specifically to the unity of science to rebut the various dualisms (such as material vs. immaterial, qualitative vs. quantitative) used by neo-Thomists to demarcate philosophy from science. He also discussed different kinds of "reductionism" to argue that they were not, as neo-Thomists charged, so mischievous or threatening to civilization. In one compelling passage, Nagel politely but firmly explained that Adler and Hutchins simply needed to grow up:

It is not wisdom but a mark of immaturity to recommend that we simply examine our hearts if we wish to discover the good life; for it is just because men rely so completely and unreflectively on their intuitive insights and passionate impulses that needless sufferings and conflicts occur among them. The point is clear: claims as to what is required by wisdom need to be adjudicated if such claims are to be warranted; and accordingly, objective methods must be instituted, on the basis of which the conditions, the consequences, and the mutual compatibility of different course of action may be established. But if such methods are introduced, we leave the miasmal swamps of supra-scientific wisdom, and are brought back again to the firm soil of scientific knowledge. (Nagel 1943, 54)

Nagel thus echoed Carnap's defense of science (as described in chapter 2) as a healthy, progressive antidote to the way people are sometimes childishly "deceived by their desires" and, more broadly, the call for scientific planning and management in modern society that Carnap, Neurath, and Hahn endorsed in the Vienna Circle's manifesto.

Hook's symposium was among the most visible items in *Partisan Review* during the first half of the 1940s. It was followed by articles, letters, editorials, and a second installment of the debate. For the most part, however, the logical empiricist émigrés did not participate directly. Frank faithfully attended future conferences on science, philosophy, and religion and apparently never lost hope that intellectuals across the disciplines might come to a more unified and accurate view of how values do, and do not, operate in science.[26] Neurath, living in Europe and, later, England, watched it from afar, asked about it, and most likely compared and contrasted this North American species of metaphysics with those he struggled against in Europe.[27]

Morris, however, was stationed in enemy territory in Chicago. Shortly after Adler's speech, Hook wrote to Morris and encouraged him to attack. Reactionary social forces only grow stronger when they are not opposed, he explained, and he added that Morris's position at the University of Chicago could well be in danger.[28] Yet Morris did not heed Hook's call to arms. He had been trying for six years to find common ground with Hutchins, and, indeed, his professional situation at Hutchins's university was never fully secured. Unlike Hook, more importantly, Morris was not a fighter. Instead of taking sides in a debate, he preferred to synthesize valuable insights he detected in the competing points of view.

---

[26] Frank wrote to Morris about his participation in the conference: "I was the only one (besides H[arlow] Shapley) among so many people who upheld the viewpoint of positivism and empiricism. I had the opportunity of getting in contact with your colleague Mortimer Adler. I attacked him very often. However we got along personally quite well and he honored me by saying: 'you are arguing very cleverly. It would be worthwhile to convert you to Thomism'" (Frank to Morris, undated but probably 1941, CMP).

[27] "What happens in the university. How goes Thomism on, and what happens with Logical Empiricism?" (Neurath to Morris, March 31, 1944, CMP).

[28] Hook to Morris, December 19, 1940, CMP.

The Russell Affair

Another conflict erupted in 1940 that further united the New York philosophers and the logical empiricists in their opposition to conservative social and religious forces and beliefs. Bertrand Russell, whose work in logic and epistemology helped to make logical empiricism possible, decided to re-enter academic life by way of the department of philosophy at City College of New York. But in 1940, shortly after being appointed, and before Russell started teaching the next year, an avalanche of protesting letters from Catholic leaders and a lawsuit filed against the City's Board of Higher Education succeeded in overturning the appointment. One of the more verbally creative prosecuting attorneys called Russell's writings "lecherous, salacious, libidinous, lustful, venerous, erotomaniac, aphrodisiac, atheistic, irreverent, narrow-minded, untruthful, and bereft of moral fiber" (in Dewey and Kallen 1941, 20). John E. McGeehan, the presiding Catholic judge, concurred with this assessment. Citing some of Russell's infamous comments, mostly from *Marriage and Morals* (Russell 1929), about adultery, homosexuality, and masturbation, he ruled that Russell's appointment was in fact invalid and "an insult to the people of the City of New York" (Dewey and Kallen 1941, 225).

The New York philosophers were enraged, but they could not save Russell's position. Dewey and Kallen assembled and edited a book, *The Bertrand Russell Case* (1941) in which educators and philosophers vented their anger in essays with titles such as "Social Realities versus Police Court Fictions" (by Dewey), "A Scandalous Denial of Justice" (by Morris Cohen), and "Trial by Ordeal, New Style" (by Walton Hamilton). Hook's late 1930s militancy against "totalitarianism" and Moscow-style authoritarianism was given pride of place at the end of the book. Under the title "The General Pattern," Hook insisted that the episode was merely one campaign in a larger war against secularism in education and culture (Hook 1941, 188). The "spearpoint" of the attack on secularism was "the Catholic Church." But its spokesmen were Adler and Hutchins, who sought to "prescribe" Thomist metaphysics in higher education and to

persuade the American people that the basic values and attitudes of our democratic way of life may not be able to withstand the attacks of totalitarianism, from without and within, unless they are fortified by supernatural sanctions. (ibid., pp. 197, 198, 204)

In the wake of Adler's incendiary attack on science and positivism and the attack on Russell, Hook was ready for war. He ended his essay with a call to arms: "The issue has been joined. The battle for a free American Culture, at the home front as well as at our frontiers, is on" (p. 210).

The Unifying Agenda

As we see in chapter 9, by 1939 the alliance between the New York philosophers and the logical empiricists had begun to sour in some quarters. In particular, the anti-Stalinist campaign against all things "totalitarian" in leftist politics by Horace Kallen involved a sustained attack against logical empiricism and the Unity of Science movement that could only have contributed to the decline of politically engaged philosophy of science. But in the science wars surrounding neo-Thomism and the Russell affair, these tensions remained contained within a broader, shared agenda to defend and promote science, both inside and outside the academy, as a means to help understand and control the world and our beliefs about it.

No mere alignment of doctrine, the two groups stimulated and nourished each other, providing resources, credibility, and epistemological muscle to fight their battles. For example, Dewey's sleepy and relentlessly general discussions promoting science as social, organized intelligence in the service of social and cultural progress were empowered by Hook's and Nagel's polemics that drew on logical empiricism's bold dichotomies between metaphysics and science, antiquated rationalism and modern empiricism, and its main conception of science as a unified whole. Similarly, Carnap, Neurath, or Morris could themselves have argued honestly against neo-Thomist hopes that "ethics and politics depend upon a metaphysical insight which is direct, immediate, and *certain*" (Hook 1941, 204), but only Hook had the nerve and theatrical sense to further compare neo-Thomism to "Hitlerism" (ibid., p. 205) in *Partisan Review*. Had Nagel never joined ranks with the Unity of Science movement, finally, he could have ably disarmed neo-Thomist arguments for the autonomy of values and ethics by presenting, say, traditional arguments against mind-body dualism. But given his central place among and his admiration for the new logical empiricist émigrés, it would seem he was likely emboldened to argue that antinaturalism was not just philosophically suspect but, in the context of the Unity of Science movement and its success, plainly anachronistic and reactionary.

4

"Doomed in Advance to Defeat"?

John Dewey on Reductionism, Values, and the International Encyclopedia of Unified Science

Dewey's willingness to participate in Neurath's encyclopedia project may seem puzzling. Dewey's lifelong concerns with values and the development of culture seem out of place in the enduring reputation of logical empiricism as a technical, value-free enterprise that took value statements, and ethical theories about them, to be empty noise. In part that reputation derives from Rudolf Carnap, whose early writings, especially, matched Neurath's in their claims for the emptiness of metaphysics and the end of traditional philosophy (see, e.g., Carnap 1959a). While Dewey accepted the movement's rejection of all things unscientific and unintelligent (or "unintelligible," as he once put it), he was very worried that the empirical, scientific study of values would be mistakenly swept away if logical empiricism came to dominate philosophy and intellectual life. Dewey therefore chose to work with Neurath and the Unity of Science movement in order, he hoped, to prevent such a catastrophe.

In his correspondence with Neurath, Carnap, and Morris about his two contributions to the *Encyclopedia*, and in those contributions themselves, we can see some of the complexities of this alliance between America's leading philosopher and the new leading philosophers of science. On the one hand, Dewey believed that logical empiricism suffered from certain philosophical faults, but came to learn, it would appear, that his critique was mistakenly based on Ayer's influential *Language, Truth and Logic*. On the other hand, though Dewey came to think more highly of logical empiricism though his participation in the *Encyclopedia*, he continued to worry about logical empiricism from a "tactical" point of view. He used his two contributions to help further the ongoing war against the neo-Thomists and similar enemies of science, and he urged his readers and his logical empiricist editors to do the same.

Morris's and Neurath's Overtures

When Morris first began helping Neurath outline his plans for the new encyclopedia, he urged Neurath to enlist Dewey for the cause. He was the most important intellectual in America, after all, and to Morris he was a personal hero.[1] Dewey's approach to values in science, moreover, was one that Morris would explicitly champion when he attempted to persuade his colleagues actively to embrace "socio-humanistic" issues as core issues in the Unity of Science movement (Morris 1938b). For Morris, it is safe to say, the collaboration among Dewey, Neurath, Carnap, and himself concerning Dewey's contributions to the *Encyclopedia* was a pinnacle in his career. For he found himself successfully brokering the very dialogue and collaboration between leading philosophical currents that he believed was required to meet the cultural and scientific challenges facing the West.

At first, however, Dewey was not eager to be involved. He was then in his seventies and did not even respond to Neurath's invitation to join the planning committee for the movement's international congresses. When Morris wrote to check up on the situation, Dewey replied, "I have not answered Dr. Neurath's communication because I did not see how I could be of any help on the Committee in question." Dewey hinted that he was not available for any such involvement, but he wished the movement well: "Of course, I hope the movement will be successful."[2]

Neurath "was accustomed to getting his ideas carried out," Morris (1973, 64) once said, and he wanted Dewey involved in the encyclopedia project. If Dewey was actively resisting involvement in 1935, Neurath soon prevailed. After meeting with Neurath in New York, Dewey agreed to contribute to the first monograph of the *Encyclopedia* and stand on its official Advisory Committee. One account holds that Neurath won Dewey's participation by meeting him in his Morningside Heights home and pledging, "I swear we do not believe in atomic propositions" (Hollinger n.d., 1).

---

[1] Morris once asked Dewey to send him a portrait photograph that he planned to include with a picture of William James in a display of those who "have come to exert a deep influence on my own development" (Morris to Dewey, July 4, 1937, CMP).

[2] Dewey to Morris, March 6, 1935, CMP.

Abraham Edel, who was present, recalls that Neurath pledged something different, that "he was interested in values, but thought that there was nothing more to be said about them other than that we held them."[3] Perhaps what was said was less impressive on Dewey than Neurath's famously infective zeal for science and its use in modern society. In either case, Dewey was now on board and aware that the new encyclopedia could be an effective ally in the ongoing science wars. Months later, when Dewey accepted Morris's invitation to contribute a second time by writing a monograph on the theory of value, Dewey's reply came with some reporting about the ongoing war: "The a priorists were out in force at the Cambridge Philosophical meeting Christmas week – all centering on the alleged impossibility of empirical normative judgments of value." Hook proposed that "something should be done to check their ravages," Dewey continued, with the result that "Hook and Kallen are thinking of having a conference here . . . in the general interests of empiricism" (Dewey to Morris, March 27, 1937, CMP).

The monograph that Dewey would deliver to his editors in about two years specifically aimed to defend what these a priorists denied: the possibility and, for the cultural goals of the Unity of Science movement, the necessity of developing empirical theories of value and normativity. Dewey warned Morris, however, that his forthcoming monograph would be at odds particularly with some of Carnap's views:

I am glad to do the axiology pamphlet; I don't see how I can do it without getting into ethics more or less, nor how I can do it without cutting across Carnap's theory, but Neurath told me to go ahead on my own, as long as I "built bridges" – or rather indicated where there are some.[4]

As we shall see, Dewey did manage to offend Carnap in both of his contributions to the *Encyclopedia*, but he was less knowledgeable about Carnap's views (as well as Neurath's) than he believed. Carnap was not so dismissive of ethics and metaphysics, and Neurath, Dewey found, had very sensible suggestions for the monograph on values.

The Unity of Science as a Social Problem

In his first contribution, Dewey's anti-Thomistic concerns were made plain in its title, "Unity of Science as a Social Problem" (Dewey 1938).

---

[3] Personal communication, January 15, 2001.
[4] Dewey to Morris, March 27, 1937, CMP.

The "social problem" was nothing like gambling or poverty. Dewey meant that aside from questions about theoretical relations among the sciences, the movement had to see itself as a response to science's enemies. Besides those whose opposition to science and a scientific outlook resulted merely from "ignorance," there was

> active opposition to the scientific attitude on the part of those influenced by prejudice, dogma, class interest, external authority, nationalistic and racial sentiment, and similar powerful agencies. Viewed in this light, the problem of the unity of science constitutes a fundamentally important social problem. (ibid., pp. 32–33)

With enemies like the neo-Thomists and antiscientific fascists in Europe, science was at "a critical juncture," Dewey warned. To respond most effectively to this situation, he emphasized, the Unity of Science movement had to remain flexible, open, and democratic. It "need not and should not lay down in advance a platform to be accepted." Rather, "detailed and specific common standpoints and ideas must emerge out of the very processes of co-operation" (pp. 33, 34).

When Dewey's manuscript circulated among the *Encyclopedia*'s editors, Neurath was in full agreement with this point. Always eager to fend off traditional philosophy and its pretensions to be queen of the sciences, Neurath himself emphasized an open, democratic posture for the movement. Like the sailors in his famous boat metaphor, scientists would chart a direction freely and without pretending that they could appeal to metaphysical foundations or some superscientific theory of science. Neurath later concluded his own monograph with a presentation of this metaphor and its implication that the movement was no attempt to legislate the future of science. That future, in fact, was unpredictable: "The whole business will go on in a way we cannot even anticipate today. That is our fate" (Neurath 1944, 47). Neurath was also pleased that Dewey employed some of Neurath's other favorite metaphors for describing the task of unifying the sciences. The goal, Dewey wrote, was "to build bridges from one science to another. There are many gaps to be spanned" (Dewey 1938, 34). Privately, Dewey acknowledged Neurath's influence. "I accepted many of the suggestions you made in our conversation on my article."[5]

Morris and Carnap were also happy with Dewey's manuscript. Yet there was one statement that irked them all, especially Carnap. For in issuing his

---

[5] Dewey to Neurath, August 17, 1938, ONN.

warnings about preconceived platforms, Dewey trampled unnecessarily on some ideas Carnap had been working out for several years. Dewey wrote,

But the needed work of co-ordination [of the sciences] cannot be done mechanically or from without. It, too, can only be the fruit of cooperation among those animated by the scientific spirit. Convergence to a common center will be effected most readily and most vitally through the reciprocal exchange which attends genuine co-operative effort. *The attempt to secure unity by defining the terms of all the sciences in terms of some one science is doomed in advance to defeat.* (ibid., emphasis added)

Morris, it appears, persuaded Dewey to revise his original claim that rejected the attempt to define all scientific terms on the basis of *physics.* Morris presumed that Dewey's target was Carnap's writings on reductionism and suggested to Dewey that he make his target broader and more general.[6] Dewey replied that he was not thinking of Carnap "at all in that sentence, but rather of some psychologists and sociologists who think everything should be reduced to terms of physics."[7] Nevertheless, Dewey rewrote his claim so that "some one science" replaced "physics."

Carnap also supposed that readers would take the remark to be directed against his views about the definability of terms. His work on the theoretical unity of science was highly visible, in part because his article "Testability and Meaning" had recently appeared in *Philosophy of Science.*[8] Here, Carnap specified different kinds and degrees of empirical meaningfulness and intertheoretic unity that claims and theories could exhibit. In particular, he distinguished between *defining* concepts of one theory in terms of those in another (a prospect that was foiled by so-called disposition terms, such as "soluble," by the logic of the material conditional) and the weaker, more practicable condition of *reducing* terms by devices he called "reduction sentences." We can aim for this kind of unity of science, Carnap explained, even though the sciences may never evolve to a point where central concepts in biology or psychology become fully definable within physics.

Carnap wrote to Dewey and sent along a copy of "Testability and Meaning." He wanted Dewey to understand his distinction between definitions

[6] Morris to Dewey, December 4, 1937.
[7] Dewey to Morris, December 7, 1939.
[8] Besides "Testability and Meaning," Carnap's writings on this topic include Carnap 1934b, 1936b, 1937a, and his reply to Herbert Feigl in 1963b.

and reduction sentences and, more importantly, to acknowledge the difference between present and future relations among the sciences:

I distinguish reducibility from definability; it is at the present time not possible to define terms of biology in terms of physics or terms of psychology in terms of biology and physics. But, on the other hand, I do not see a reason for assuming that the present impossibility of such definitions should hold in all future.

Indeed, Carnap must have been baffled by Dewey's remark, for his larger point was that no preconceived or a priori intuitions or expectations should be brought to the task of unifying the sciences. On what grounds, then, could Dewey confidently assume that such definitions will be forever impossible to achieve? Agreement was at hand if only Dewey would redirect his claim to the "present state" of the sciences. In that case, Carnap explained, "there would not be a discrepancy between our views."[9] Dewey had already adjusted his sentence to satisfy Morris, however, and he was unwilling to make further changes. He thought his meaning was sufficiently clear and correct:

My belief that the categories of sociology and biology cannot be "reduced" in the sense in which the English reader naturally understands the word to physical categories (i.e. categories of physical science) is so firm that I do not see how I can alter my revised statement.[10]

Neurath found the statement puzzling, as well. He wrote to Dewey some months later and was less concerned than Carnap about the distinction between definitions and reductions. He recommended a different distinction to Dewey, the distinction between unifying the sciences by relating them all to physics, on the one hand, and by employing natural, "physicalist" language as a unifying jargon or "universal slang" for all of science, on the other. Neurath favored the second approach and, by comparing his "universal slang" to Carnap's observational "thing language," noted that his physicalism was not unconnected to Carnap's.[11] Indeed, Neurath had to defend his physicalist program to Dewey because it was plainly a program and Dewey had urged in his essay that unified science proceed without any restrictive program. As Neurath saw it, however, this kind of physicalism was necessary for unified science and would indeed make possible the kind of democratic, collective collaboration that Dewey called for. Neurath therefore asked Dewey to distinguish these kinds of

---

[9] Carnap to Dewey, December 28, 1937, ASP RC 102-39-06.
[10] Dewey to Carnap, December 30, 1937.
[11] On Carnap's "thing language" see Carnap 1936/37, 467; 1937a, 52; 1938, 52–53.

physicalism and, as Carnap had, to back off his excessive claim about certain kinds of definitions being "doomed in advance to defeat":

I should appreciate it if, if you could let [*sic*] open the answer whether it might be possible to reduce the scientific terms of the different sciences to the terms of this universal slang ("thing language" according to Carnap) in concordance with the "program" of physicalism or not. This program is not exactly identical with "defining the terms of all the sciences in terms of some or one science" and I cannot see that such a program even in the narrow sense you explain "is doomed in advance to defeat."[12]

Had Neurath seen Dewey's response to Carnap and his distinction between technical concepts (namely, Carnap's "reduction") and their interpretation in ordinary language, he could have told Dewey that he was in fact no stranger to Neurath's preferred kind of physicalism. The task of consciously managing terminology in science by connecting technical concepts to ordinary, empirical language was a cornerstone of Neurath's conception of unified science.

But this possibility for mutual understanding and agreement was missed as confusions and disagreements began to pile up between Dewey and the *Encyclopedia*'s editors. Dewey, to begin with, was puzzled by Neurath's physicalism and Carnap's "thing language." He did not see the importance of a "thing" language for science, he explained to Neurath, because science crucially involves "operations" and behaviors. Appealing again to "the ordinary use of the English language," Dewey objected that these operations and behaviors "are not 'things' nor yet 'objects.'" More importantly, these operations and behaviors in science crucially involve valuations and value propositions:

In a strict sense of thing language there can be no genuine evaluation propositions or sentences. In terms of a behavior or operational language I think the case can be clearly made out in behalf of the genuinely logical character of some – though not all – value-expressions.[13]

As he had told Morris earlier, Dewey told Neurath that this was the concern that lay behind "the sentence in my contribution that Carnap didn't like." The fashion for social scientists to mimic natural science as much as possible in their language was "a fundamental mistake," Dewey explained, and doomed to defeat because the social sciences *require* and make crucial use of value terms.

[12] Neurath to Dewey, August 3, 1938, ONN.
[13] Dewey to Neurath, August 17, 1938, ONN.

Neurath was probably not happy to hear this, for Dewey seemed to be relying on traditional distinctions between the natural and social sciences that the Unity of Science movement aimed to leave in the historical past. And he did so while leaning on the term "value" that Neurath worried was dangerously metaphysical. For Dewey, however, "value" and value terms were crucial to any science or philosophy of science that was going to assume a leading role in charting a future for humanity. The movement must therefore address values and matters "socio-moral." As he wrote to Neurath,

> in virtually leaving out a large field – the socio-moral – I think that in the end this course will produce a reaction to the a priori, while an operational language in enabling the field to be brought under the empirical cuts completely under the a priori.[14]

Dewey could have been clearer. This "reaction to the a priori" was not primarily some philosophical or logical maneuver, but rather a *social* reaction of the kind that Hutchins, Adler, and other a priorists would be happy to lead. It was a reaction against modernity, science, and progress of the sort that Dewey, Nagel, and Hook would in a few years criticize as the "new failure of nerve."[15]

Dewey's *Theory of Valuation*

Several months later, when Dewey's second contribution to the *Encyclopedia* circulated among the editors, this chain of events nearly repeated themselves. The monograph defended the view that normative judgments of value must be understood (as he earlier told Neurath) behaviorally and operationally – as versions of "liking and disliking" – so that they can be treated empirically and scientifically. Dewey argued against any sharp and robust distinction between ends and means and insisted that a "continuum of ends-means" must instead be embraced by any valid theory of value. Our distinctions between ends and means are in fact shifting and "temporal and relational" (Dewey 1939, 423), he argued: Means toward an end are, as means, desired as ends; while ends achieved typically become means for yet further and different goals.[16]

---

[14] Ibid.

[15] See chapter 3 for more on Dewey's, Nagel's, and Hook's campaign.

[16] "Every condition that has to be brought into existence in order to serve as a means is, *in that connection*, an object of desire and an end-in-view, while the end actually reached is a means to future ends as well as a test of valuations previously made" (Dewey 1939, 423).

What stood in the way of any such future theory of value, however, was the dominant "ejaculatory" view of value statements. It holds, Dewey wrote, that "value-expressions cannot be constituents of propositions, that is, of sentences which affirm or deny, because they are purely ejaculatory" (ibid., pp. 386–87) and contain at most an expression of subjective "feelings" or mood. Dewey cited several examples of this view from a source he did not identify. One reads,

In saying "tolerance is a virtue" I should not be making a statement about my own feelings or about anything else. I should simply be evincing my own feelings, which is not at all the same thing as saying that I have them. (p. 387)

Since sentences that do not say anything cannot be inconsistent with each other, it follows, the anonymous quote continues, that "it is impossible to dispute about questions of value" (ibid.). On this view, a legitimate theory of value becomes impossible, and Dewey devoted about one-tenth of his essay to attacking it.

Carnap read Dewey's manuscript and reported back to Dewey that he liked it very much. Again, however, he suggested to Dewey that he was unfairly painting logical empiricist views with a broad brush. For Carnap accepted parts of the "ejaculatory" view and took those parts to be representative of the movement. But in the interests of fairness and accuracy, he urged Dewey to disentangle some things. "There is only one small remark I should like to make," he wrote:

I suppose that your criticism of the "ejaculatory" view is especially meant against Schlick. But Schlick and the others of us do not mean to say that value expressions have no meaning at all, but only that they have no cognitive content. You say yourself that value statements are not derivable from factual statements. Therefore, I suppose that you agree with us in the view that there is a non-cognitive component.... We certainly agree with you that besides this non-cognitive component there is also a cognitive factual component, if the value statements are interpreted in your way.... Certainly we do not deny, but rather admit explicitly the great psychological and historical effect of metaphysical statements.

Carnap's remark was not so small, but it was careful and polite. "You understand that it is not at all my intention to censor your ms.... My concern is only to prevent the reader from getting a not quite adequate picture of the views which you criticize."[17] Again, Carnap believed that

---

[17] Carnap to Dewey, March 11, 1939, ASP RC 102-39-03.

agreement was in easy reach, as long as the logical empiricist views Dewey did not like were specified in sufficient detail.

Dewey responded twice to Carnap on this point. The first time, he said he didn't recall whose views he had quoted, but they weren't Schlick's. In any case, he offered to add "a qualification in the proof" to make this clear.[18] A week later, Dewey wrote to say that the quotations were taken from A. J. Ayer and that he did not give "reference to the sources of my quotations because I wanted to avoid the appearance of any personal controversy."[19] Though Dewey maintained his policy of not identifying the source of his quotations, he found Carnap's comments compelling and decided to add a qualification to his argument.

Once again, Neurath was generally pleased with the monograph, too. He objected to some of Dewey's terms and formulations, but these objections were less important than the larger alliance. "I am so glad that we are cooperating that I do not much object to some formulations and some criticism in your manuscript. I think you will change certain phrases in the second edition."[20] Neurath and Dewey corresponded in 1938 and 1939 about Dewey's monograph and were often at odds over particular terms, including "value."[21] But Neurath was busy and left the major editing of Dewey's essay to Carnap and Morris. He was also willing to relax his editorial grip somewhat because he did persuade Dewey to accept the monograph's title, "Theory of Valuation," and the key term "valuation." "You see our agreement concerning the term 'valuation' seems to me to be an example of common understanding, goodwill and cooperation."[22]

---

[18] Dewey to Carnap, March 17, 1939, ASP RC 102-39-01.

[19] Dewey to Carnap, March 24, 1939, ASP RC 102-39-02. Dewey took his quotations from Ayer's *Language, Truth and Logic* (1936, 107–10).

[20] Neurath to Dewey, March 24, 1939, ONN. There were no further editions of the monograph. Neurath probably meant the second *version* of the manuscript, and the presentation here assumes that reading. Note the effect here of Neurath's Germanic English. "I think you will change certain phrases" sounds like a command; but later in the same letter he conversely envisions himself and others in the movement changing "our expressions" to better harmonize with "your ideas and the ideas of your scientific friends." His thought was that the evolution of terminology would be collective and all participants' favorite expressions would evolve in time as the movement progressed.

[21] "Value" is one of the "doubtful nouns" that appears in Neurath's infamous *Index Verborum Prohibitorum*, or Index of Prohibited Words (Neurath to Dewey, October 12, 1938; and January 2, 1939, ONN). For more on Neurath's index, see Reisch 1997a.

[22] Neurath to Dewey, March 24, 1939, ONN. In other cases, as discussed in chapter 10, Neurath would greatly annoy his encyclopedia writers with his strong views about the proper titles for monographs.

About Dewey's attack on Ayer's noncognitive view of value statements, however, Neurath was as emphatic as Carnap that the movement as a whole did not uphold Ayer's view. "Maybe it is our fault that you got the impression we underestimate the importance of verbal expressions which are not 'statements' – we do not, not at all." To prove it, Neurath quoted four remarks by Philipp Frank and one by Joergen Joergensen made at International Congresses for the Unity of Science concerning, for example, "the large role of metaphysical statements" in "practical life."[23]

But Dewey had already begun drafting the "qualification" he had promised Carnap. It takes the form of a long footnote. It is the only substantive note in the monograph and effectively capitulates to Carnap's (as well as Neurath's) protests. Though his focus had broadened from "value statements" to "metaphysical sentences," Dewey clearly qualified his argument to distinguish between different kinds of meaning and different kinds of practical effect that metaphysical language can have. The note reads,

The statement, sometimes made, that metaphysical sentences are "meaningless" usually fails to take account of the fact that culturally speaking they are very far from being devoid of meaning, in the sense of having significant cultural effects. Indeed, they are so far from being meaningless in this respect that there is no short dialectic cut to their elimination, since the latter can be accomplished only by concrete applications of scientific method which modify cultural conditions. The view that sentences having a nonempirical reference are meaningless, is sound in the sense that what they purport to mean cannot be given intelligibility, and this fact is presumably what is intended by those who hold this view. Interpreted as symptoms or signs of actually existent conditions, they may be and usually are highly significant, and the most effective criticism of them is disclosure of the conditions of which they are evidential. (Dewey 1939, 444)

Dewey thus came to agree, as Carnap insisted he must ("You say yourself that value statements are not derivable from factual statements"), that noncognitivists about value statements and metaphysical statements do not necessarily dismiss them as entirely meaningless.

However much the note signals agreement and understanding, however, it also indicates Dewey's continuing frustration with logical empiricism. Well into his correspondence with Neurath and Carnap, Dewey told Morris that he could not accept the way that the "l.p.s." (or logical

[23] Ibid. Neurath cited Frank's afterword in *Erkenntnis* 6: 448 ("...im praktischen Leben des Einzelnen wie der Völker eine grosse Rolle spielen...") and Joergensen in *Erkenntnis* 7: 288.

positivists) viewed relations between the "moral" and the "physical."[24] The footnote suggests that Dewey continued to find other aspects of logical empiricism objectionable, as well. Logical empiricists may take themselves to distinguish different kinds of meaning embedded in metaphysical statements, the note implied, but it is irresponsible to use the word "meaningless" when so much meaning – especially meaning embedded in, or perhaps expressive of, cultural and social "conditions" – is in play. In addition, Dewey's comment that metaphysical statements cannot be simply eliminated by philosophical means ("there is no dialectic short cut to their elimination") is perhaps a nod to Carnap's essay "The Elimination of Metaphysics through Logical Analysis of Language" (in which Carnap infamously dismissed metaphysicians as "musicians without musical ability" (Carnap 1959a, 80)). If so, Dewey hinted that this is another project "doomed in advance to defeat" since linguistic therapy would hardly affect these broader conditions giving rise to metaphysical language and their meanings.

Dewey's enduring reservations are also suggested by the note's location. Instead of inserting it near his discussion of Ayer's view, Dewey placed it in the final section of the paper, "Valuation and the Conditions of Social Theory." Here, he objected that contemporary scholarship reinforces "unexamined traditions, conventions, and institutionalized customs" that have effectively prevented "valuation phenomena" and valuational behaviors from receiving the scientific study that they deserve. The footnote appears after the words "institutionalized customs" as if Dewey saw Ayer's noncognitivism becoming strong and entrenched, something like an "institutionalized custom." If so, Dewey's point is less directed at logical empiricism and more directed at stereotypes about logical empiricism. Some degree of unity had been achieved: However much Ayer's *Language, Truth and Logic* did come to shape perceptions of and beliefs about logical empiricism, Dewey had joined Carnap and Neurath in trying to avoid its misleading generalizations.

Science, Values, and Tactics

In Dewey's eyes, both of these run-ins with his logical empiricist editors involved his concerns about science's enemies and how to keep them at bay. The dispute over "doomed in advance to defeat" was persistent because its real significance far outstripped technical questions about

---

[24] Dewey to Morris, March 24, 1939, CMP.

whether and how concepts in social science related to those in physics or chemistry. From Dewey's point of view, it was about the scope of philosophy of science and whether, as he insisted it should, the movement embraced values as a core component of unified science. Only then could it arm itself to defuse neo-Thomism's critique of "positivism" and find a credible, influential voice in intellectual and popular debate about the course of contemporary culture.

Dewey's response to Carnap's objections about Ayer's noncognitivism rested on the same underlying purpose. Dewey told Morris that the purpose of his long footnote was to clarify and emphasize that the logical empiricist critique of "meaningless" metaphysics was in part, if not primarily, a *strategic*, "tactical" mistake:

Of course I agree that "metaphysical" statements in the sense of non- or anti-empirical are unverifiable. But I think the attempt to dismiss them entirely at one swoop by calling them "meaningless" is a serious tactical mistake.[25]

If the Unity of Science movement was to accept its role in the "social problem" of the unity of science, that is, then it had to balance its narrower philosophical concerns with broader "tactical" postures and maneuvers. These included recognizing the importance and different kinds of meaning that otherwise cognitively empty statements can have and recognizing the centrality of valuations in science itself. Otherwise, Dewey feared, defenders and advocates of science risked losing leadership and influence in the quest to manage modern life scientifically and intelligently. If Adler and Hutchins successfully fooled the world into believing that science was technical and value-free, that is, then they could more easily persuade the world that Thomism (or some other nonscientific, rationalistic system) had to be embraced as a source of values and guidance for contemporary life. And in that case, both the New York philosophers and the logical empiricists would be on the losing side in the war over science.

[25] Ibid.

5

Red Philosophy of Science

Blumberg, Malisoff, Somerville, and Early Philosophy of Science

While the New York philosophers enlisted logical empiricism and the Unity of Science movement in their battles where they could, they were also welcomed by other left-leaning philosophers of science of the 1930s. Some lived or circulated outside New York City and remind us, therefore, that leftist philosophy of science was not a historical phenomenon isolated within New York. They were also farther to the left than the liberal pragmatists. In terms of the survey of the philosophical left sketched in chapter 3, those discussed here belonged to the Socialist left and admired logical empiricism and the Unity of Science movement as allies in their various campaigns for international socialism.

Albert Blumberg (1906–1997)

Albert Blumberg played a very early role in the reception of logical empiricism. Born in 1906 to Russian parents in Baltimore, Blumberg grew up to study at City Colleges in New York, Johns Hopkins, Yale, and the Sorbonne. He then received his doctorate under Schlick at the University of Vienna. From 1931 to 1937 he taught philosophy at Johns Hopkins and helped to introduce logical empiricism to readers of the *Journal of Philosophy*. With Herbert Feigl, who arrived at Harvard from Vienna in 1930, he co-wrote "Logical Positivism: A New Movement in European Philosophy" (1931). They described the new movement as a synthesis of empiricist and rationalist traditions that was made possible by Russell and Whitehead's new logic and that promised unprecedented analytic power for making sense of the recent and startling developments in physics. Charles Morris, Ernest Nagel, and W. V. O. Quine all traveled to Europe in the first half of the 1930s and had undoubtedly read and studied Feigl and Blumberg's paper before seeing the European philosophers they were eager to meet.

Blumberg was also an ardent Communist and, for a short while at least, his political and intellectual work overlapped. In the mid-1930s, for example, he helped to found the Marxist journal *Science & Society*, where his name appears as an editor in the first volume from 1936.[1] Though he later wrote articles in the journal (e.g., Blumberg 1958), Blumberg's name does not appear as editor after 1936 probably because he had redirected his career into the Communist Party. Judging from how Blumberg was once introduced at a party banquet in 1942–

It is your good fortune here in Maryland to have such a man, who is not only a Doctor of Philosophy and a good Party leader, but a fighter and an organizer and a political leader of whom you can all be proud. Greetings, Comrade Blumberg[2] –

Blumberg was effective and well liked by his comrades. He rose quickly to a position of national party leadership. He was the party's mayoral candidate in Baltimore in 1938; chairman of the Maryland branch of the Communist Party U.S.A. in 1940; one of five vice presidents of the Maryland-D.C. District of the party in 1944 and, by August 1945, the district vice chairman.

Though Blumberg left academic life, he remained a teacher in his political career. As he busily shuttled back and forth between Baltimore and New York City in the late 1930s and '40s, he taught courses at the Communist Party's New Worker's School in Manhattan with titles such as "What Is Philosophy?," "World Politics," and "The Negro in America." He also organized and spoke at various rallies and reached outside the party to popularize the Communist cause. He ran for the U.S. Senate in 1940,[3] wrote articles for Communist magazine *The Clarion*, and delivered radio addresses from Baltimore radio stations. In one address from 1942, as Hitler rampaged through Europe and the Soviet Union, Blumberg called for U.S. aid to the Soviets and Great Britain. "The closest cooperation between the U.S.A., Great Britain and the U.S.S.R. and

---

[1] The editors listed along with Blumberg are Edwin Berry Burgum, V. J. McGill, Margaret Schlauch, and Bernhard Stern.

[2] Quoted by the FBI from the *Baltimore Morning Sun*, November 11, 1942. Unless otherwise noted, the following information on the life and activities of Blumberg comes from his FBI file, parts of which obtained by the author under a Freedom of Information Act request.

[3] Albert E. Blumberg, obituary, *Baltimore Sun*, October 14, 1997. I thank Gary Hardcastle for providing me with this document.

also active collaboration with China" was required to halt Nazism and Japanese fascism, he explained. Well aware of communism's enemies at home, however, Blumberg reminded his listeners that this war against fascism was domestic, as well. "In Great Britain and the United States, there are many powerfully situated reactionaries who would sell out their own people, the Soviet Union and world democracy to Hitlerism at the first opportunity."[4]

Blumberg's greatest effect on the American political landscape concerns the national elections of 1948 in which much of the American left put its hopes in the candidacy of Henry Wallace, Truman's former vice president. Wallace's resounding defeat, as discussed in chapter 12, marks the rising tide of anticommunism that would dominate the Cold War landscape. Wallace and the new Progressive Party that sponsored him was red-baited and criticized as overly sympathetic to communism. According to FBI investigations, what alliance there was between the Communist Party and the Progressive Party was implemented by Blumberg, who, according to an unknown informant, "traveled through the nation and from coast to coast, contacting Party functionaries and Party sympathizers and supporters in connection with the Henry Wallace campaign."

Whether or not it was because of Blumberg's role directing the "third party campaign" and Wallace's overwhelming defeat, or because of his ongoing surveillance by the FBI, Blumberg retreated from public life and went underground after the election of 1948. Blumberg knew he had enemies worth hiding from. In 1940 he was called before the House Un-American Activities Committee, led by Martin Dies, and cited for contempt on refusing to answer questions put to him by the committee. He was fined and given a suspended sentenced of thirty days in prison. During the McCarthy years, in 1956, he was indicted in Philadelphia under the anticommunist Smith Act, a cornerstone of anticommunist legislation that made it illegal to advocate violent overthrow of the U.S. government.[5] As Cedric Belfrage recalled after sympathetically covering the case for the *National Guardian* (which Belfrage edited), prosecution witnesses included one Mary Markward, a teacher who had testified that Blumberg advocated violence (as Belfrage dryly put it) "over lunch in 1941" (Belfrage 1973, 250). Blumberg's case was one of 150 filed

---

4 The address was delivered August 17, 1941, when Blumberg was state secretary of the Communist Party of Maryland.

5 Albert E. Blumberg, obituary. *Baltimore Sun*, October 14, 1997.

by federal prosecutors attempting to break the Communist Party (Klingaman 1996, 341). At one point, Hoover himself paid close attention to Blumberg. When agents finally nabbed him, Hoover proudly dispatched the headline that Blumberg was "seized . . . on a Manhattan street corner as an underground red" (cited in Belfrage 1973, 250).

Blumberg and many others never served time for their convictions because in 1957 the Supreme Court ruled that Smith Act convictions failed to distinguish between advocacy of abstract doctrines and violent political action. However, Blumberg's wife, Dorothy Blumberg, was earlier indicted under provisions of the Smith Act and served a three-year jail sentence. The Blumbergs left their radical political life in 1958 and, as Albert's fate was still being appealed and negotiated in federal courtrooms, he began to resume his academic and philosophical career. He reached out to his former co-author, Herbert Feigl, who coached Blumberg by mail about recent developments in philosophy of science and, in particular, Carnap's work in semantics. Blumberg also undertook some scholarly translations, including Schlick's *General Theory of Knowledge* (1985), and by the end of the decade had become a popular professor of philosophy at Rutgers University. In 1976, he published the successful textbook *Logic: A First Course* (1976). As Blumberg completed his "re-entry" from politics to professional philosophy, however, he did not abandon his political beliefs.[6] He ardently defended student protests against American involvement in Vietnam and, though he became district leader of the Democratic Party in Manhattan, continued to write and lecture on dialectical materialism.[7] Blumberg died in 1997.

William Malisoff (1895–1947)

While Blumberg circulated primarily around Baltimore and New York City, William Marias Malisoff was a research biochemist based at different times in Philadelphia, Baltimore, and New York City. His interest in science and its philosophical aspects led him to found the journal *Philosophy of Science*, which he edited from its first issue in 1934 until his sudden death on November 15, 1947. Like Blumberg, Malisoff was a founder of professional philosophy of science in the United States.

---

[6] Blumberg to Feigl, June 18, 1959, HFP 02-40-04.

[7] Blumberg to Feigl, May 20, 1961, HFP 02-40-08; Blumberg to Feigl, May 26, 1969, HFP 02-40-12. Blumberg defended "dialectics" against criticisms from philosophers including Hook and Nagel in Blumberg 1958.

The subscription list for his journal grew to become today's Philosophy of Science Association.

Ten years older than Blumberg, Malisoff was also Russian. He was born in the Ukraine in 1895 and brought by his parents to New York City in 1905. Following tradition in his family of educators, he attended Columbia University and then earned a Ph.D. in chemistry from NYU. No ordinary chemist, Malisoff then worked in a variety of academic, medical, and industrial settings. His publications, editorials for *Philosophy of Science*, and letters to the editor of the *New York Times* indicate a wide range of affiliations and titles. He taught philosophy of science at the University of Pennsylvania, taught chemistry at the New School for Social Research, and conducted research for the Atlantic Refining Company and also at Montefiore Hospital, New York. Malisoff also founded his own company, Unified Laboratories, Inc.[8] By 1937, Malisoff had also become a paid consultant to "two of President Roosevelt's 'brain-trusters,'" as he described it to Neurath.[9]

In light of Malisoff's early publications, he was bound to be interested in the Unity of Science movement and scientific philosophy. His book *A Calendar of Doubts and Faiths* (1930) contains curiously personal ruminations on the evolutionary history of humans and the future of human culture and knowledge. Critical but often undisciplined in his thinking and writing, Malisoff argued for scientific research to increase the human life span. As suggested by his occasional fascinations with vague or pseudoscientific topics (such as when he posed "for research" the question "How is will related to time?"(Malisoff 1930, 190)), Malisoff was not a philosopher of science by the standards that the logical empiricists would soon put in place. Still, he was a capable scientist, patent-holding inventor, and intellectual who read widely in science, philosophy, and art. He was also well acquainted with the intellectual soil from which logical empiricism grew. In this book, he alluded to Mach, to the logic of Russell and Whitehead, to psychological theories of all sorts, and to debates over mechanism versus vitalism. He criticized theology as a "morass of malobservation, primitive speculation and sheer superstition" (ibid., p. 277) and did so without succumbing to a naïve positivism holding that empirical science was an easy path to epistemic certainty.[10] As most logical

[8] These affiliations are given in Kaempffert 1937a, in Malisoff's letters to the *New York Times*, and in his patent applications.
[9] Malisoff to Neurath, July 14, 1937, ONN.
[10] See, e.g., Malisoff 1930, 272.

empiricists would agree, he believed that science was "thoroughly rooted in philosophical assumptions" and properly "subject to philosophical criticism" (p. 148).

Malisoff's interest in the unity and classification of the sciences, however, most directly connected his projects to Neurath's. "One of the crying needs of the day," Malisoff wrote, "is to get away from over-specialization, and it is this factor that tends to bring the sciences together" (p. 249). Though the sciences were divided according to convention, Malisoff explained, their subject matter is a unified nature (pp. 249, 253). Too few scientists understood this, Malisoff lamented, as he complained about rivalries among different areas of specialization and "clannishness that makes a certain type of mathematician look upon a physicist or chemist as a most inaccurate amateur" (p. 262).

Scientists needed to work together, after all, if science were to be used as a tool to change the world. Malisoff urged his readers to respect science's "efforts at the intellectual reconstruction of the world" (p. 263) and the "value of science in the reconstruction, or construction, of a practical world" (p. 290). Like Neurath and others in the Unity of Science movement, Malisoff was an advocate of rational planning on a national scale: "We need badly a Congress of Strategy for the location of industries" (p. 297), he explained, to make transportation of goods and people more rational and efficient. Written a year after Neurath, Carnap, and Hahn published their pamphlet *Wissenschaftliche Weltauffassung*, Malisoff's *Calendar* includes many of the same modernist themes as it called for cooperation among the sciences. Malisoff even anticipated Neurath's argument that a unified science was necessary because "no single man can be all-knowing in even one industry, men shall have to come together."[11] As unity increased, therefore, science would be better able to study and to serve life in all its social and intellectual complexity.

Malisoff and the Unity of Science Movement

Malisoff's *Calendar of Doubts and Faiths* and his next book, *Meet the Sciences* (1932), established him as a popularizer of science and its unity. Though it is not clear when and what Malisoff first learned about Neurath, he was

---

[11] Ibid., 303. Neurath wrote that "since it takes a great number of people to carry the totality of knowledge of an epoch, it is understandable that through a series of successive assimilations and rejections uniform manners of thinking, in short, are formed.... The powers of an entire generation of scholars are hardly sufficient to perceive all the consequences of a single theory" (Neurath 1936b, 157).

very eager to make contact with him. Morris brought about their meeting. He told Neurath to contact Malisoff in part because Malisoff, then making plans for "organizing an American *Philosophy of Science Organization,*" was central to North American philosophy of science. "The soil is very fertile for such things in this country," Morris proudly told Neurath.[12] True, Malisoff was not an "eminent" philosopher, Morris explained in a different letter, but he would be an important contact for Neurath to cultivate. As editor of "our new *Philosophy of Science*," he would have a most useful list of potential subscribers to the new encyclopedia.[13]

Malisoff was thrilled when he received Neurath's letter. He replied,

I had been wanting to communicate with you for the longest time. Indirectly rumors have reached me via Feigl, Morris, Carnap, of the plans for an international meeting in Paris to be devoted to the glory of philosophy of science. May I be allowed to cooperate to the best of my ability?

Though perhaps merely accidental, Malisoff's letter is peppered with terms that are suggestive of Malisoff's leftism. Here, he echoed the famous socialist maxim, "from each according to his ability; to each according to his need." When describing his journal's list of subscribers, Malisoff explained that the 700 individual subscribers and libraries did not reflect the "many 'sympathizers,' i.e. people who cannot or will not spare the subscription price." *Philosophy of Science*, he hinted, was a journal with considerable proletarian appeal.

As he ended his first letter to Neurath, Malisoff raised the subject of communist politics and, in particular, Neurath's proclivities and travels:

By way of personalities, – you are described to me as "a brilliant mind, outstanding organizing genius, communist, sociologist, scientist, positivist, Marxian, etc." I may be going to the Physiological Congress in Leningrad and Moscow in August. How can I manage to meet you even if I cannot come to Paris? Will you be going to Russia?[14]

Malisoff did not say whom he was quoting, nor did Neurath ask. Neurath responded by inviting Malisoff to the Paris congress and adjusted his profile via a friendly lesson in Neurathian language management:

1. For "Brilliant" we have no precise definition. 2. "Mind" is a suspicious term, but admissible in a wider sense. 3. I was not and I am not member of the communist party. 4. "Marxian" is not a generally intersubjective term, because this term is

---

[12] Morris to Neurath, May 25, 1935, ONN.
[13] Morris to Neurath, March 16, 1935, USMP, box 1, folder 15.
[14] Malisoff to Neurath, May 22, 1935, ONN.

used by different Marxian groups only for their own adherents. I do not know if in this moment any group names me thus. 5. "Positivist" only in the widest sense, better "scientist" in a special sense, as a propagator of the "New Scientism" or of the "Scientific Empiricism" – as Morris names this tendency. As a thus reduced person I hope to meet you in August. I myself shall not go to the Soviet union in this year.[15]

Neurath was no official communist, nor was he a "positivist" if that implied any kind of epistemic foundationalism. Though he "reduced" himself in Malisoff's eyes, he did not deny Malisoff's recognition of him as a fellow leftist intellectual or, depending on the definition, a Marxist.

Neurath took full advantage of Malisoff's enthusiasm. In the last half of the 1930s, Malisoff ran articles and book reviews that Neurath suggested and agreed to dedicate a few pages in each issue to a "Symposium of Unified Science" presenting news and short articles concerning the movement.[16] Malisoff was not so generous to a mere stranger, however. He and Neurath met socially in New York and appeared to enjoy each other's company. Neurath regularly sent his regards to Malisoff's family and his "clever children," and he invited Malisoff to join the Advisory Committee of the new *Encyclopedia*. Morris and Neurath also discussed asking Malisoff to write a monograph for volume 3 of the *Encyclopedia* (which was never realized).[17]

Neurath's science-writing cousin at the *New York Times*, Waldemar Kaempffert, also contributed to the friendship between Malisoff and Neurath. Malisoff contributed many book reviews and features to the newspaper in the 1930s and '40s and appears to have been personally and intellectually friendly with Kaempffert. In 1937, Malisoff wrote a letter to the *New York Times* taking Kaempffert's side in a debate over the merits of organization in science. Malisoff defended "the cooperation of the sciences, sometimes called the 'unity of the sciences'" and defended it against critics who invalidly reasoned that "all organization must be regimentation, because sometimes it is." Instead, "organization, properly conceived, is unity in variety and not the grayness of regimentation." Malisoff enlisted his favored bio-chemical metaphor for the unity of science and the coordination of different areas of science: "Just think of

---

[15] Neurath to Malisoff, June 5, 1935, ONN.
[16] Neurath to Malisoff, January 18, 1936; March 8, and June 25, 1937, ONN. See also Morris to Malisoff, July 28, 1937, CMP.
[17] Neurath to Malisoff, January 16, June 3, June 25, 1937; May 24, 1938, ONN; Morris to Neurath, August 4, 1938, USMP, box 1, folder 17.

our own bodies as an illustration. Strange that students of such marvels of organization should talk against it!"[18]

Months later Kaempffert favorably reviewed Malisoff's book *The Span of Life* (1937) in which Malisoff argued for scientific organization and cooperation toward the goal of extending human life. Kaempffert emphasized that Malisoff's plan would "strike the social note" and require scientists to be aware of the "social circumstances" of their research, "to think socially and philosophically."[19] Kaempffert and Malisoff apparently agreed not only about the importance of broad, social thinking on the part of scientists, but about their shared affection for Neurath. "Kaempffert has been telling me about you," Malisoff wrote to Neurath in Holland. "We all wish you were here permanently."[20]

Philosophy of Science versus Erkenntnis

The alliance and camaraderie between Malisoff and the Unity of Science movement was not without rivalry and competition. Most of it concerned Neurath, Carnap, and Morris's plan to rescue *Erkenntnis* and bring it to the University of Chicago Press under the name the *Journal of Unified Science*. As of 1938, Neurath had not told Malisoff about these plans, plans that Malisoff would not have welcomed. The new journal would compete with *Philosophy of Science* and it would be marketed probably using the subscription list Malisoff had been willing to share with Neurath for marketing the *Encyclopedia*.[21]

The issue surfaced when Neurath wrote to ask why an article of his, long since submitted to Malisoff, had not yet appeared.[22] Malisoff explained that he had been talking with Carnap, who was then leading efforts to secure independent funding for the new journal, and had learned that the movement "was to have an organ of its own published by the University of Chicago Press." Malisoff therefore put Neurath's article on hold, apparently figuring that Neurath would want the article published in the movement's new journal. "I gathered the impression that your group was acting as a unit."[23]

[18] Malisoff to the Editor, *New York Times*, April 28, 1937.
[19] Kaempffert 1937a.
[20] Malisoff to Neurath, May 31, 1938, ONN.
[21] Malisoff to Neurath, May 22, 1935, ONN. Malisoff told Neurath that he did not have a subscriber list but would ask his publisher to send one to Neurath.
[22] Neurath to Malisoff, May 24, 1938, ONN. The article was Neurath 1938a.
[23] Malisoff to Neurath, May 31, 1938, ONN.

Malisoff and the logical empiricists never fully embraced each other. In part, the logical empiricists were not comfortable with Malisoff's editorial control. On one occasion, when Carnap asked Warren Weaver of the Rockefeller Foundation for funds to help resuscitate *Erkenntnis*, Weaver recommended that the Unity of Science movement merge *Erkenntnis* with *Philosophy of Science*. That would not be acceptable, Carnap explained: Malisoff would want control and "we want of course to be independent."[24] Years later, Philipp Frank helped to fill in the circumstances Carnap alluded to: Malisoff "suffered from great many complexes and inhibitions which made him afraid of any cooperation."[25]

If Malisoff was resisting cooperation by withholding Neurath's paper from publication, Neurath soon smoothed the situation. He told Malisoff that the fate of *Erkenntnis* had nothing to do with relations between the Unity of Science movement and *Philosophy of Science* in particular, because several journals (including *Scientia, Theoria, Analysis,* and *Synthese*) had agreed to publish communications of the movement. Regardless of the fate of *Erkenntnis*, he urged Malisoff, "our nice corner in your periodical should remain!" A week later, he wrote again and asked Malisoff for a personal biographical sketch and bibliography. Neurath had just concluded negotiations giving *Erkenntnis* to the Dutch publisher Van Stockum & Zoon and now promised Malisoff that he would arrange for a review of Malisoff's *Meet the Sciences* in the new *Journal of Unified Science* once it was up and running.[26] Malisoff published Neurath's paper, "Encyclopedism as a Pedagogical Aim: A Danish Approach" (1938), later that year in *Philosophy of Science*.

Malisoff, Soviet Communism, and the KGB

Like Blumberg, Malisoff was politically active. While Blumberg worked openly in the CPUSA, Malisoff had secret meetings and conversations with Soviet spies. Among intercepted communications between Moscow and U.S. Communist Party offices (known as Venona documents, made public in 1995) are nine messages referring to Malisoff under the code names Talent and Henry (or Genry). (Because not all Venona intercepts

[24] Carnap to Weaver, dated "Chicago, 1938," USMP, box 1, folder 4.
[25] Frank to Morris, November 28, 1947, CMP.
[26] Neurath to Malisoff, June 9 and 18, 1938, ONN. For more on the story of *Erkenntnis's* rescue and the *Journal of Unified Science*, see Reisch 1995 and Stadler 2001, 60.

were successfully decoded, and because not all messages between Moscow and the United States were intercepted, these messages almost certainly do not exhaust Soviet communications concerning Malisoff.) Dating from 1943 and 1944, the messages tell something, however impressionistic, of Malisoff's activities and his importance to his Soviet contacts.

One message reports to Moscow that Malisoff had met with his contact after being ill for three weeks. Malisoff reported that while he was sick he had been visited by a representative of the New York Board of Health who was interested in Malisoff's "work." Malisoff did not understand this reason or nature of this meeting. In the Venona document, a note connects this decrypt to another that reports that, in the next month, the operative code named "Anton" canceled his meeting with Malisoff because he was "under surveillance by two Competitors."[27] If the message intended to say that Malisoff (and not Anton) was under surveillance, it was probably correct. Malisoff was being investigated by the FBI.[28]

"Anton" was Leonid Kvasnikov, assigned to New York by Soviet foreign intelligence to coordinate spies pursuing information about atomic weapons (Romerstein and Breindel 2000, 200). Kvasnikov, a trained chemical engineer, is also credited with first bringing to the attention of Soviet intelligence in 1940 that American and British scientists had suspiciously ceased publishing articles on uranium. He entered New York under diplomatic cover and by 1948 had risen to head of KGB scientific intelligence (Haynes and Klehr 1999, 392).

According to the Venona documents, Anton met with Malisoff on twenty occasions in 1943 (ibid., p. 291). This fact and the bits of information in the decrypts suggest that Malisoff was of genuine interest to Kvasnikov, most likely because of Malisoff's expertise in chemical engineering and his contacts in science and industry. One document connects Malisoff to goings-on at the University of Chicago, where its metallurgical laboratory, site of Enrico Fermi's creation of the first atomic pile, would have been a natural focus of Soviet interest. It reports that Malisoff was able to recommend "reliable people" for a post in "Chicago University" and appears to request permission and instructions for Malisoff

[27] New York to Moscow, November 30, 1944, Venona no. 1680. This and some of the other Venona transcripts quoted here are in the public domain and made accessible by the FBI over the Internet.
[28] At the time of this writing, Malisoff's FBI files had yet to be made public under a Freedom of Information Act request filed by the author.

to meet someone (who is not specified) regarding his potential recommendation.[29]

Frank's comment that Malisoff was sometimes a difficult, temperamental collaborator finds some support in these decoded messages. Malisoff became angry with his contacts when they refused his request for funds with which he would add a manufacturing capability to his firm, United Laboratories, Inc. Malisoff believed he had been highly valuable to his contacts, yet they, in turn, would not share their wealth (or Moscow's) with him. Malisoff threatened that it would be "impossible to expect much help from him" in the future (ibid, p. 290). Since public Venona documents referring to Malisoff extend over only 1943 and 1944, it is not possible to determine whether Malisoff did sever his relationship with the Soviets during this dispute. If he did not, however, his relationship did not last much longer. Malisoff died suddenly of a heart attack in November 1947.

Malisoff's Public Profile

There is no evidence that the logical empiricists or New York philosophers knew about Malisoff's relationship with the Soviets. Decades later, Sidney Hook recalled "a certain Dr. W. M." as an ordinary (for Hook, despicable) Stalinist who naïvely praised Russian biomedical research (Hook 1987, 219–20). Shortly after Malisoff's death, a two-part obituary in *Philosophy of Science*, written by Philipp Frank and C. West Churchman (who succeeded Malisoff as editor of the journal), praised him as a progressive scientist and philosopher. He sought to popularize science "as a part of the whole spectrum of human activity which embraced also social, political, religious, and artistic activity." According to Frank, Malisoff founded his journal to fight against "the specialization and occasional hyperspecialization of the average scientist" and to create a venue for dialogue between philosophers and scientists, as well as "a channel" for scientists to speak to the educated public. Both praised Malisoff as a brave progressive among more conservative colleagues. Churchman wrote that he had a "well thought-out plan for his philosophy, and the impact of his thinking on the conservatives in the philosophical world could not help but have fruitful effects." Frank wrote that he "took the initiative which had been awaited a long time, but which no other scientist has dared to take" (Frank and Churchman 1948, 1–2).

---

[29] New York to Moscow, August 2, 1943, Venona no. 1276.

Malisoff's politics were also plainly visible in his published writings. In "A Science of the People, by the People and for the People," an essay Malisoff first read at the American Association for the Advancement of Science meeting in September of 1944, he invoked his metaphor of society as an integrated organism to explain the kind of democracy he advocated. "*The highest democracy is to be sought only at a maximum of organic integration of society.*" This integration should be realized willfully, deliberately, and scientifically:

For the realization of democracy we need a *complete people's economy*, or a planned, realizable, self-correctable economy, covering *all* the goods of existence from the most concrete to the least tangible, *from meals to ideals.* (Malisoff 1946a, 168)

To that end, scientists had a "moral responsibility"

to pass beyond the mere level of popularization, which is significant enough, and to finish the task of communication with the people by giving them a science of the people which will (1) give an adequate definition of democracy; (2) give an adequate justification of it; (3) explain and forecast the conditions of its increasing realization and (4) strengthen our desire for it.(ibid, p. 167–68)

In the democratic paradise Malisoff envisioned, a peculiarly biological version of the familiar socialist slogan would ring true: "from each organ according to its ability and to each organ according to its needs" (p. 169).

Malisoff also editorialized in *Philosophy of Science* about specific issues. During postwar debate over whether the new atomic technologies should come under civilian or military control, he weighed in with a semantic lesson on the popular talk of "atomic secrets" – secrets that, in the wake of Hiroshima and Nagasaki, many believed the United States should withhold from the Soviets and others. Malisoff pointed out that commentators and politicians typically failed to distinguish whether the "secrets" in question were merely empirical recipes for making bombs or whether they represented genuine knowledge of scientific laws of the atomic nucleus. "I say we have discovered no such law. We have wrested no such secret from nature.... In fact, we are rather thoroughly baffled by the nuclear secret, in the second sense of the word 'secret.'" To gain genuine secrets from nature, Malisoff explained, international, organized research would be necessary: "It will take the combined and organized intellectual power of the entire new 'One-World' to bring into existence and to complete the revelation of the so-called atomic secret" (Malisoff 1946b, 2).

Malisoff's leftism and his strong sense of sarcasm were clear in the book reviews he wrote for *Philosophy of Science*. In 1945, he criticized the essays collected in *The Science of Man in the World Crisis* for lumping together Communism and Nazi-fascism as "totalitarian." In this book, Malisoff wrote,

it is alarming to report that the "world crisis" remains quite unanalyzed, and where it is pecked at one can find the good old Goebbels line, e.g. the menace of Communism, disguised, of course, as a deadly (?) parallel to fascism. (Malisoff 1945, 228)

In 1947 he questioned the credentials and motivations of one of the New York philosophers' campaigns against neo-Thomism, the Conference on the Scientific Spirit and Democratic Faith, organized by Dewey, Horace Kallen, and Jerome Nathanson of the Ethical Culture Society.[30] Reviewing this conference's proceedings, Malisoff fingered its anticommunism and raised questions about its source of funds:

The contributors on the whole have clear records as liberals, although a couple have advocated serious restrictions of the liberties of unpopular minorities. One thing the present reviewer has missed – no one has given a plan how to educate the educators and to prevent their prostitution to vested interests. That would be awfully impolite. (Malisoff 1947a, 104)

Malisoff was frequently distrustful of professors and educators. On the one hand, he disdained those who shied away from ideology and politics. About Henry Wieman's *The Source of Human Good*, for example, he complained that "no concrete issues are discussed, and one gets the impression that the author has no definite position concerning the immediate problems of society, or doesn't care" (Malisoff 1947e, 173). On the other hand, he disagreed with many who took political stands: About Norman Cousin's *Modern Man Is Obsolete*, for example, he wrote, "Why do not men like Cousins have the courage to say 'capitalism is obsolete'?" (Malisoff 1947b, 171). About Lyman Bryson's *Science and Freedom*, he remarked, "The author has either read Marx not at all or with his eyes tight shut" (Malisoff 1947c, 171). Malisoff's words were sharpest, however, when he felt authors unfairly criticized aspects of Soviet life or intellectual culture.

[30] Hook 1987, 347. For more on the Ethical Culture Society and its relationship to New York philosophical life, see Hollinger 1996.

John Baker's attack on Lysenkoism in his book *Science and the Planned State* was "vicious and intellectually dishonest":

> Where others see organization the author sees "regimentation." He is not out to produce better organization, but to destroy it before it really begins to roll up successes. "Free Enterprise" has its spokesmen even in science. (Malisoff 1947d, 171–72)

For Malisoff, "free enterprise" was self-evidently noxious.

One book that Malisoff found self-evidently wonderful was John Somerville's *Soviet Philosophy* (Somerville 1946). "Hurry, get this book!" This book is a "must," he wrote, because Somerville adroitly showed what was "essential in the midst of the almost insane distortions this subject of Soviet Philosophy has received from the professional enemies of the U.S.S.R." Part of Somerville's achievement, Malisoff enthused, was to analyze "the miserable term 'totalitarianism'" (Malisoff 1947f, 172) that assimilated communism, fascism, and socialism and that would come to dominate Cold War political thought and discourse.

John Somerville (1905–1994)

John Somerville is a central, connecting figure in leftist philosophy of science. To his political left, for example, he was a defense witness for Blumberg in his Smith Act trials at a time when publicly defending a known Communist was nearly always taken as a confession of one's own radicalism (Belfrage 1973, 250). Yet, unlike Blumberg and Malisoff, Somerville supported the Soviet Union from within the academy. He taught philosophy at Hunter College and, like Dewey, William Gruen, and others, supported the Unity of Science movement and the idea that unifying the social and natural sciences would allow society to "realize the benefits of intelligent control of its own forces" (Somerville 1936, 297).

In 1936, in "The Social Ideas of the Wiener Kreis's International Congress," Somerville reported on the Unity of Science movement's first International Congress held in Paris a year before. His report was informed not only by his observations at the congress, but also by his correspondence with Neurath. Immediately after the congress, he wrote to Neurath to inquire about the status of the social sciences and social problems within the movement. (Neurath, Somerville wrote, was one of the few figures who "touched more or less on the field of social problems" (ibid.).) Neurath reassured Somerville that the movement harbored no antipathy to the social sciences. It was only that, at this early stage, "the number of scientists is very little for these special discussions." In general, Neurath explained, logical empiricism was neutral with respect to the conventional distinctions between natural and social science: "We can speak by means of our logic instruments about all problems."[31]

Somerville imbibed Neurath's optimism and suggested that the congress's dominant interests in epistemology and the natural sciences, as well as the upcoming Second Congress's emphasis on physics, were preliminary to future consideration of social issues and social science. Instead of rejecting the movement for failing to embrace his social concerns explicitly, Somerville positioned himself as a cheerleader. Like Dewey, he was concerned that American social science was becoming excessively positivistic and ignorant of its social relevance. Neurath and his movement, he hoped, would help to correct that. Thus, he wrote, "it is heartening in these times to see a fresh and vital philosophical movement at least aware of its social obligations" (ibid., p. 300). Somerville then attended the Second International Congress in Copenhagen and read papers addressing both biology and the social sciences (Stadler 2001, 376).

The book Malisoff praised, *Soviet Philosophy* (Somerville 1946), was one of Somerville's major works. A few months before it was published, he defended it in *Philosophy of Science* against the growing number of critics of the Soviet Union and Marxism in postwar intellectual culture. "Ninety-five percent of the fears and misgivings expressed by American writers in regard to this subject are quite unwarranted," he explained. These fears and misgivings are artifacts of "superficialities and distortions of the sensational press" or of faulty logic and rhetoric (Somerville 1945, 23). Somerville was in a position to know, he explained, because he spoke Russian and had spent almost two years in the U.S.S.R. researching Soviet philosophy for his book.

Somerville responded to criticisms by Kallen, Hook, and others that Soviet science was deprived of liberty by dialectical materialism, which, as the official philosophy of the Soviet Union, was increasingly perceived by Western intellectuals as a totalizing intellectual straightjacket. Not true, Somerville explained. "The principles of dialectics are not regarded as a priori or unchangeable dogmas, but as a guide to action, as working hypotheses summing up the trend of scientific observations." Dialectical theory is not a substitute for scientific method, but really a second order theory *of* scientific method, and as such is not in the position to be

---

[31] Neurath to Somerville, November 8, 1935, ONN.

a controlling overlord of science. Thus, he concluded, dialectical materialism could not "dictate to scientists what conclusions they must come to."

Hand in hand with Somerville's defense of dialectical materialism was his support for the unity of science. Properly understood, dialectical materialism was a tool in the service of unified science. It offered "broad working hypotheses which go beyond any one field of specialty, and which stress what the different specialities have in common" (ibid., pp. 25, 26). Its goal was "to link up specialties with one another, and to link up science consciously and fruitfully, rather than unconsciously and chaotically, with the whole field of social problems" (p. 26). It was also an indispensable tool for social planning. In fact, Soviet planning, Somerville explained, "has not retarded or discouraged science, but has greatly increased its range and scope" (p. 29).

Somerville was indeed an apologist for most things Soviet, even in the face of the Lysenko affair. His main purpose in 1945, however, was not to convert his readers but to defend a kind of political realism in philosophy. It was important to examine Soviet philosophy carefully, he believed, if only because it was part of Soviet culture. "It is not necessary to agree with the Soviet system," he explained, "but it is very necessary to cooperate with it, to meet it half way in a friendly spirit, just as in the case of the British or the Chinese." Above all, Soviet and American philosophy shared a respect for science that could make "a very powerful contribution" toward attaining "that peace and harmony in the world which we all desire" (ibid., p. 29).

Rautenstrauch, Schrickel, and Planning

Blumberg, Malisoff, and Somerville were not unique in their sympathy with socialism and Soviet Communism. The cluster of causes and ideas they embraced – such as the need to unify the sciences, their use in social and economic planning, and the duty of philosophers and scientists to contribute toward solving social problems – appeared often in *Philosophy of Science* under Malisoff. H. G. Schrickel, for example, defended "social philosophizing" as scientific on the grounds that, first, the social philosopher aimed to be "objective" and free of "personal biases and prejudices." Second, the social philosopher engaged science critically to make it a better tool for realizing "some societal plan of living that will make possible the greater and more widespread enjoyment of human values" (Schrickel 1943, 212). For Schrickel, the enemy of the admirable social philosopher was overspecialization and disunity of the sciences:

Instead of being unified and comprehensive, [our knowledge of human relations] is broken up into more or less isolated bodies of knowledge; and . . . does not readily lend itself to application in any endeavors to understand completely present social problems and their relations. (ibid., pp. 208–9)

Another obstacle was the infamous dichotomy between values and facts. Both scientists and philosophers, he urged, should "recognize the necessarily cooperative nature of a progressive knowledge of social values and facts" (p. 212).

Two years later, Walter Rautenstrauch of Columbia University contributed the article "What Is Scientific Planning?" Like Malisoff, Rautenstrauch promoted large-scale social and economic planning in United States and his task was to outline how the process would work. At its core, planning required "setting up a framework of relationships among the processes of civilization which has a high probability of attaining optimum social satisfactions" (Rautenstrauch 1945, 8). As Rautenschrauch detailed his proposals, he did not hesitate to editorialize about social and political concerns. He argued for management of natural resources ("the foundation upon which rests all of our material welfare") and added that this concern expressed genuine patriotism, as opposed to "mere flag waving and other ceremonials of like character" (ibid., p. 11). In explaining that different economic plans can have vivid powerful social effects, he urged that future planners emulate some aspects of Chinese civilization (which he found admirably progressive) and avoid those that historically gave too much power to social reactionaries, such as Rome's enemies of Galileo. In the present, Rautenstrauch wrote, this problem took the form of Martin Dies's House Un-American Activities Committee. Several years before McCarthy found his anticommunist voice, Dies crusaded to eradicate "subversive" ideas (including "unholy beliefs" such as the "spread of communism"). Rautenstrauch told readers of *Philosophy of Science* that this was just silly (pp. 17, 18).

A Decade of *Philosophy of Science*

Under Malisoff, *Philosophy of Science* was as an intellectual journal with a unified leftist voice. Malisoff, Somerville, Rautenstrauch, and Schrickel positioned themselves in similar ways vis-à-vis communism, social and economic planning, the importance of the unity of science, and the evils of fascism. They criticized philosophers who avoided engagement with social issues as much as they criticized politicians and institutions that obscured the progressive, socialistic vision they shared for the United States and the world. Both agendas are highlighted in Malisoff's editorial of 1944 celebrating the journal's ten-year anniversary. Excited about the impending defeat of Nazi Germany, Malisoff described the journal as nothing less than a progressive war effort against fascism:

[The journal] started publication in 1934 when German Fascism was already in the saddle and was beginning its systematic abuse of science and philosophy. Many of us had hoped to carry on its pages a fight against the rising tide of obscurantism and its not too unfriendly appeasers.

Like Dewey and Morris, Malisoff regarded the military fight against fascism as one part of a cultural fight for the survival of science. In that regard he appreciated the joining of forces with logical empiricism: "we were fortunate in the adherence of distinguished refugees to our circle" (Malisoff 1944, 1).

Yet the logical empiricists did not join this circle empty-handed. Implicitly, Malisoff's editorial seemed to acknowledge their high intellectual standards. Malisoff admitted that in the course of ten years much progress had been made toward improving the "quality of our papers," many of which were "unnecessarily difficult, confusing and involved, even when the authors had a substantial thought" (ibid., p. 2). Sharing Neurath's recommendation that language in science (and philosophy of science) remain rooted in everyday language (his "universal slang" of physicalism"[32]), Malisoff urged the journal's contributors

to make a real effort to make their essays direct, simple, clear, and to avoid the top-heavy trade jargon of certain schools of philosophy which survive immediate annihilation by sound fact and theory by throwing up the barriers of a "secret language." It is almost impossible to edit a paper full of such weasel words and expressions. The paper should not be written that way in the first place. (ibid.)

He also nodded in the direction of Carnap's syntax-era prohibitions against corrupt pseudostatements that only appear to have cognitive content (e.g., Carnap 1935). Some papers that he had received, Malisoff lamented, were "just barely intelligible." "The Editor frequently has been forced to say to prospective contributors, 'I do not know whether you are right or wrong. I can't understand what you are saying'"(Malisoff 1944, 2).

[32] See, e.g., Neurath 1941.

Perhaps the main reason to read Malisoff's editorial as a testament to his collaboration with the logical empiricists is his central contention that the journal deserved to be the centerpiece of a community dedicated to unifying the sciences and, in turn, unifying the world:

> The Editor does not tire of repeating that in order for us to fall in step with the hard won victories of the day, we must make *Philosophy of Science* itself a model of democracy, an integrating, dynamic influence for establishing a universal *amity of the sciences*. Then, and only then, can we expect the emergence of a supreme *science of amity* in a warless world. (ibid.)

As Blumberg emphasized in his anti-Nazi radio address, Malisoff noted that the enemy was not only overseas. The journal also faced opposition from "domestic fascism" as evidenced by "poison pen communications" that Malisoff had received from various quarters within "the legions of crankism." Whether amused, angry, or thinking of his secret conversations with Soviet agents, Malisoff found it remarkable that "*Philosophy of Science* was red-baited!" (p. 1).

Epilogue: Hook versus Somerville

Malisoff did not live to see the extent to which anticommunist politics affected his journal and some of his colleagues. Philipp Frank and Rudolf Carnap, whose "On the Character of Philosophic Problems" (Carnap 1934a) was the journal's premier essay, were investigated by the FBI as potential subversives. Walter Rautenstrauch earned a stigmatizing place in the Senate's "Handbook for Americans" (U.S. Senate 1956) in a list of about a hundred "typical sponsors of [Communist] Front organizations."[33]

Malisoff did, however, witness John Somerville's dispute with Sidney Hook and Freda Kirchwey, editor in the 1940s of *The Nation*, in which Hook skewered Somerville and his opus *Soviet Philosophy*. Hook described the book as an apology for "the party character of philosophy and science"(Hook 1947, 189) in the Soviet Union and described Somerville personally as a substandard intellectual whose love for Communism led him into logical and factual errors. Somerville struggled to defend himself against Hook in a way that illustrates how difficult life would soon become in the postwar world for leftist philosophers and intellectuals.

---

[33] The senate printed 75,000 of these booklets in 1956 for distribution throughout the United States.

First, Somerville wrote a letter to *The Nation* explaining that his book sought neither to apologize for, nor condemn, Soviet philosophy. Its goal was to highlight differences between Soviet and Western philosophy and thus assist international intellectual understanding. Kirchwey printed Somerville's letter but, to Somerville's surprise, in a highly edited form. Somerville's attempts to rebut Hook's claims were shortened, if not removed, in a way that played into Hook's charges. Additionally, Kirchwey then permitted Hook to respond and gave him about 50 percent more space than Somerville's (shortened) letter. Hook sustained his attack vigorously, writing that Somerville's book was "a shabby piece of apologetics" full of "deceptions peddled by fellow-travelers of the Communist Party line."[34] Somerville had no direct evidence that Hook had something to do with how his letter was edited, but he was surely suspicious.

Seeing himself skewered by Hook for a second time, Somerville was understandably furious and soon incredulous that Kirchwey refused to print anything further in the exchange. Somerville then turned to Malisoff, who printed his account of the attack and some further reflections on its significance in *Philosophy of Science*. One of the most remarkable features of Somerville's account is that he pitched it to *future* historians studying "the cultural history of our times." He seemed aware, that is, that if he was right about what Hook and Kirchwey had done to him, then he was indeed powerless to defuse their attack. Only in the future, he suggested, after the day's anticommunist and anti-Soviet hysteria had become a historical curiosity, would Hook's attack be seen as the contradiction is really was: an example of bullying, if not censorship, masquerading as a defense of free inquiry; or, as Somerville put it, "a sharp contrast between a pompous claim to moral principle, and an actual shabbiness of practice" (Somerville 1947, 345).

Somerville alluded to the damage that Hook's attack might do to his "professional standing." But his main concern was the fact that Soviet-American relations continued to spiral downward. International understanding, he feared, was becoming impossible because honest and fair-minded intellectual dialogue was becoming impossible:

The real insult [of Hook's and *The Nation*'s attack] is that offered to the public, which must sooner or later wake up to the fact, if only the hard way, as I did, that the standards still assumed to be present have in reality vanished. If we cannot take elementary honesty for granted, what can we take for granted when we try

---

[34] Somerville's letter and Hook's in response appear in *The Nation*, May 19, 1947, pp. 555–56.

to apply the results of serious professional research to a problem of the gravest public concern?

Somerville was witness to a phenomenon that would become characteristic of the postwar campaigns against Communist or radical intellectuals: the transformation of intellectual differences about Soviet culture and politics into robustly perceived differences of national loyalty. Somerville seemed not quite sure what to make of this lapse of "elementary honesty" and goodwill, which is partly why he offered his account as "data" for future interpretation (ibid., p. 345). But he was sure that in light of tensions between the postwar superpowers and the new reality of atomic weaponry, this deterioration of standards and discourse could imperil literally everything worth caring about:

We live in times when we should remind ourselves every hour of the day that every bit of social knowledge capable of strengthening the chances of peace must play its part. Otherwise, there may be no world left for us to study, nor any philosopher to explain its passing. (p. 347)

Somerville turned out to be wrong, of course. The world remained and the Cold War's antipathies to all things Soviet and "totalitarian" eventually passed. Somerville himself survived and helped to establish and maintain the philosophical study of Marxism as a niche in North America. But neither the kind of profession-wide social engagement he promoted nor the Unity of Science movement he encouraged in the 1930s survived the Cold War.

6

The View from the Left

Logical Empiricism and Radical Philosophers

At the height of the so-called hysteria over communism in the 1950s, the logic of anticommunist accusation and persecution was intricate and subtle. With respect to one's political essence, the logic was binary. As presupposed by Senator McCarthy's famous inquiry – "Are you now or have you ever been a member of the Communist Party?" – one either was or was not loyal to the United States at the time in question. With respect to outward appearances, however, the markers of one's political essence were believed to be often obscured or hidden. For communists within the party, one could be a public communist, such as a candidate in an election or a party officer, or one could work hidden in the party's underground. Albert Blumberg, for example, took both these paths at different times of his career. Outside the party, anticommunist investigators targeted "fellow-travelers" who supported or even participated in communist causes but refrained from officially joining the party. They, too, could be more public or less public about their motives and activities. Some openly supported Moscow or some form of communism, while others were believed to hide their support or disguise it as support for general, populist, and even pro-American causes or institutions. Organizations or events known as "communist fronts" outwardly promoted peace, the arts, or social and economic justice while actively promoting Moscow's interests in the Cold War by recruiting party members or cultivating spies. William Malisoff and John Somerville were typical fellow-travelers insofar as they defended Soviet politics and culture as individuals, not as party members. Given the tone of some of Malisoff's editorials in *Philosophy of Science*, it is likely that some of the red-baiting, "poison pen" letters he noted in his editorial of 1944 accused his journal of being a "communist front" (Malisoff 1944).

Since the markers of one's political loyalty could be so hidden or obscure, the skill of an anticommunist investigator often lay in scrutinizing a subject for clues revealing his or her true relationship to communism or to the party. Here, a logic of association and proximity often came into play. Were a subject observed by the FBI to attend parties, rallies, or correspond via mail or telephone with a known communist or fellow-traveler, or were their beliefs or public statements about policy observed to track and always to agree with Moscow's, investigators would find their suspicions confirmed and continue looking for more direct and incriminating clues. In the resulting climate, an individual or institution whose writings or beliefs were similar merely in some respect to those of known Communists or who engaged known Communists publicly in collegial debate or dialogue was more likely to become a candidate for investigation than one who avoided such appearances and associations altogether.

When the Cold War began in the late 1940s, the Unity of Science movement was ripe for investigation and suspicion. Malisoff and Blumberg, though they were not central leaders, were already being investigated by the FBI. As discussed in chapter 13, a rumor would soon place the name of Philipp Frank and later Rudolf Carnap on J. Edgar Hoover's desk. In addition, logical empiricism and the Unity of Science movement had been publicly engaged in various kinds of public debate with moderately socialistic intellectuals (such as the New York Intellectuals) as well as with more aggressive leftists writing for journals such as *Science & Society* (discussed in this chapter), and the party's official journal, *The Communist* (discussed in the next). With few exceptions, we see below, this debate was one-sided. Given this anticommunist logic of association and proximity, it would have been foolhardy for any intellectual to spend too much time publicly arguing the finer points of metaphysics, the unification of the sciences, or the political implications of physicalist language with others who openly crusaded for communism in North America. Hook and other former leftists usually issued only denunciations of communists and Stalinists in the intellectual left, while most logical empiricists did not respond to these criticisms. Philipp Frank did, however, as we see in chapter 15. And when Frank responded, as this and the next chapter show, he was not wrong to underscore that there were substantive points of agreement, such as promoting the unity of science, alongside points of disagreement, such as the value of metaphysics and dialectical materialism between the Unity of Science movement and philosophers of science on the extreme left.

Lenin versus Mach

Nearly all criticisms of logical empiricism and the Unity of Science movement from the extreme philosophical left bore some imprint of Vladimir Lenin's earlier attack on Ernst Mach. In his influential book *Materialism and Empirio-criticism*, first published in 1908, Lenin laid groundwork for attacking logical empiricism by attacking Mach and Russian and European Marxists who sought to synthesize Marxism with Machism in any form. Taking Mach's phenomenalism to be a metaphysical thesis, Lenin saw Mach as nothing but an agent of regressive, reactionary philosophy that repackaged Berkeley's subjective idealism:

> No evasions, no sophisms (a multitude of which we shall yet encounter [in this book]) can remove the clear and indisputable fact that Ernst Mach's doctrine that things are complexes of sensations is subjective idealism and a simple rehash of Berkeleianism. If bodies are "complexes of sensations," as Mach says, or "combinations of sensations," as Berkeley said, it inevitably follows that the whole world is but my idea. Starting from such a premise it is impossible to arrive at the existence of other people besides oneself: it is the purest solipsism.... (Lenin 1908, 34)

For Lenin and his followers, this cluster of metaphysical views was heretical for at least three reasons. It violated Marx's point (and famous slogan) that the purpose of philosophy was not merely to study the world but to change it.[1] Though a robust idealism might seem a path of lesser resistance for this goal, Lenin insisted that Marxism and ontological realism were inseparable. Unless scientific knowledge in some sense illuminates or reflects an objective and independently existing material world, Lenin believed, then it could not be of any use as a tool for collective social action. Second, solipsism obviously threatened ideals of social solidarity among the proletariat who stood to change and inherit the world on the foretold collapse of capitalism. Third, idealist metaphysics aided Marxism's enemies. Idealist and nonmaterialist philosophies obscured the revolutionary tasks at hand in the social and economic here and now.

Lenin convinced many that Marxist materialism was automatically victorious over its idealist and reactionary enemies. Only Marxism was

---

[1] The last of Marx's theses on Feuerbach announces, "The philosophers have only interpreted the world, in various ways; the point, however, is to change it."

conscious of the socio-economic conditions of all criticism and philosophy – conditions that allowed Marxists to explain away anti-Marxist critique. Lenin explained, for instance, that despite the popularity of Machism and "empirio-criticism" (which, for Lenin, included all forms of positivism and conventionalism),

> recent philosophy is as partisan as was philosophy two thousand years ago. The contending parties are essentially, although it is concealed by a pseudo-erudite quackery of new terms or by a feeble-minded non-partisanship, materialism and idealism. The latter is merely a subtle, refined form of fideism, which stands fully armed, commands vast organisations and steadily continues to exercise influence on the masses, turning the slightest vacillation in philosophical thought to its own advantage. The objective, class role of empirio-criticism consists entirely in rendering faithful service to the fideists in their struggle against materialism in general and historical materialism in particular. (ibid., pp. 434–35)

Lenin's attack on empirio-criticism could move easily between questions about ideas, methods, and their integrity, on the one hand, and different questions about the socio-economic status, interests, and biases of philosophers who might question or criticize its logic or foundations. In the hands of the doctrinaire, Marxism could become in this way self-justifying and unquestionably true: Its critics were simply "fideists" who could not see that they were puppets of a doomed-to-fail socio-economic system. For critics and less doctrinaire Marxists, such as Sidney Hook, this view was understandably maddening and helped to lay the foundation for the Cold War's popular representations of communists as puppets, "brainwashed" persons, or mechanical robots who seemed intellectually inflexible and incapable of criticizing Marx or Moscow.

Horkheimer's and Marcuse's Attack

After fleeing Frankfurt and a brief stay in Geneva, Max Horkheimer and his Institute for Social Research were welcomed by Columbia University to Manhattan's Morningside Heights in 1935. There Horkheimer published his *Zeitschrift für Sozialforschung*, in which he soon published his own essay, "The Latest Attack on Metaphysics" (1937). The essay was itself a vigorous attack on logical empiricism and suggests how deeply Horkheimer felt the sting of logical empiricism's rejection of metaphysics and its contention that only empirical science provided genuine information about the world. If this was true, Horkheimer's intellectual project, Critical Theory, had no platform on which to stand and issue its critical, objective analyses of all aspects of modern life – of science as well as economics, politics, and society. When the logical empiricist critique of metaphysics gained currency in the mid- to late 1930s, and the logical empiricists themselves were celebrated as important new intellectual emigrants in New York, Horkheimer and his associate Herbert Marcuse fought back.[2]

Horkheimer acknowledged that logical empiricism had now "established itself as the most thoroughgoing antimetaphysical school" partly because it appealed to leftists eager to criticize the metaphysics and confusions of fascism. But logical empiricism was in this regard a fraud, "for this philosophy in its present form is securely bound as metaphysics to the established order" (Horkheimer 1937, 140). Logical empiricists could not claim to be independent of politics in the first place, Horkheimer argued, for it was inevitably (as Lenin would have agreed) caught in a web of social and economic interests that sustained it and that it was therefore unable to criticize effectively.

The problem for logical empiricism was its empiricism. By insisting that genuine knowledge issued only from empirical science, and by promoting unified science (which Horkheimer, like other critics, mistakenly took to be a kind of universal physics: "the correct form of knowledge [for logical empiricism] is identical with physics" (ibid., p. 146)) logical empiricism lacked the creativity and imagination required for genuine reform. As Horkheimer put it,

New forms of being, especially those arising from the historical activity of man, lie beyond empiricist theory. Thoughts which are not simply carried over from the prevailing pattern of consciousness, but arise from the aims and resolves of the individual . . . do not belong to the domain of science. (p. 144)

Empiricism, for Horkheimer, reduced science and philosophy to mere chroniclers or portraitists of the sensible surface of an unjust world:

If science as a whole follows the lead of empiricism and the intellect renounces its insistent and confident probing of the tangled brush of observations in order to unearth more about the world than even our well-meaning daily press, it will be participating passively in the maintenance of universal injustice. (p. 151)

Empiricism could never be radical, even reformist, for Horkheimer because it has no extra-scientific facilities to critique scientific, much less journalistic, observations.

---

[2] For an extended account of the debates between representatives of Critical Theory and logical empiricism and pragmatism, see Dahms 1994.

Another fundamental mistake, according to Horkheimer, was logical empiricism's neglect of human subjectivity and "the relation [of all knowledge] to a knowing subject" (p. 142). "Empiricism rejects the notion of the subject *in toto*" (p. 149), he wrote, and helps to maintain an illusion that scientific or "technical rationality" exhausted or epitomized human rationality. Drawing on the Kantian distinction between reason (*Vernunft*) and the narrower concept of understanding (*Verstand*), Horkheimer insisted that human subjectivity provided rational resources connected to but also exceeding those of science.[3] While logical empiricism "reduces the thinking subject to the role of subsuming protocol sentences under general propositions and deducing other sentences from them" (ibid.), critical theory held that thinking subjects could be more than mere scientists and thus attain legitimate power to "criticize the conceptual forms and structural pattern of science," to criticize "a branch of technical science from outside," or to convince "a specialist" that the course of science requires "direction and meaning" from a global, historical point of view (p. 145).

Horkheimer was relentless in his attack. He liberally quoted Neurath, Carnap, and Russell and (less liberally) Schlick and Hans Hahn to argue that, with the exception of some technical scientific matters, these philosophers had nothing correct to say about knowledge, consciousness, history, politics, and the role of science in society. He ridiculed them for trying to "dispose of all problems by . . . dubious purifications of language," for attempting to isolate epistemology from problems of history and sociology, and, joining Dewey's critique, for pretending to have achieved "freedom from value judgments (*Wertfreiheit*)" (pp. 155, 159, 164).

Two years later, Herbert Marcuse issued a similar indictment as he reviewed the first few monographs of the *International Encyclopedia*. Writing in the institute's journal, Marcuse reported that monographs by Neurath, Carnap, Morris, Victor Lenzen, and others vividly illustrated how positivism's attention to given facts blinded it to possibilities for alternate arrangements of facts: "That 'real possibilities' and their realization belong to reality [*Wirklichkeit*] is a truth lost on modern positivists" (Marcuse 1939, 229). Anticipating his popular critique of "one-dimensionality" in his *One-Dimensional Man* (1964), Marcuse chastised these unified scientists for attending only to the one-dimensional realm of "Fact" and

---

[3] See Carus 2004 for a lucid account of this distinction.

ignoring the two-dimensional space of "Fact" and "Being" and the mani-
fold possibilities in that space recognized by Critical Theory.[4]

Lewis Feuer (d. 2002)

Only a few years after Horkheimer's and Marcuse's attacks, several of their
criticisms were articulated by Lewis Feuer, one of logical empiricism's
strongest American critics. Like Ernest Nagel and Sidney Hook, some
twenty years his senior, Feuer was a student of Morris Cohen at the College
of the City of New York. He then received a doctorate from Harvard in
1935 and returned to teach at City College. There, he worked urgently
to save truly revolutionary philosophy from the mistakes he saw lurking
in logical empiricism.

Feuer's 1941 article "The Development of Logical Empiricism" in *Science & Society* was occasioned by the appearance of Frank's collection
of essays *Between Physics and Philosophy* (Frank 1941). In light of Lenin's
brief dismissal of Frank as a "Kantian" (Lenin 1908, 190), Feuer took
Frank's book to signal that Machism had not only survived but thrived
after Lenin's attack. It had "matured into its well known variant, logical
empiricism" and achieved "academic respectability in American universities" (Feuer 1941, 222). Since 1939, after all, Frank had been teaching
at Harvard.

Feuer admired some aspects of logical empiricism. Regarding Frank's
new book, he liked his call for compromise and cooperation between
Marxists and logical empiricists and Frank's "excellent use of the method
of historical materialism." He also noted with some admiration Neurath's
administrative work in Bavaria on behalf of socialism (ibid., pp. 223,
224, 228–29). Still, much was amiss with logical empiricism. First, Feuer
attacked the popular view that logical empiricism was politically neutral.
Taking aim at Carnap, Schlick, Reichenbach, and Neurath, he scoffed
at their various claims that logical empiricism's "formulations are solely
those of the logic of science":

In fact . . . its distinctive doctrines are not the product of a dispassionate analysis
of scientific practice. They are rather to be regarded as ideological expressions,
the ideology of a group of bourgeois scholars living under unique historical
conditions, the expression of their varying moods and anxieties. (p. 222)

---

4 "Sie bewegen sich nicht in der Zweidimensionalität von Wesen und Tatsache, welche der
eigentliche Ort der Wahrheit ist, sondern nur in der Eindimensionalität der Tatsachen"
(Marcuse 1939, 231).

Even worse, he argued, for all logical empiricism's pretensions about remaining aloof from politics, it actually positioned itself on the *wrong* side of the revolutionary battle. Its emphasis on language and logic and its disdain of metaphysical disputes functioned as a distraction from, if not denial of, the all-important "world of social relationships." Marx knew all about this kind of evasive maneuvering by intellectuals, Feuer explained:

Marx used to speak of the bourgeois economist who could only see economic relations through the mist of a "fetishism of commodities," and who lost sight of the world of social relationships. Likewise, the bourgeois empiricist, preoccupied with his language, comes by a kind of occupational affliction to elevate his symbols into a self-contained substitute for the world. His ideology reflects his social isolation; his theory, separated from practice, becomes an autonomous realm of theory. (p. 224)

Where Horkheimer complained that a logical empiricist was inevitably a student of the status quo, Feuer complained that he or she must completely disengage from the world:

The limits of his empiricism are unconsciously dictated by the narrow confines of his own activities, by his withdrawal from active scientific participation and confrontation with facts. The "empiricist" wishes to reject a world which he cannot manipulate like his symbols, at will. (p. 224)

Unless they accepted Marxism, Feuer believed, logical empiricists would remain powerless to break out of the solipsistic, linguistic, and idealistic prisons they had built for themselves. Even Neurath's otherwise laudable efforts to plan and administrate a socialized economy in Bavaria, Feuer wrote, had been doomed to fail because of this escapist "fetishism of propositions."[5]

A year later, Feuer attacked again. Writing once more in *Science & Society*, he attacked logical empiricism for its noncognitive view of ethical propositions. As it was with Dewey, Feuer's target was A. J. Ayer's presentation in *Language, Truth and Logic* (Ayer 1936) and, additionally, Carnap's "Philosophy and Logical Syntax" (Carnap 1935). Feuer dismissed arguments for noncognitivism and a value-free picture of scientific theory by arguing that logical empiricism was itself a historical

---

5 Feuer wrote, "Neurath's pamphlet [on transforming a war economy into a peacetime economy-in-kind (Neurath 1919)] enjoyed a large success among the Social Democrats for whom the endless discussions of the Socialization Commission were an adequate substitute for the taking of a single, practical step. The fetishism of propositions is a reformist ailment of long standing" (Feuer 1941, 228).

expression of class interests. Members of the Vienna Circle, he wrote, belonged to an "academic class" that after the First World War "endeavored to preserve its social status by insisting on the irrelevance of its 'science' to social issues" (Feuer 1942, 252). Though Neurath, Carnap, and Hahn in their pamphlet *Wissenschaftliche Weltauffassung* (1929) had specifically upheld the *relevance* of the new scientific philosophy to modern social movements, Feuer insisted on the contrary that the movement was happily and intentionally isolated from real ideological struggle:

Logical empiricism has a tendency also to make formal logic serve as an ideology. A fetishism of rigor is pursued, a "faith in formal logic." . . . Pure logic served as a kind of facade moreover for the inner conflicts within the academic mind. The "critical method," as thus employed, is a device for guaranteeing that, whatever the questions, the philosopher will end up without taking sides. (Feuer 1942, 254 n. 35)

In Feuer's eyes, most logical empiricists were apologists for false consciousness: "The empiricist does not try to penetrate beneath the level of the 'false consciousness.' He takes the ideological statements in the form in which they are uttered on the conscious level, and shows they are meaningless" (Feuer 1941, 253). From a Marxist perspective, however, the proper task of philosophy is "to go much further." Tacitly nodding to Horkheimer's and Marcuse's critiques and possibly also to Dewey's point that otherwise meaningless statements point to the "cultural conditions" that nourish them, Feuer praised Marxists who employ scientific and "sociological methods" to find the genuine significance (about "class preferences and repressions," typically) of statements that logical empiricism disregards as empty (ibid., pp. 253–54).

One reason Feuer so bluntly attacked logical empiricism and all other non-Marxist approaches to ethics in this essay was his own devotion to Marxism and, it would appear, to Moscow. These early essays exhibit an unquestioning regard for the writings of Marx and Engels and a defensiveness about the Soviet Union. At a time when no reader would fail to think of rumors of deadly mischief behind Stalin's show trials, purges, and the disasters of forced collectivization, Feuer defended Marx's theory of a postrevolutionary dictatorship of the proletariat on the grounds that dictatorial maneuvers are sometimes justified by the ends they bring about. He reached awkwardly for Cromwellian England to reject "a dire picture of the Soviet Union" that Dewey and others had painted as their anti-Stalinism had grown in recent years. "Social relations which are founded historically through dictatorial means," Feuer apologized, "are not themselves irrevocably committed to a dictatorial superstructure" (Feuer 1942, 265).

Like many other philosophical radicals, however, Feuer eventually moved to the political right. Yet he did so much later than the so-called old left of the 1920s and '30s that had already begun to separate itself from communism and Stalin. Through the 1950s and into the 1960s, Feuer supported the New Left and the student movements at the University of California at Berkeley where he taught from 1957 to 1966. Alongside his specialty in Spinoza, he continued to write about Marxism and social philosophy. In 1960, Nathan Glazer recalled, Feuer defended student activists who had demonstrated against the House Un-American Activities Committee in San Francisco (Glazer 1969).

By the mid-1960s, however, Feuer had become highly critical of the student movement. Student movements around the world, he now argued in his popular book *The Conflict of Generations* (1969), were best understood as Freudian generational revolts of youth against their parents – revolts that were, moreover, doomed to degenerate into irrationality and violence. As Feuer felt increasingly isolated at Berkeley, he and his wife Kathryn, a specialist in Russian literature, accepted offers from the University of Toronto, where Feuer joined the department of sociology (see Feuer 1969, 497 n. 85). In subsequent decades, and partly on the reputation of *The Conflict of Generations*, Feuer was adopted by right-wing intellectuals and students critical of postwar academic leftism.[6]

Margaret Schlauch (1898–1986)

Feuer's criticism that logical empiricism was an escape from the social world were seconded and developed by Margaret Schlauch, who taught medieval literature at NYU. When Morton White and Albert Wohlstetter surveyed the popular general semantics movement for the readers of *Partisan Review*, they found that within this field dominated by second- and third-rate scholars the real "friends of semantics" were the logical empiricists of the Unity of Science movement (Wohlstetter and White

---

[6] Criticism of American culture from academics and, mainly, English professors, Feuer wrote, "is a smug, unexamined, and unchallenged consensus that dismisses dissent as a rude intrusion" (quoted in Sullivan 2000).

1939). That might be so, Schlauch implied some three years later as she surveyed the field of semantics in *Science & Society*, but only if semantics was a reactionary project peddling (as her title announced) "Semantics as Social Evasion."

Schlauch explained that one of the most popular books in the general semantics movement, Alfred Korzybski's *Science and Sanity* (1933), purported to trace all our individual and social ills (such as crime, mental illness, poverty) to misleading logic embedded within our ordinary language. But Korzybski's project was bound to fail, in Schlauch's eyes, because it paid no attention to the all-important dynamics of class struggle. Only

rarely, and then in a most perfunctory manner, does he indicate the basic importance of class conflicts and the rivalry of empires. Only the vaguest indications are made concerning eradicable concrete maladjustments in the objective world, such as unequal distribution of income, exportation of capital to helpless "backward" countries, international trade competition, and recurring crises and periods of unemployment, which are the causes of insecurity leading to psychological ills. Exorcism seems easier when these problems are treated as primarily linguistic. (Schlauch 1942, 321)

Stuart Chase's *The Tyranny of Words* and other bestsellers reproduced this mistake, Schlauch explained, by confusing linguistic effects with their social, nonlinguistic causes. Regarding racial stereotypes, for instance, Chase appears not even to consider

the *causes* of [emotion-laden stereotypes] and . . . the need, therefore, for something more than language therapy to solve a situation involving class interest, the nature of the state, exploitation, and a legacy of fear on the part of rulers inherited from the days of slavery. (ibid., p. 325)

As far as society's ills were concerned, semantic linguistic therapy was obscurantist snake oil.

In Schlauch's eyes, logical empiricism did little better. Following Feuer's charge that logical empiricism succumbed to a fetishism of propositions, she denounced Carnap's program of logical syntax for holding that "philosophical positions such as idealism and naturalism" were "not about the world, but only the use of words" (p. 329). Even though the logical empiricists had made

contributions to the analysis of language, it is difficult to escape the conviction that they too have greatly exaggerated its therapeutic values. And their approach, like that of the semanticists, may be used as an avenue of escape from active concern with an external world which daily presses the most urgent problems upon our attention.[7]

If Carnap had provided a potential "avenue of escape" from the world and class struggle, however, Schlauch reserved her main criticism for Harvard physicist and philosopher of science Percy Bridgman. Bridgman, she charged, forged that path of escape, step by step, as he created a reactionary, individualist ideology.

Bridgman's operationism held that the meaning of a concept consists in the set of procedures or operations with which the concept is employed in scientific practice. For natural science, Schlauch suggested, operationism does well, offering plausible, pragmatic analyses of key scientific concepts. But when applied to social and ethical concepts, operationism led to ideological disaster. "The quest for social meaning via operations ends in a blank," she wrote, because collective ideals and goals cannot find home in a semantic theory whose focus is "the individual (as Bridgman conceives him)."

A Marxist could point out . . . that Bridgman's approach is a fundamentally mistaken one, since he posits a deep division between his intelligent individual and society. He sets them as it were in opposition to each other instead of viewing them in intimate dialectical relationship. (pp. 328, 329)

Bridgman's operationism was reactionary because it threatened to obscure the properly collective, dialectical aspects of meaning in modern, progressive life.

In 1942, when Schlauch published her critique, her optimism that collectivism would prevail over Bridgman's individualism was neither naïve or unique. The United States was officially allied with the Soviet Union in the battle against Nazi Germany, and in that light, Schlauch took Bridgman's semantics to be "outmoded":

In the few years since [Bridgman's] book was written, however, its whole message might be felt to be already outmoded by many who were once impressed by it. Joint effort in a hopeful, ultimately constructive war effort is making collective thinking habitual to many who formerly held themselves consciously isolated, as Bridgman's intelligent individual was supposed to do. (p. 329)

---

[7] Schlauch 1942, 330. Here Schlauch referred her readers to V. J. McGill's criticisms in his article "Logical Positivism and the Unity of Science" (McGill 1937), described in chapter 7.

Nor would readers have found out of place her triumphant final call for semantics to join forces with social philosophy and strengthen ties between academic philosophy, materialism, and social progress:

> The authentic investigation [in semantics] will be made, we may be sure, by persons actuated by a social philosophy more generous than any evinced by current writers on the subject. Such a study must be made ... by those who are not seeking escape from the actuality of non-verbal problems. It must be made by students who accept the existence of our material world, with everything in it that is unpleasing to us now; and who are willing to fuse theory and practice in realistic efforts toward its amelioration. For such students, the lore of semantics will become an instrument of social progress instead of technique of social evasion. (p. 330)

Like Feuer, Schlauch remained critical of logical empiricism and radical in her politics through the 1940s. In 1947, she added Charles Morris to her pantheon of reactionary semanticists in her review of his *Signs, Language, and Behavior* in the magazine *New Masses*. Schlauch disliked the book because Morris had not heeded her (or anyone else's) call to cultivate semantics in a Marxist direction. Like other critics of *Signs, Language, and Behavior*, she especially disliked its behaviorism. Morris's "account of ... abstract logical elements of discourse is based entirely on the conditioned behavior of an animal in a laboratory learning to locate food." This reductionism was plainly "unsatisfactory to a Marxist scientist" who acknowledged the different "levels of phenomena with which scientific discourse deals: the physio-chemical, the psycho-biological, and (for the study of human beings) the social-historical" (Schlauch 1947, 17).

Schlauch's disdain may seem misplaced. Far from an eliminative, reductionistic project, semiotic was a tool, Morris explained, to relate, compare, and possibly unify different "material which has been approached in isolation" by scientists in different fields. These included the languages of natural science, biological science, "humanistic studies," and psychology. "The sense in which semiotic itself is a phase in the unification of knowledge should be evident from all that precedes." This needs, he assumed, "no elaboration" (Morris 1946b, 225). Morris's assumption was wrong, of course. Schlauch illustrates how intellectuals on the left could have shared Morris's goal of unifying the sciences partly for the sake of progressive social reform but could abide only one theoretical program, namely Marxism, with which to work toward unification. For Schlauch, a good scientist was a "Marxist scientist" to whom this "social-historical" domain of inquiry was primary.

Unlike Feuer, Schlauch never turned to the right or renounced her political radicalism. In 1951, during the height of McCarthyism and the Red scare in America, she used her sabbatical year as an opportunity to leave the United States and to begin a career as a medievalist in Soviet-controlled Eastern Europe. Instead of returning to New York, she settled in Poland, where her sister and her sister's husband, Leopold Infeld, Einstein's colleague and co-author, had fled after his deportation from Canada (Schrecker 1986, 294–95). Schlauch quickly established herself as a highly respected medievalist at Adam Mickiewicz University in Poznan, her arrival later celebrated as "a breakthrough in the studies of medieval English" in Poland (Fisiak 1984, 1).

Maurice Cornforth (1909–1980)

These criticisms of logical empiricism and the Unity of Science movement from Horkheimer, Marcuse, Feuer and Schlauch in the late 1930s and early '40s were echoed years later in the writings of the British philosopher Maurice Cornforth. Cornforth was a student of C. D. Broad at Trinity College at Cambridge and subsequently moved in British communist circles as a philosophical defender of dialectical materialism and critic of logical empiricism.

Cornforth was assigned the topic of logical empiricism in the volume *Philosophy for the Future: The Quest of Modern Materialism.* Since this volume appeared in 1949, just as anticommunist investigations at the University of Washington set an ominous tone for academic philosophical radicalism, the volume reads today as a swan song of radical philosophy.[8] Besides Cornforth, its contributors include the mathematician Dirk Struik, Leopold Infeld, the Dutch philosopher H. J. Pos, J. D. Bernal, and the volume's editors Roy Wood Sellars, V. J. McGill, and Marvin Farber. Their collective goal was to chart a future for philosophical materialism. Curiously, however, the editors acknowledged the "enormous influence" of dialectical materialism but did not include an essay dedicated to it. It was, they explained, "a matter of regret to the Editors that an article on this subject could not be obtained in time" for publication. If, as this disclaimer might suggest, the editors chose not to include such an

---

[8] Sometime before 1968, one borrower of this volume from the University of Chicago Library warned subsequent readers by writing in the margin: "N.B.: This author is a dialectical materialist, apparently a Marxist."

essay, it was not in order to conceal their sympathies with socialism and its egalitarian economic values. Those sympathies were in full view:

> Because modern materialism recognizes that cultural values must, in general, wait upon the servicing of vital needs, it favors forms of social organization which release the productive forces of the economy, so that men, living in some leisure and dignity, can express their genius, their intellectual and artistic bent. It demands a society which organizes full production for the maximum benefit of all its members. (Sellars, McGill, and Farber 1949, viii–ix)

These values required a more substantial foundation than the "naturalism" defended by Dewey, Hook, and Nagel in their campaign against neo-Thomism. Their naturalism, the editors explained, consisted mainly in criticisms and rejections of other programs "rather than any positive tenets of its own about the cosmos." A properly articulated materialist philosophy would articulate a concrete world picture – "a synoptic view of man and the universe implicit in the sciences at their present stage of development" (Sellars et al. 1949 ix–x).

Cornforth began his review of logical empiricism with a fair and generous account of its historical development within the empiricist tradition and its practical culmination in the Unity of Science movement. But when he considered whether logical empiricism "serves the interests of the advance of scientific knowledge and of social progress," Cornforth described a programmatic train wreck. Following Lenin, he charged that rejecting the idealism-realism question prevented logical empiricists from offering any account of scientific objectivity and put them "in the same camp as the older empiricists, the subjective idealists, although they protest that they reject subjective idealism as well" (Cornforth 1949, 506). Carnap's proposal, for example, that both physicalistic and phenomenalistic languages are viable for reconstructing science was not acceptable to Cornforth: "the difference between saying that a thing is 'a complex of atoms' and saying that it is 'a complex of sense data' is merely a difference in choice of language." With Carnap, he argued, "every essential point has been conceded to the side of subjective idealism" (ibid., p. 507).

Cornforth also joined Schlauch in attacking Morris's *Signs, Language, and Behavior*,[9] and repeated charges against logical empiricism filed by

---

[9] Morris motivated his semiotic project in that book partly on the grounds that semiotic sophistication could help "to protect the individual against the exploitation of himself by others" (Morris 1946b, 240). Cornforth, however, charged that Morris had provided tools to the enemy. Semiotic techniques and terminology might just as well be used by "powerful interests antagonistic to progress" in order "to deceive and hoodwink people" (Cornforth 1949, 520).

Schlauch, Feuer, and Horkheimer. These include the complaint that logical empiricism neither provided nor pursued a positive theory of knowledge that anchored scientific thought in "the actual material social foundations of the thought process which is expressed in language" (p. 507); the charge that it refused to recognize the transformative possibilities of science by fixing attention on empirical facts and sense data (p. 514); and the claim that a mature unified science, as Carnap and Neurath conceived it, was merely a linguistic web, an interconnected system of sentences that need make no contact with material realities (p. 510).

By the late 1940s, Carnap had embraced semantic theory as the study of relations between language and the external world. But Cornforth found this project just as problematic. Both Carnap's and Reichenbach's then recent views of scientific theories, in which theories had empirical meaning as long as elementary statements derived from theoretical statements could be tested, were yet more concessions to subjective idealism. For Cornforth, all Carnap's and Reichenbach's technical sophistication hardly exceeded Arthur Eddington's quasi-mystical pronouncement that, behind the visible "pointer readings and similar indications" involved in laboratory science, "something unknown is doing we don't know what – that is what our theory amounts to."[10]

Semantics could not save logical empiricism from political mischief and irresponsibility, Cornforth concluded, because its focus on *language* would forever prevent it from engaging the material and social conditions in the world. Those conditions extended, of course, into the lives of both science's experimenters and theorists. Singular observations of a meter stick or a track in a cloud chamber are made possible, Cornforth insisted, by surrounding conditions and circumstances in social, economic, and political life. Logical empiricism's attempts to interpret science via an analysis of language neglects the fact that science is a "social activity." Materialism, on the other hand, properly "sees the observational data of science as obtained in the course of men's struggle to understand and master natural and social forces" (Cornforth 1949, 517).

In fact, Cornforth all but provided, on his own terms, a defense of logical empiricism against his charges. For he admitted that, besides making useful contributions to "the study of language and signs" and to "formal and mathematical logic" (ibid., p. 520), logical empiricism was poised to make substantive contributions to science. Especially in light of the excessive specialization and isolation of scientific languages, he wrote, "the

---

[10] Cornforth 1949, 513–14. Cornforth quoted Eddington's *The Nature of the Physical World.*

logical empiricists are undoubtedly right in stressing the importance of the logical study of this language [of science]." It is important

not only for the philosophical understanding of science and its significance for us today, but also for the development of science itself, since theoretical difficulties arise which can be shown to be associated precisely with the use of the instrument of language. (p. 517)

If logical empiricism was poised to engage scientific practice, however, it would appear no longer incapacitated by subjective idealism, as Cornforth charged. It could engage social and economic realities as much as other sciences. But Cornforth did not follow such a line of reasoning as he returned, again and again, to a picture of logical empiricism trapped in "the shackles of the traditional empiricist standpoint," unable to see science as more than "a means of referring to observational data and organizing expectations of future observational data" (pp. 518, 517) and thus forever wedded to the presocialist status quo.

The Critique from the Left

From Lenin's Russia, Horkheimer's, Feuer's, and Schlauch's New York, and Cornforth's England, logical empiricism and its Unity of Science movement were seen as a pair of connected projects that commanded the attention of the intellectual, philosophical left. As powerful, creative and iconoclastic as these projects were, however, and as much as they joined the radical left in promoting the ideal of a unified science and rejecting popular opiates such as theology and superstition, they were consistently seen as impaired by a cascade of mistakes triggered by their refusal to endorse metaphysical realism (or, specifically, dialectical materialism). First, since they did not accept metaphysical realism, they could not provide an acceptable account of scientific objectivity. In addition, their formalism and exclusive attention to logic and language deprived them of the power to situate and to critique scientific knowledge in its socio-economic contexts. Nor could they provide a positive theory of discovery that embedded scientific ideas, again, in these contexts. At the same time, as Horkheimer emphasized, they promoted an unflattering, incomplete portrait of human beings as mere symbol manipulators and rational calculators who, from a radical perspective, lacked the power and imagination to remake the world anew. Aside from Cornforth, who wrote somewhat later, what seemed most maddening to these critics in the late 1930s and early 1940s was the popularity of logical empiricism and the

Unity of Science movement. For the antimetaphysical arguments in logical empiricism and the movement at least threatened the metaphysics of dialectical materialism embraced by most radical philosophers.

In part, this critique was shared by the other, more moderate leftist philosophers who nonetheless admired logical empiricism and the Unity of Science movement or called for cooperation between them and Marxist philosophers and critics. John Somerville and William Gruen called for such collaboration, while Morris and especially Dewey (as discussed in chapter 4) contributed to Neurath's *International Encyclopedia* in order to inject a broad social and pragmatic approach to meaning and value into the movement's program. Without sharing or, at least, without relying on Marxist orthodoxies about class struggle or dialectical materialism, these critics joined Horkheimer, Schlauch, Feuer, and Cornforth in believing that logical empiricism's interests in logical and linguistic analysis were too narrow to defend science against its enemies effectively and to promote science into the leading creative, social, and cultural role it deserved to have.

As this critique from the left took shape, however, another more radical critique issued from philosophers who explicitly identified themselves and their work as "communist." From this point of view, explored in the next chapter, some important ground shared by logical empiricism and their leftist critics comes into view, namely a distinctly intellectual conception of philosophy (or logic of science). Radical leftist philosophers may have variously charged that logical empiricists did not understand language, values, or the substance of the realism-idealism controversy. But they did not believe, as these communist philosophers did, that logical empiricists and their supporters fundamentally misunderstood the point and function of philosophy itself.

7

The View from the Far Left

Logical Empiricism and Communist Philosophers

The most radical camp of philosophers that observed and criticized logical empiricism and the Unity of Science movement upheld Marxism as both a true and comprehensive theory of the world and a foundation of Communist Party policy. The editors of *The Communist* acknowledged no legitimate division between intellectual theory and revolutionary practice. Its subtitle was "A Magazine of the Theory and Practice of Marxism-Leninism Published Monthly by the Communist Party of the U.S.A." Symbolizing this identity of intellect and party, the journal's chief editor in the late 1930s and early '40s was party head Earl Browder.

Browder was assisted by V. J. Jerome, who operated within the party as a kind of cultural and intellectual commissar.[1] Jerome and book reviewer Philip Carter together wrote about philosophy and science for *The Communist* and they did so in a way that set them apart from the other camps on the philosophical left of the 1930s and '40s. For them, the relationship and priority between philosophy and politics understood by the others was inverted. Recall that even for emphatic leftists such as John Somerville, Margaret Schlauch, the young Lewis Feuer, Maurice Cornforth, and arguably Lenin himself – who surely considered, at least, that their own philosophical pronouncements were as much propped up by economic determinants as those of their philosophical targets – philosophical criticism was an independent tool that one could use freely

---

[1] Jerome, whose real name was Isaac Jerome Romaine, was later named by Elia Kazan (in his official testimony against the so-called Hollywood Ten) as a main contact between Hollywood and the Communist Party. Jerome was later imprisoned under the anticommunist Smith Act (Fast 1990, 272; Folsom 1994, 297).

and reliably to navigate the world of theory and practice. For Jerome, Carter, and Browder, however, philosophy and philosophy of science had no prior, independent position in intellectual affairs. Instead of using philosophy to articulate, refine, and popularize the tenets of dialectical materialism (as the editors of *Science & Society* pledged to do in their new journal), Jerome, Carter, and Browder took the fundamental writings of Marx, Engels, and Lenin as absolute yardsticks by which the adequacy and value of philosophy and philosophy of science were to be measured.

Philosophy Is the Party

One of Maurice Cornforth's arguments against logical empiricism helps to introduce a distinctive feature of the philosophy practiced in *The Communist*. At the end of his essay in *Philosophy for the Future*, he wrote,

All philosophers who work in universities operate in factories of ideas which go forth into the world and play a part quite independently of the intentions of their creators. The ideas propagated in the discussion and writings of logical empiricists certainly do not help people to form a scientific picture of the world and man's place and prospects in it, or to understand how science can be utilized in the service of the common man. But this very scholastic negativity, this very failure to present a scientific philosophy comprehensible to the common man, means that the way is left open for the deception of the people by supernatural, idealistic and antiscientific illusions. If the scientific philosophers put it out that science is concerned with nothing but pointer readings, the common man will agree ... that in that case science does not tell us very much. It is vain for philosophers to pretend that they are energetically combating idealism and irrationalism and supernaturalism as meaningless metaphysics, when they have nothing to put in their place, and when they just as energetically combat the scientific materialism which is the only practical alternative. (Cornforth 1949, 519–20)

Cornforth demanded that logical empiricism be a philosophy of science that was not only coherent, but also comprehensive and popular. For the revolutionary philosopher, the task was not only to understand science or knowledge properly, but further to provide appropriate intellectual tools for "the common man" engaged in the realities of social life. Once logical empiricists had deflated the pretensions of superstition, religion, and ineffective philosophy, Cornforth worried, what would they provide "to put in their place?" Some kind of "scientific materialism" was the only "practical alternative" Cornforth could see.

Cornforth's point suggests a common feature of all prewar leftist philosophy of science: the view that criticism of reactionary or illegitimate ideas was not enough. Society further required an outlook or world-view that somehow effectively substituted for what was lost. In the case of Dewey, Nagel, and Hook, for instance, the replacement proposed was "naturalism." The logical empiricists of the Vienna Circle proposed their *Wissenschaftliche Weltauffassung*, which Morris complemented and expanded into his semiotically structured "scientific empiricism." Farther to the left, philosophers writing for *Science & Society* or Sellars's *Philosophy for the Future* proposed versions of materialism, dialectical or other. At the extreme left, however, Browder, Jerome, and Carter answered Cornforth's demand somewhat differently: The "common man" should be supplied with not only a theory or theoretical outlook on the world, but a form of life defined by various practices, institutions, and other embodiments of that theory. That form of life was the Communist Party.

For these communist philosophers, the party would provide all the intellectual tools, perspectives, and beliefs that the masses required. A sidebar in a 1939 issue of *The Communist* quoted Browder to make this point:

The greatest contribution of all, which Marxist-Leninist theory has given to the masses, is *the Party*. The Communist Party is the organized theory, embodied in growing tens and hundreds of thousands of men and women, preserving and transmitting the experience and wisdom of past generations, enriching it by the experience of the present, transmitting it to the broadest masses, providing thereby the illumination, the guidance, the leadership, which will organize victory for the masses in their age-long struggle against the forces of darkness and reaction. (vol. 18, p. 169)

Jerome made a similar point several issues earlier. The party became the crucible of Marxist theory in revolutionary Russia, he explained, when "the revolutionary content of Marx and Engels was restored and developed only there where Bolshevism as a party came into being in the historic split with Menshevism" (Jerome 1938, 90). Where Menshevism required parliamentary debate and analysis as prolegomena to gradual socialist reform, Bolshevism dispensed with the need for independent intellectual analysis and fused theory and practice in revolutionary action. The result, often insufficiently understood, Jerome explained, was "the Party nature of philosophy." "Implicit in Marxism is the vanguard party of the proletariat" (ibid.).

Trotsky: Heretic

Communist criticism of logical empiricism is connected to its criticism of pragmatism. In the pages of *The Communist*, criticism of pragmatism, in turn, is inseparable from anger toward Dewey, Hook, and others for their defense of Leon Trotsky. Consider Philip Carter's review of Dewey's *Logic: A Theory of Inquiry*. Carter first criticizes Dewey on philosophical grounds familiar to the philosophical left. He complains that Dewey's pragmatism is not sufficiently friendly to metaphysical realism and thus obscures our view of "a correspondence of knowledge with reality" that is "progressively attained through the dialectical materialist advance of science." For Dewey, propositions are "mere means to an end" so that "truth, which Marxism has shown to be a reflection of reality, drops out of the picture." Thus, Dewey's book "totters continually on the brink of subjectivism" (Carter 1939b, 164, 165).

As he neared the end of his review, Carter noted that this philosophical mischief created social mischief. In particular, it was an obstacle to social planning. For Dewey's (as well as William James's) pragmatism, he explained, celebrated a pluralism through which "all the doors of the universe, or of human possibility, open, so that crackpots and even fascists and Trotskyites were allowed to enter to disrupt any unified plan for social progress" (ibid., p. 168). Though resentment against Trotsky seems to percolate throughout the review, once Carter had mentioned Trotsky, it boiled over in an ad hominem attack – against not Trotsky but Dewey:

A present-day liberal who brings unjust charges against the Soviet Union and gives aid to its enemies, cannot continue in the character of a liberal. The man who in 1928 warmly praised the Soviet Union, ten years later joined the preposterous Trotsky "commission," a set of men who publicly advertised their frantic enmity to the Soviet Union, which they propose to judge "impartially." (ibid.)

Dewey's defense of Trotsky was "the great blunder of his career" and it was directly caused by his inadequate philosophy that relentlessly relativizes and subjectivizes truth and forever hesitates to take disciplined, principled revolutionary action. Dewey's pragmatism "cannot, by its very nature, accept the consequences of far-sighted, relentless struggle for democracy and socialism" (p. 169). The Moscow Trials and the resulting exiles and executions of Stalin's enemies, Carter seemed to suggest, were merely consequences of that noble, far-sighted struggle.

Dewey was also attacked by Jerome in a review of Theodore Brameld's article "American Education and the Class Struggle," which appeared in the first issue of *Science & Society* (Jerome 1937). Jerome dismissed Brameld's article as thoroughly corrupted by Deweyan pragmatism, a charge that led Brameld to reply (in a letter to the editor) that Jerome's comments were perhaps not very objective, in light of the fact that "professor Dewey is not popular with the Communist Party at the present time." Brameld, who was devoted to Dewey and Communism alike (and had just published his book *A Philosophic Approach to Communism*) tried to move debate away from Trotsky and toward appreciation of Dewey. Dewey's work, especially its "frequently dialectical approach" to problems, was a valuable tool for progressive, Marxist philosophy, Brameld argued. Because Dewey's methods were appropriate for introducing Marxism to an academic audience that was often prejudiced "against anything Marxian," they were also an asset to the anti-Fascist popular front, the success of which "depends upon the principle that, despite basic differences between liberals and Marxists, they can agree enough to work together in an effective program against fascism and for democracy" (Brameld 1938, 381).

Jerome would have nothing to do with Brameld's compromises. Pragmatic methods for fighting fascism were no justification for philosophical pragmatism. He replied that there was no acceptable middle ground between pragmatism and dialectical materialism and that every committed Marxist philosopher ought "to wean a pragmatist from his philosophy and gradually . . . win him over to *dialectical materialist philosophy* in order to make him a good colleague in the struggle for peace and democracy."[2] Pragmatists coming to the revolution would have to leave their pragmatism at the door. Dewey, of course, was not invited.

Jerome's Survey of *Science & Society*

There was a typical format for philosophical criticism in *The Communist,* one that illustrates the delicate job that Browder, Jerome, and Carter performed as intellectual critics. On the one hand, they had to uphold ideological and intellectual orthodoxy. On the other, they had to cultivate solidarity and avoid offending intellectuals who, stung by their criticism, might turn away from the party. Often, Jerome and Carter skewered their targets in the body of their articles, but then moderated their tone and became more conciliatory and collegial as they neared their final sentences. Heretics were scolded, and then warmly encouraged to pursue Marxist-Leninist truths more aggressively in their future writings.

---

[2] Jerome responded to Brameld's letter in *The Communist* 18: 382–84.

This balancing act took the spotlight in an extended, two-part review by Jerome of the new Marxist journal *Science & Society* (Jerome 1937; 1938). Jerome examined its first year's issues and declared that the results were mixed. When it came to the essays concerning philosophy and semantics, he praised some articles outright (including Margaret Schlauch's "Social Basis of Linguistics" and Dirk Struik's "Concerning Mathematics") but qualified his approval of V. J. McGill's "An Evaluation of Logical Positivism." McGill, who had met Neurath personally at the New York reception described by Nagel, wrote about logical empiricism twice in the first year of the new journal. Though Jerome cited only the first article, he detected that McGill, unlike his colleagues Schlauch and Feuer, and despite his clear criticisms of logical empiricism, was nonetheless attracted to Neurath's movement.

In his first article, for instance, McGill noted that "logical positivism is usually at variance with dialectical materialism at many points" (McGill 1936, 77). Along with Schlauch and Feuer, he depicted the project as having retreated from the social class struggle and into a "tower of symbols" (ibid., p. 78). But that did not mean, he emphasized, that logical empiricists were "reactionary philosophers. They have not yet turned to religion as have so many other schools, nor have they offered their services to reaction; and most of them are liberals and some even claim to be Marxists of a sort" (McGill 1937, 77). With Neurath in mind as a logical empiricist Marxist "of a sort," McGill was happy to see Marxism as a broad, progressive program that far exceeded questions about metaphysical orthodoxy. In light of Carnap's *Logical Syntax of Language* and the principle of tolerance, he added, "the newest phase of logical positivism is, so far as it goes, almost as realistic as the Marxist could desire" (McGill 1936, 77).

By the fourth issue of *Science & Society*, McGill was even more sanguine about relations between Marxism and logical empiricism. In "Logical Positivism and the Unity of Science" (McGill 1937), he called for outright cooperation: "Marxists are also committed to the unification of science," he noted, and suggested that the Soviet Union would probably be a natural proving ground for the movement. There, "communication between theorists, planners, laboratories, factories and schools should increase enormously, and . . . the alienation of workers of hand and brain, so prejudicial to the unity of science, should be resolutely combated" (ibid., pp. 552–53). Again, McGill used Carnap's "extraordinary" book, *The Logical Syntax of Language*, to make a case for organized collaboration involving not only logical positivism and Marxism, but also working scientists.

Referring to the upcoming Fifth International Congress to be held at Harvard, McGill wrote,

> It occurs to me that there might be a special advantage in admitting realists and Marxian materialists to the discussions of the Congress for the Unity of Science. Any principle of tolerance which would admit them would also admit hard working scientists who are uninstructed in philosophy and ignorant, in particular, of such refined doctrines as positivism and idealism, doctrines which scientific investigation by itself does not seem to suggest or to encourage. (p. 561)

McGill took Carnap's principle of tolerance to be a big tent capable of uniting not only Marxists and logical empiricists, but also philosophers and scientific workers, in common action.

For Jerome, all this admiration simply showed that McGill failed to understand what he was talking about. He explained away McGill's sympathy and interest, "the obvious abatement of McGill's attack," by arguing that McGill simply failed to undertake "an adequate analysis of the class basis of logical positivism":

> Were this basis made clear, the social implications and dangers of this philosophic trend would be brought out more explicitly. The high-titled metaphysical medley [of logical empiricism] would be laid bare as the embroidered poverty of bourgeois philosophy. The pretensions made in its behalf would be exposed as the rationalizing logistics of a social order that has long lost its reason for being. (Jerome 1937, 1151)

The future convergence that McGill envisioned between logical empiricism and dialectical materialism was for Jerome as unimaginable as a convergence between Marxism and Deweyan pragmatism.

For all his criticism of McGill and others, however, Jerome had to sound the right note of collaboration between *Science & Society* and the party. He knew, after all, that the new journal's "unfortunate silence" on the question of Trotsky's heresy suggested that it was a Trotskyite "camp organ" (Jerome 1938, 88). Nonetheless, he called on the editors (including McGill) to remain true to Marxist-Leninism. He especially urged them to remain vigilant against Trotskyite anti-Stalinists who were now ascendant in colleges and universities (where "imposters like Sidney Hook are given chairs from which to teach 'Marxism'"(Jerome 1937, 1147)). Indeed, *Science & Society* "has its work to do," Jerome concluded at the end of his review. For that reason, despite its flaws, it deserved "the fullest support of the Communist Party and of all progressives" (Jerome 1938, 90).

Unity of Science in *The Communist*

Jerome's resentment about "diluters and distorters" of Marxism and "pseudo-Marxist academicians" (Jerome 1937, 1146, 1147) was matched by his journal's indictment of the U.S. educational system. Richard Frank's "The Schools and the People's Front" took for granted that public schools in America operated for "the *training of efficient and docile wage slaves*" and, thus, "the perpetuation of capitalism" (Frank 1937, 433). It was true, Frank admitted, that colleges and universities were less beholden to "the state machinery" of capitalism. Yet as long as Marxism-Leninism was withheld from students, the educational system might produce, at best, "a cynic or an ineffectual reformist rather than a revolutionary" (ibid., pp. 444–45).

To remedy this situation, Frank – not to be confused here with Philipp Frank – proposed educational changes that were quite sympathetic to the Unity of Science movement. Using arguments and observations similar to some of Neurath's, Frank called for something like a parallel movement in colleges and universities:

Scientific training is offered as a commodity in the higher schools, but in accordance with the general planlessness of capitalism, no effort is made to care for the well-rounded intellectual advance of students. Subjects are offered haphazardly to be chosen at the student's discretion. Thus even in college it is almost impossible to acquire a genuinely scientific outlook. (p. 444)

Neurath would have agreed. A "genuinely scientific outlook" was one that emphasized the unities, interconnections, and practical compatibility among different areas of study. "A world view is ultimately a consistent whole," Neurath wrote in his early Marxist essay "Personal Life and Class Struggle" (1928, 297). Marxism offered "a kind of solid skeleton that serves as support [for] a powerful body of thought," unlike bourgeois intellectual life, Neurath explained, which dabbles in disconnected fields with contradictory methods and presuppositions:

Today it sends out feelers in all directions, proceeding here in an anthroposophic way, there mathematicising, here psychologising, there pursuing the idea of fate, here in a technical manner, there in an occultist one. The wealth of scientific detail is no longer held together by a unitary approach, and in a certain sense it is left to chance whether a man thinks about some linguistic formations in Chinese or about a medieval legal text, about African beetles or about wind conditions at the North Pole. (ibid., pp. 294–95)

Neurath joined Richard Frank and the editors of *The Communist* in endorsing unified science as a jointly political and intellectual goal. He shared – as Jerome put it – "a conscious purpose to transform science into an instrument for re-fashioning society" (Jerome 1938, 91).

But Neurath stopped well short of their confidence in dialectical materialism. On the one hand, Neurath and the Unity of Science movement distinguished a scientific approach or attitude toward the world from specific and detailed theories about the world used to explain and predict events. Unified science was the task of a scientific approach and attitude to the world, a *Wissenschaftliche Weltauffassung*, which was not itself a body of articulated scientific, much less metaphysical, doctrine. For true believers, moreover, dialectical materialism was regarded as a complete and true theory of the world of just the sort that Neurath's pseudorationalist strictures consistently denied was available to us. One would never detect any such skepticism in the pages of *The Communist*, however, where Marxist-Leninist writings and alleged laws of dialectical materialism were regularly regarded with pseudorationalistic confidence, if not awe.

Dialectical Materialism

There are three laws of dialectical materialism. Despite being vague and qualitative, these laws were widely regarded by Marxists as a basis for understanding, explaining, and predicting the structure and evolution of nature and human society. Originally articulated by Marx and Engels, and developed further by Lenin, Plekhanov, and other theorists, the three laws were summarized by John Somerville in Dagobert Runes's *Dictionary of Philosophy*. They are: (1) the "law of interpenetration, unity and strife of opposites," (2) the "law of transformation of quantity into quality and vice versa," and (3) the "law of the negation of negation" (Runes 1960, 79). Against a metaphysical canvas depicting the world as matter or material in constant motion, the first law depicts existing material things as composites of opposites, the tensions among which keep them continually in motion and flux. The second addresses qualitative properties of things or systems that emerge as they increase their complexity in some quantitative sense. The third refers to the unending, dialectical sequence through which quantitative and qualitative changes yield new "syntheses" or phases of development that, in turn, are subject to opposition and further dialectical development.

In the pages of *The Communist*, these laws were taken for granted. For example, Carter recommended J. B. S. Haldane's book *The Marxist*

Philosophy and the Sciences as a useful primer in technical dialectical materialism:

The statement of the principles, the unity of theory and practice, materialism, and dialectical materialism with its three laws, follows the account of Engels very closely and should prove especially helpful to scientists who are making their first approach to Marxism. (Carter 1939a, 572)

The three laws are also treated as the foundation of the unity of the sciences and a tool to help realize that unity in practice. In the book under review, Haldane himself wrote:

I am convinced that Marxism proves the greatest value in studying the develop-ment of science and the relationship of the different sciences to one another, particularly the relation of chemistry to physics, and of biology to chemistry. And it is particularly useful in those branches of science which are themselves con-cerned with change, for example, in the theory of evolution. (in ibid., p. 573)

Carter agreed and approved Haldane's account of how "recent achieve-ments in mathematics, astronomy, physics, chemistry, biology, psychol-ogy and sociology exemplify and embody these principles" of Marxism (p. 572).

On the other hand, Carter criticized British philosopher H. Levy's *A Philosophy for Modern Man* as a catalog of mistakes about dialectical materialism. In particular, Levy failed to appreciated properly how the laws of dialectics unify and relate the sciences to each other. He failed, Carter explained,

to describe with sufficient concreteness the specific dialectical interplay between particular sciences. This is due in part to his almost exclusive preoccupation with the passage of quantity into quality [i.e., the second law] and his neglect of the other two laws of dialectics. (Carter 1938, 668)

Levy broke ranks with orthodoxy because he was *too* eager to explicate the unity of the sciences. He argued that the second law of dialectics was prior to the others and itself a unifying basis to which they (in some sense) reduced. Carter disapproved and was confident that Engels and Lenin would, as well:

In his effort to fuse together into one law the passage of quantity into quality, the interpenetration of opposites and negation of negation, Levy has weakened and distorted the latter two principles and confined their operation to a single phase, which he wrongly calls, "the dialectical point." Every point is a dialectical point. Every phase of motion is a unity of opposites. This principle is as old as Heraclitus (sixth century B.C.) and it is a pity that Levy did not make full use of it. Engels and Lenin put the principle very clearly. (ibid., p. 670)

For Carter, these laws contained nothing less than the wisdom of humanity and were not to be toyed with. Still, at the end of his review, Carter extended an encouraging hand of comradeship to Levy across the Atlantic: "It is highly likely that the present trend toward dialectical materialism on the part of distinguished British (and American) scientists will continue and that theoretical shortcomings will be ironed out in the process" (ibid.).

Another belief commonly held by dialectical materialists was that these laws promised a theory of discovery, an account of the origin of scientific ideas. Feuer's and Schlauch's criticism of logical empiricism moved in this direction, insofar as they criticized it for ignoring the embeddedness of scientific research in social and economic realities that shaped and guided that research. In his review of Haldane, Carter complained that the book was "too short" specifically because Haldane did not explain in enough detail how the laws of dialectics inform and guide scientific discovery. Had Haldane taken the time to explain these dynamics, "these principles [of dialectics] would have emerged as internal and integral factors, as the constant and inevitable accomplices of the whole process" (p. 573). Too often, Carter observed elsewhere, "the living process of inquiry, experiment and discovery" was taken to be some inexplicable "miracle" or the work of inscrutable "genius" (Carter 1939b, 163). But Marx had made clear in his theses on Feuerbach "the dependence of theoretical knowledge upon the practical activity of men, including of course economic and political activity" (ibid., p. 165). It remained, therefore, the responsibility of philosophers to refine our understanding of the laws of dialectics as tools to help us to understand and to advance the course of science.

The Landscape of the Left and the Reputation of the Unity of Science Movement

Since party philosophers writing for *The Communist* and their less doctrinaire comrades writing for *Science & Society* equally looked back to Lenin's *Materialism and Empirio-Criticism* for guidance in matters epistemological, metaphysical, and scientific, their critiques of logical empiricism and the Unity of Science movement had a family resemblance. Still, there are important differences. Broadly speaking, Schlauch, Feuer, and Cornforth (who criticized logical empiricism and the movement) as well as McGill, Somerville, and Gruen (who were much more sympathetic) were committed to Marxism and Marxist ideas because they found them intellectually and politically agreeable. For Jerome and Carter, however, these ideas were more than merely agreeable. They had a sacred, unquestionable quality that, joined with idea that party life was a kind of embodiment of proper philosophy, encouraged anti-Stalinist criticism that communist intellectuals had traded away their intellectual freedom.

Most important, these portraits of the philosophical left provide context for understanding how the Unity of Science movement and its leaders could come to be seen by various colleagues and authorities as dangerously leftist (if not actively seeking to import communism into the United States). As this and earlier chapters show, logical empiricism and the movement occupied roughly a middle (or, perhaps, center-middle) position in the spectrum of the left. In venues such as the *New York Times* and *Partisan Review*, they were praised as friendly to socialism. Farther to the left, however, most writers for *Science & Society* urged their reform, while philosophers of *The Communist* urged that they be discarded as reactionary and bourgeois. In the big picture of leftist philosophical life before and during the war, logical empiricism and the Unity of Science movement were in the center of the frame.

As subsequent chapters show, however, the moderate leftism of logical empiricism and the movement of the late 1930s became, relatively speaking, more extreme during the Cold War. As intellectual, popular, and political cultures moved to the right in the years and decades after the Second World War, those individuals or projects that remained on the left had fewer neighbors, appeared to stand closer to those populating the extreme left, and thus became more visible as targets for criticism and suspicion. In some cases, Frank, Morris, Neurath, and Carnap in fact took, or had taken, positions on intellectual as well as popular issues that were shared with radical and communist philosophers of the 1930s and early '40s. Besides their devotion to the cause of unifying the sciences, for example, Neurath and Frank shared the view that logical empiricism was too abstract, formal, and "scholastic," just as Feuer, Schlauch, Cornforth, and others had argued. Frank, in addition, effectively agreed with the far left when he planned to make the new Institute for the Unity of Science a sponsor of collective, coordinated research in the sociology of science. As if answering to McGill's, Somerville's, and his own calls for cooperation between dialectical materialists and logical empiricists, Frank's sociological ambitions easily appeared designed to join logical empiricism's "scholastic" and abstract studies of science to Marxist-friendly research in the social conditions in which scientists worked and the effects of those conditions on science itself. Along with Carnap's habit of supporting workers' causes and international peace in the pages of the *Daily Worker*, these and other circumstances help to explain how Neurath and the Unity of Science movement, despite their antitotalitarian and antimetaphysical credentials, would nonetheless come to have a reputation as "communistic" by the end of the 1950s.

8

Postwar Disillusionment, Anti-Intellectualism, and the Values Debate

American culture's tolerance for communists and "fellow-traveling" sympathizers fluctuated according to both domestic and international circumstances. Communism was broadly acceptable during the early 1930s when federally run projects such as the WPA were benefiting many and when the Great Depression vividly posed the question (and nodded toward the negative) of whether capitalism was viable. It was also possible to be a communist during the last half of the 1930s (when most logical empiricists arrived in America) because Moscow and the Communist Party reached out to progressive and liberal organizations to form a united, "popular front" against Nazism and fascism. During the last years of the war, America and the Soviet Union were official allies, and thus not communism but *criticism* of communism could be deemed unpatriotic.

Always in the background, however, and filling the gaps between these interludes were events and circumstances that eroded communism's popularity and acceptability among intellectuals and the broader public. Beginning with Stalin's rise to power in the late 1920s, increasing numbers loss respect either for the Communist Party or for the idea of Communism. For the many intellectuals who were devotees of Leon Trotsky, Stalin's persecution of Trotsky and other enemies and rivals was dismaying enough, even without the insulting show trials through which Moscow asked the West to pretend that Stalin's consolidation of power was altruistic and justifiable. Equally damaging was the debacle of Soviet agrarian reforms in the early 1930s (known as "five-year plans") that led to widespread starvation and death among Soviet peasants. News of these items sometimes traveled slowly and was often discounted by the faithful as capitalist or fascist propaganda. For true believers, the USSR was encountering predictable bumps on its way to a utopian future. For others, the evidence began to pile up that Stalin, Marxist communism, or both were neither what they claimed to be nor what most leftist intellectuals had dearly hoped them to be.

For many, the greatest blow to their faith came in August 1939 with news of Stalin's nonaggression pact with Hitler. If there was a universal plank within the political left, it was opposition to fascism. Even with doubts and worries about Stalin's motives, many continued to support Russia as a valuable bulwark against the spread of fascism through Europe. It was therefore unthinkable to many that Stalin would make such a huge concession to Hitler. Though contemporary historians still debate the extent of its impact (party membership, for instance, did not precipitously and immediately decline), contemporary literature such as the testimonies of six prominent intellectuals in the book *The God That Failed* (Crossman 1949) shows that it was for many a decisive turning point.

The God That Failed began with an essay by Arthur Koestler in which he used the metaphor of a high wire or tightrope:

At no time and in no country have more revolutionaries been killed and reduced to slavery than in Soviet Russia. To one who himself for seven years found excuses for every stupidity and crime committed under the Marxist Banner, the spectacle of these dialectical tight-rope acts of self-deception, performed by men of good will and intelligence, is more disheartening than the barbarities committed by the simple spirit. Having experienced the almost unlimited possibilities of mental acrobatics on that tight-rope stretched across one's conscience, I know how much stretching it takes to make that elastic rope snap. (Koestler 1949, 71–72)

Koestler's final break occurred not after his imprisonment in Franco's Spain, nor after close friends in the party suffered through baseless and paranoid denunciations as spies for Hitler. It happened instead on "the day when the swastika was hoisted on Moscow Airport in honor of Ribbentrop's arrival and the Red Army band broke into the *Horst Wessell Leid.* That was the end" (ibid., pp. 67, 72, 74). Some of the New York intellectuals experienced their "snap" earlier, partly because they were such close observers of the Soviets and the Communist Party.

The Conversion of Sidney Hook

Before the mid-1930s, Sidney Hook was among the leading defenders and expositors of Marxist socialism, his reputation secured by his book *Towards the Understanding of Karl Marx* (1933). A telling snapshot of mid-decade Marxism among Hook, Dewey, and others is the collection *The Meaning of Marx, a Symposium* (Russell et al. 1934) published the following year. Intended mainly for students and teachers, the book presents a symposium in which Hook passionately defended Marxism against its critics. He praised Marxism for being comprehensive and respecting the "culture of a society" as a "living whole" (in ibid., p., 56). It was properly naturalistic and scientific, philosophically anti–a priori, and politically progressive and socialistic. "This bare outline of the meaning of Marx," Hook added, "is intended only as an introduction. . . . [It] cannot suggest the wealth of insight [his writings] contain nor reveal the perspective which the Marxian approach to culture opens up" (in ibid., p. 81).

Despite Hook's enthusiasm for Marxism, his turn against communism had begun. His main worry was the Communist Party and Stalin, whom he believed had begun to pervert, misunderstand, or simply abandon its responsibilities as the standard bearer of socialist society. Instead of representing the socialist masses, he complained, the party had become bullying and capricious. The 1930s were witnessing the "substitution of the *dictatorship of the party* for the dictatorship of the proletariat" (in ibid., p. 80).

Dewey, Bertrand Russell, and Morris Cohen responded to Hook, each of them writing according to the theme, "Why I Am Not a Communist." They largely agreed with Hook's criticisms of the Soviet party leadership, but they were more critical and suspicious of revolutionary Marxism than Hook was willing to allow. All three, and especially Cohen, rejected the Marxist view that prosocialist change must be violent and revolutionary. "The program of civil war," Cohen wrote, "may bring more miseries than those the communists seek to remove" (in ibid., p. 92). They also rejected the view that the future of America was inevitably socialist. They denied that communism and fascism were the only available alternatives and questioned why liberal, nonrevolutionary reform of capitalism would not be possible (ibid., pp. 85, 89, 100).

For Hook, however, the alternatives could be ruled out to show that communism was more or less inevitable. Fascism was transparently noxious; capitalism was "wasteful and destructive." There was therefore no practical good to be had by dwelling on the problems of "the communist position":

[I]t is the absence of a realistic alternative program and path of action which makes the criticism of the communist position – justified as it may appear to be from an abstract ideal position – irrelevant to the pressing tasks of combating capitalism, fascism, and war. (in ibid., pp. 104, 105)

Yet because the communist leadership in Moscow had proved itself "irrelevant" to these "pressing tasks," Hook called for a new American leadership to lead the way effectively, and without corruption, to a Marxist future:

It seems to me that only communism can save the world from its social evils; it seems to me to be just as evident that the official Communist Party or any of its subsidiary organizations cannot be regarded as a Marxist, critical or revolutionary party today. The conclusion is, therefore, clear: *the time has now come to build a new revolutionary party in America and a new revolutionary international.* (in ibid., p. 144)

The "new revolutionary party" Hook had in mind was the short-lived American Workers Party, whose platform he had helped to draft at its founding in late 1933. Hook was indeed a socialist radical at this time, but he would no longer take orders from Moscow or the Communist Party.

In a few years, Hook's radicalism was greatly eroded by the Moscow trials of 1936 and 1937. In the wake of the Dewey Commission's determination that Moscow's charges against Trotsky were empty, Hook was dismayed that this news was not received by leftists as proof of Moscow's decline into corruption and dishonesty, as "a deadly indictment of the Soviet Union's pretension to have developed a free, democratic, and Socialist culture" (Hook 1987, 234). Instead, as Hook wrote in his autobiography, many of his fellow intellectuals seemed suddenly blind, stupid, or hypocritical because they could not admit that their God, or its representatives in the Soviet Union, had failed:

Most of those who passionately upheld the verdicts of the Trials were secularists who were wont to scoff at the myths and superstitions of traditional religions. When it came to evaluating events in the Soviet Union, however, the will to believe – perhaps it would be more accurate to call it the will to illusion – prevailed over ascertainable fact and rational analysis. (ibid., p. 241)

The trials, Hook wrote, were "a decisive turning point in my own intellectual and political development" (p. 218). By the late 1930s, his fiery defenses of Marxism had become fiery indictments of those who failed to denounce Stalin, Moscow, or the Communist Party. And through the 1940s, '50s and '60s, Hook remained an influential anti-Stalinist who not only denounced his enemies (domestic and Russian) in his writings, but also organized conferences and organizations dedicated to pro-American and anti-Soviet cultural matters. These included his Committee for Cultural Freedom (for which he enlisted Carnap in 1939) and the long-running Congress for Cultural Freedom, which first convened in Berlin in 1950 and thrived into the 1960s as a mainstay of Cold War intellectual life in Europe and Asia.[1] Hook was awarded the U.S. Medal of Freedom by President Ronald Reagan in 1983.

Disillusionment and Dualism in the Postwar Intellectual World

Among former leftist intellectuals, Hook was possibly unique in the speed of his swing to the right and the wide arc between his youthful radicalism and his mature anticommunism. Other intellectuals moved more slowly or, unsure of themselves, became skeptical if not agnostic about politics in general. It was a confusing and confused time for intellectuals. Along with *The God That Failed*, books and articles routinely highlighted themes such as uncertainty, disenchantment, and disillusionment about Stalin or socialism itself. *Witness*, the autobiography of Whittaker Chambers that was a national best-seller in 1952, recounted in minute detail Chambers's self-searching journey from underground communist operative to anti-communist informer and prosecutorial witness against State Department official Alger Hiss. In the late 1940s and early '50s, "disillusionment" was in the air.[2]

The series "The Future of Socialism" dominated *Partisan Review* in 1947 with articles by Hook, Arthur Koestler, George Orwell, and others. The editor's note introducing the new series claimed that "the entire socialist perspective" had been thrown into question by world history since 1917. Working classes in Germany and Italy failed to realize socialism, and widespread confusion remained over the distinction between genuine socialism and "its Stalinist perversion." Now that Stalin had created "a totalitarian system with a dynamism of its own that throttles the development

---

[1] For more on the Congress for Cultural Freedom and its covert CIA sponsorship, see chapter 15.

[2] During the Cold War, a genre of literature emerged examining the (usually former) emotional, psychological, and pseudo-religious allure of communism or Marxism. Besides *The God That Failed*, this genre included the book *Appeals of Communism* (Almond 1954), the articles described below concerning "disillusionment" of former fellow-travelers, and the classic treatment of totalitarian's appeal, Eric Hoffer's *The True Believer* (Hoffer 1951). The view that communists were emotionally needy and maladjusted was common and took root in the diagnosis of "neurotic susceptibility" (Herberg 1954, 11).

of socialist thought and democratic socialist movements," the left was in "a state of intellectual disorientation and political impotence."[3]

A year later, editor Philipp Rahv wrote about the varying degrees of "disillusionment" that leftist writers and philosophers had experienced and the fact that many former leftists were moving into the conservative opposition. His editorial, "Disillusionment and Partial Answers," was, in part, a pep talk for those keeping the faith: "The perspective of a democratic socialism, that is, of a planned and socialized economy combined with the fullest political and cultural liberty, has by no means been annulled by historical events." In fact, it was "still the only possible perspective," Rahv explained, "in a world dominated by Soviet Totalitarianism on the one hand and American Capitalism on the other" (Rahv 1948, 521).

The dichotomy that Rahv described would come to be the conceptual skeleton for Cold War culture in subsequent decades. Compared with the 1930s, when the spectrum from left to right was populated by varieties of socialists, communists, libertarians, and capitalists, the Cold War had little interest in continua and subtle differences. By the late 1940s, events such as the Soviet clampdown in Czechoslovakia and the Berlin airlift had fueled mutual suspicions between Washington and Moscow and structured the international political landscape in Manichean terms: Capitalistic, North American democracy, on the one hand, stood opposed to "totalitarianism" of all kinds, on the other. As Malisoff and Somerville complained in *Philosophy of Science*, programs and political systems as diverse as Italian fascism and Soviet communism were routinely lumped together as "totalitarianism."

Most important, however, this dualistic opposition between American democracy and Soviet communism was anything but academic or theoretical. Fears of nuclear war added urgency to the debate and reinforced the simplistic dichotomy between democracy and totalitarianism. Since the Cold War could heat up at any moment and easily become the most destructive war in history, loyalty and patriotism were demanded, almost as if one's insecurities and worries were mitigated by knowing that neighbors, coworkers, and friends were uniformly united against communism. Those who objected to the simplistic dualisms in play were viewed with suspicion. "The current self-preening smugness of the non-communist socialist has a phony ring," quipped one columnist (Davis 1952, 20 n.). Another

---

[3] Editor's note, in *Partisan Review* 14 (1947): 23.

distinguished the 1930s from the postwar climate and apologized for the current disdain for hair-splitting intellectualism. "The necessity for choosing sides irrevocably in a conflict of two hostile worlds and ways of life was not as urgent and inescapable then as now" (Chamberlin 1950, 20).

Diagnoses of the Intellectuals

As this Manichean political landscape took shape in the late 1940s, one popular theme concerned intellectuals themselves. Titles concerning "The Intellectuals" or "The Intelligentsia" appearing in *Partisan Review, Commentary, The Nation,* and the *New Leader* were either critical of leftist intellectuals (in conservative journals) or self-critical and navel-gazing (in the leftist press). The main question that columnists and essayists set out to answer was this: In the 1930s, Stalin and Moscow were seen by intellectuals as harbingers of utopia. But by the late 1940s, most had come to see their true, evil colors. How, then, could America's best and brightest have been so egregiously wrong?

Early explanations came from early converts. Arthur Koestler published "The Intelligentsia" in *Partisan Review* in 1944, about six years after he left the Communist Party. He reflected on the meaning of "intelligentsia" and its role in society from the Enlightenment to its present death throes. One of Koestler's claims was that intellectuals of the 1930s and early '40s lacked engagement with any ascending social class and thus lost the ability to perform the intelligentsia's traditional task of forming public opinion. The revolution had long since failed, leaving the intelligentsia directionless and mired in some sort of Freudian neurosis:

An intelligentsia deprived of the prop of an alliance with an ascending class must turn against itself and develop that hot-house atmosphere, that climate of intellectual masturbation and incest, which characterized it during the last decade. (Koestler 1944, 275)

Aside from questions about the merit of Koestler's suggestion, its tone and language illustrates his vitriol and anger at the unreconstructed left.

Another genre of essays about intellectuals was the essay-apology. Martin Gardner (who would later edit Carnap's book (1966)) confessed in the *New Leader* that he was a fellow-traveler who in his socialist youth was frustrated by H. G. Wells's "long and vigorous opposition to communism." Now, however, Gardner praised Wells as an unsung hero who helped to usher "the American Liberal's fast growing disenchantment with the Soviet myth":

There is a tendency among recently disenchanted Communists and fellow-travelers to rationalize their period of domination by the Soviet Mystique. Not until the last few years, you hear them say, has the behavior of Russia forced upon us a realization of her corruption and tyranny. . . . I think there is a more honest attitude. There were those who saw more clearly than we, and we did not listen to them.[4]

While many leftist intellectuals alternately licked their wounds and attacked themselves from within, others attacked them from without. Atomic scientist Edward Condon complained that intellectuals in America were "accorded neither the social prestige nor the remuneration" that they enjoyed in Europe. Now, scientists especially were suffering from "the post-war loyalty hysteria" that placed them under immediate suspicion for passing so-called atomic secrets to communist countries (on the popular model of Julius and Ethel Rosenberg). The result was nothing less than "the abridgement of scientific freedom in this country" (Condon 1950, 267). Leonard Engel, a columnist for *The Nation* who defended Condon when he was "accused of communist sympathies," put out the warning for scientists as clearly as he could in the title of his columns: "Warning All Scientists" (Engel 1947) and "Fear in Our Laboratories" (Engel 1948). "If the attack on Condon succeeds," Engel reasoned, "no civilian scientist will be safe" (Engel 1947, 117, 119).

If scientists and intellectuals were not considered security threats, they were considered fools. Even the *New Leader*, which regularly published articles by intellectuals, ran one titled, "Who Listens to the Intellectuals?" Writer Wallace Martin Davis mocked "the Intellectual's past wrongness, ranging from the cosmic to the merely asinine" and happily chided "the intellectuals" for fully deserving what has happened to them. Because the political realities of the world turned out to be "much more complex, subtle and unpredictable . . . than the Intellectual[s] had imagined," he suggested, they should pull back, retreat from politics, and develop their expertise in small, technical domains where it could be trusted:

To the Intellectuals who weep for their lost estate, I can only say this: It is your competence to speak, not your freedom to speak, that should be concerning you. We consumers will not shut you up, but we don't have to pay any attention to you. Your toaster burns my toast, gentlemen, and you would do well to stay away from sales conferences at the Waldorf and get going in the engineering department.

4 Gardner 1950a, 20, 21.

"Economists, political scientists, educators, psychologists, sociologists, historians" had produced merely a "cemetery of dead ideas" and a record of Red politics symbolized by the Waldorf conference (discussed in chapter 13). No one listens to the intellectuals anymore, Davis concluded, because they had lost legitimate "claim for leadership" in the postwar, anticommunist culture (Davis 1952, 20, 21).

Public Bullying and Confrontation

Davis broadly attacked "the intellectuals." Other writers and editors targeted individuals. Few targets had a higher profile than Albert Einstein – or were easier to criticize given his pacifist sentiments and proclamations in the 1950s. In a review of his autobiography, *Out of My Later Years*, one Jim Cork verged into schadenfreude as he reflected on this great intellectual's declining reputation: "What is particularly discouraging, to friends of intellectual freedom everywhere, is Einstein's 'soft' attitude toward Russia." The book contained "scattered . . . criticisms of Russia's authoritarian methods in education" and other abuses, but the book was "offset by the complete silence in regard to the enormous magnitude of her *major* crimes against the material well-being, dignity and freedom of millions." For Cork, "the so-called puzzles and paradoxes of Einsteinian Relativity are as nothing compared to the profound paradox of Einstein's reticence here." He could only lament that Einstein's book "left the inescapable impression of a pro-Soviet attitude" (Cork 1950, 20).

Bertrand Russell was another frequent target. In the early 1950s, over a decade after his dismissal from City College, he elicited the wrath of many for his frequent letters to the editors of magazines and newspapers denouncing intellectual conformity and McCarthyite, anticommunist hysteria in the United States. To *The Nation*, for example (and in support of Einstein's public comments against McCarthyism), he wrote

I am astonished that there is not more objection in America to the inquisition by your new Holy Office into the lives and opinions of American citizens and eminent aliens. . . . It is somewhat ironic that in America, which was once regarded as the land of liberty, the most resounding voice in favor of freedom should come from a German. (1953, in Russell 2002, 175)

Few editors could tolerate the fact that Russell aggressively criticized both the USSR and the United States during the Cold War and did not simply take one side or the other. That he expressed his criticisms with flair and wit, moreover, made him an even more inviting target for Red-baiters and the conservative right.

As illustrated by John Somerville's exchange with Sidney Hook in *The Nation* (described in chapter 3), journals and magazines used their editorial authority to stage conflicts or to bully individuals into public stances. Russell noted that

> it is extremely difficult to get letters accepted by most American newspapers except such as I should consider intolerably pro-Russian. And when I do succeed in getting published in America, I am exposed to editorial distortions without my knowledge or consent.

Russell once found that an article he had sent to *Look* magazine appeared with its title most provocatively changed from "What Is Wrong with Anglo-American Relations" to "What's Wrong with Americans?" (Russell 2002, 174). On another occasion, the editors of the *New Leader*, angry at Russell's comments about America, demanded that he clarify his views. Russell sent them a statement that they printed along with their own rebuttal, roughly twice as long. Russell's comments, they urged, "serve[d] not the cause of freedom and peace but the cause of the Kremlin" (Russell et al. 1952, 4). The operative dualism of Cold War debate was clearly in play: If one did not praise the United States, then one must be a servant of Soviet communism.

Towering intellectuals such as Einstein and Russell were obvious targets within the anti-Stalinist left. But there were many from which to choose. In January 1953, the editors of the *New Leader* assembled a feature titled "Rosenberg Clemency Backers Comment on Red Anti-Semitism." Their targets were thirty-five of the hundreds of public figures who one week earlier had signed a statement appearing in the Communist *Daily Worker* in support of clemency for Julius and Ethel Rosenberg.[5] Since one common claim by Rosenberg supporters was that they were persecuted unfairly and overzealously because they were Jewish, the editors seized an opportunity to expose what they suspected was a double standard. In the wake of Kremlin-sponsored trials in Prague that targeted Jewish doctors, intellectuals, and party officials, would these thirty-five suspiciously pro-Soviet Americans publicly condemn *Soviet* anti-Semitism?

---

[5] Among the twenty "academicians" joining these writers and artists who lent their names to the cause were Carnap and Morris. See Daily Worker, "Cry of 'Save Rosenbergs' Sweeps World of Art, Science, Literary Circles," January 14, 1953. In Carnap's case, his public support of the Rosenbergs prompted the FBI to investigate him, as discussed in chapter 13.

The feature did not politely invite these figures to comment. Rather, they wired each of them a telegram that read:

Note your support Rosenberg clemency. In name human rights we ask you make equally forthright condemnation antisemitic Prague trial and imminent execution of soviet Jewish doctors. Please wire collect protest up to one hundred words.("Rosenberg Clemency Backers Comment on Red Anti-Semitism" 1953, 4)

They then printed the responses that they received along with their own commentary. Some were complemented for condemning the trials. Those that did not outright condemn them were lampooned and criticized, including Harvard's Harlow Shapley, the University of Chicago's Robert Morss Lovett, and the Reverend Willard Uphaus. On this occasion, Einstein fared well, for he denounced the trials. As for those that had not responded within in six days, the editors assembled and printed their eighteen names (including well-known leftists Nelson Algren and singer Paul Robeson) under the bold-face heading, "No Reply" ("Rosenberg Clemency Backers . . . " 1953, 4).

The Rediscovery of Values and the Conference on Science, Philosophy and Religion

One intellectual who sustained a career writing breezy, partisan attacks on leftist intellectuals was Wellesley professor and poet Peter Viereck. His best known book, *The Shame and Glory of the Intellectuals* (1953) began with four words – "The Rediscovery of Values" – that signaled both the decline of scientific socialism in North America and the alleged confusions and missteps of the formerly influential intellectuals who once promoted it. Viereck and others believed that the leftist intellectuals of the 1930s guaranteed their marginalization by embracing science or a scientific naturalism upholding scientific methods, Marxist or other, as the best tool for understanding society, history, and international relations.

Mortimer Adler and Robert M. Hutchins had argued similarly as they promoted neo-Thomism in the 1930s. In the wake of Adler's infamous speech, "God and the Professors" at the first Conference on Science, Philosophy and Religion (CSPR) in 1940, the CSPR continued to be a forum where the values debates and this critique of scientism was articulated. Nearly all leading figures in the humanities participated at one time or another in the CSPR, the most notable exceptions being Dewey, Hook, and Kallen, who set up a rival conference, the Conference on Methods in Science and Philosophy, to oppose the neo-Thomism and, Hook later commented, the anti-Semitism he felt he detected at the CSPR (Hook 1987, 337).

Alongside debate about science and values, as we see below, the conference specifically wrestled with questions about the roles of intellectuals in cultural and political life. As editors Finkelstein, Lyman Bryson, and Robert MacIver explained in one of the conference's first published proceedings, the conference was under attack by "fellow specialists in different fields" who urged intellectual specialization and disengagement from world affairs. They believed that "the business of human relations ought to be left to professionals, technicians and statesmen" (Bryson, Finkelstein, and MacIver 1947, v). From the other side, it was attacked for quietism and failing to contribute to "the preservation of peace, mutual understanding, and co-operation among men" (ibid.).

One of the leading activists was Northwestern University's Paul Schilpp, original editor of the series *The Library of Living Philosophers*. At the conference's eighth meeting in 1947, dedicated to the question, "How can scholarship contribute to the relief of international tensions?" Schilpp addressed "the task of philosophy in an age of crisis." The cliche that humanity stands "at the crossroads" no longer applied, Schilpp wrote. Instead, "it stands before the Abyss." Nuclear annihilation seemed inevitable. "We can no longer speak in terms of generations or even of decades but – probably – only in years below two digit figures" before all major cities become Hiroshimas and Nagasakis. Schilpp was therefore exasperated that "leaders of church, science, and the university insist, in general, upon continuing to act as if nothing had happened. True, we do talk about 'The Atomic Age,' but . . . we are not *doing* anything" (Schilpp 1948, 300).

On the one hand, Schilpp explained this quietism by appealing to the personalities involved and "the practical impossibility of getting the 'scholar' out of his scholarly rut. It seems to be in the very nature of scholarship to cause its devotee to shrink from action." He also attributed it to the mood of a country that was heading toward McCarthyism and hysterical anticommunism:

[The scholar] shuns public speaking on "controversial issues." And, for this very reason, he is ineffective. In his blind devotion to scholarship he has lost his devotion to the cause of mankind. Even though he may not actually be "fiddling while Rome is burning," he is quite likely to be counting commas while the atomic bomb is rubbing him out of existence. (ibid., p. 301)

Despite social pressures to avoid controversy, Schilpp urged his fellow intellectuals at the conference to maintain engagement in two ways. The first was to contribute to the understanding of other cultures and "the understanding of human nature" as it plays out in foreign nations. The other recalled the original popularity of logical empiricism among progressive intellectuals in the 1930s. "The paramount need of this hour is *thoughtfulness*," Schilpp wrote (p. 302):

Here, in helping both leaders and the masses of people to learn to think clearly, honestly, logically and consistently, and in the light of ascertainable facts, lies certainly one of the major tasks of philosophy in this human crisis. (p. 304)

In the CSPR, Schilpp was an ally of logical empiricism and the Unity of Science movement, for it had promoted these goals and values since the mid-1930s. He echoed Carnap's comments in his Harvard Tercentenary radio interview and followed Morris's interests in studies designed to enhance intercultural understanding. Within two years, Philipp Frank (who attended CSPR meetings regularly) would issue his own call for "active positivism" as president of the new Institute for the Unity of Science.

As he closed his talk, Schilpp seemed to know that his call for activism was not going to be persuasive. He ended on "the frankly pessimistic note required by the subject" and quoted Arthur Lovejoy's indictment of humanist intellectuals: "In view of the mess man has made of this planet, a suitable costume for humanists is sackcloth." However compelling that view may be, Schilpp wrote, philosophers should ignore it. If philosophy were to help reduce international tensions, it would require

more than penitence for the past; it will need an unswerving resolution today and tomorrow to do those things which the exigencies of the present world situation require and demand of us as philosophers and thinkers. Are we men and thinkers enough to accept this challenge? (p. 310)

Values and the Critique of Scientism

One reason Schilpp was pessimistic about a scientifically informed philosophical activism was that a compelling argument (or family of arguments) against scientism was well articulated at the CSPR: Scientism overlooked not only the importance of values, the argument went, but also the determinative, often irrational forces lurking within human nature. Recent history, and the Second World War, in particular, had revealed human nature to be a power or agency that was beyond the reach of scientific understanding.

Several in Schilpp's audience at the CSPR took the podium and endorsed this point of view. Paul Kecskemeti, a friend and student of Morris's, shared Schilpp's alarm about impending war. But he questioned whether philosophy was an appropriate tool for understanding and controlling humanity's destructive impulses:

> While his primary concern is with the determination of behavior by reason, insight and responsible, free decision, the philosopher cannot overlook the vast extent to which human action is determined by such things as instinct or fear or compulsion or convention or routine.

Social science and international law, not philosophy, he argued, would be the best way to approach one of the pressing questions driving the values debate: the question of "how the human agent comes to embrace his ends" (Kecskemeti 1948, 323).

Others suggested that neither philosophy nor science could make sense of world affairs. David Baumgardt, billed as "consultant in Philosophy, Library of Congress," spoke against myths of the Enlightenment – "that, by allowing poise and cool, sober reflection to get the upper hand, all international and class conflicts could be satisfactorily overcome" – and criticized intellectuals for failing "to pay full regard to the strength of emotions in fields of thought where emotions legitimately has to play a major role" (Baumgardt 1948, 366). Similarly, Ralph Flewelling took it to be obvious that science and materialism are to blame for our "ever deepening embranglement of discontent":

> The disadvantages have arisen largely from the erection of scientific knowledge into an all sufficient end, to the disparagement of the artistic, intuitional, and spiritual forces necessary to the complete man. (Flewelling 1948, 373)

The complete, objective reality of human beings, Flewelling and these others argued, involved more than any scientific approach could grasp.

Popular Antiscientism

This argument that science and scientific methods were incapable of understanding or predicting matters of society, history, and politics joined the more popular critique of leftist intellectuals and their various Marxist and scientific over-enthusiasms of earlier years. The same year that Schilpp spoke, James Schlesinger, Jr., had written in *Partisan*

Review that Enlightenment liberalism was powerless against "the dark and subterranean forces in human nature" now exposed by the war and the holocaust. Liberals were mistaken to have "dispensed with the absurd Christian myths of sin and damnation and believed that what shortcomings man might have were to be redeemed, not by Jesus on the cross, but the benevolent unfolding of history." For history had not unfolded benevolently. Tapping into our dark and sinister natures, Schlesinger explained, "practical men, like Hitler, Stalin, [and] Mussolini, transformed depravity into a way of life" (Schlesinger 1947, 235).

Though the Unity of Science movement was not in Schlesinger's sights, it did not need to be. The "rediscovery of values" and the dark, inscrutable forces driving human beings flatly opposed the Promethean optimism that Neurath, Carnap, and Hahn described in their manifesto, *Wissenschaftliche Weltauffassung*. In their empirical, scientific, and modern approach to the world, they explained,

neatness and clarity are striven for, and dark distances and unfathomable depths rejected. In science there are no "depths"; there is surface everywhere.... Everything is accessible to man; and man is the measure of all things. (Neurath et al. 1929, 306)

Not so, anti-Stalinist intellectuals and commentators now claimed. At least some of the forces driving history and human events are not accessible by examining the empirical surface of the world. Science, and therefore any scientific philosophy, cannot therefore understand these forces, much less predict and consciously manage their effects.

In one influential conservative venue, William F. Buckley's magazine *National Review*, this critique of scientism was fused to that magazine's anticommunism by writer Frank Meyer. Like Schlesinger, he saw scientism as the Achilles' heel of Enlightenment liberalism:

This is the bigotry of science: the demand that all activities of the intellect which do not follow the methods of the sciences, and all intellectual conclusions which do not square with the conclusions of that methodology, be cast into outer darkness as infantile and/or superstitious.

The trouble with this "bigotry" was practical and political:

The effects of this science-worship are reflected in the cry that arises as the Soviet threat mounts: "Science will save us" – when our deepest problems are defects of moral understanding and will, defects which the scientific methodology can do nothing to correct.

Science, according to Meyer, was effective at predicting the trajectories of colliding billiard balls and "the behavior of physical matter," but it shed no light on international relations and politics precisely because they involved human beings "endowed with consciousness, will, and the faculty of moral understanding" (Meyer 1958, 234). Thus some literary critics, social scientists, behavioral psychologists, as well as "logical positivists and the analytical philosophers" were unequipped to understand or shape public opinion about politics and world affairs.[6]

One does not need Meyer's accusatory nod to "logical positivists" to confirm the growing marginalization of scientific philosophy in Cold War social thought. As anti-Stalinism's influence increasingly dominated both popular and intellectual life, not only the regular appeal to super-scientific values but the metaphysical and theologically tinged language of the day would have made it difficult for the Unity of Science movement to join these public discussions. When Hook recalled events decades later, he wrote that he "discovered the face of radical evil" in those who disagreed with him that "any conception of socialism that rejected the centrality of moral values was only an ideological disguise for totalitarianism" (Hook 1987, 218). In 1951, Harvard's president James Bryan Conant, reputed and revered as a defender of science and science education, spoke of "the forces of good [and] evil" when speaking to the American Chemical Society about the future of science and the world's political landscape (Conant 1951, 2). During years when intellectuals in general were being chastised, often by themselves, for not recognizing that the God that had failed them was "evil" or inspired moral or metaphysical "evil" in the world, it is hard to imagine Neurath, Carnap, Frank, and other logical empiricists gaining much prestige and influence in this dialogue. They could have argued that "evil" was an obfuscatory (if not intellectually juvenile) concept or they could have claimed, as Frank set out to convince his audiences at the CSPR (and as discussed in chapter 11), that it was logical empiricists who were perhaps best equipped to elucidate the complex interplay of facts and values in modern science and its historical development. Neither approach, it would appear, would have been effective in this antiscientistic intellectual climate of the early 1950s.

---

[6] Meyer knew something about the "logical positivists and the analytical philosophers" he dismissed, as well as the Unity of Science movement and its *Encyclopedia*. A year before, he reviewed Morris's *The Varieties of Human Value* (Morris 1956), and dismissed Morris as "an outstanding leader of the collegium of analytical philosophers who have in this century systematically sucked the substance from thought" (Meyer 1957, 118).

The End of Ideology

Consensus grew during the 1950s that whatever lay behind the shocking behavior of the Nazis, Stalin, and other proponents of "totalitarianism," "human nature" and "the complete man" was something far more complex (and dangerous) than most of the optimistic and ambitious intellectuals of the 1930s had ever suspected. Humanity, it turned out, was not something that could be reasoned with or understood in simple, clean scientific ways. This contributed to a general and widespread pulling-back of intellectuals – especially those scientifically oriented – from politics and social issues. Leadership in political affairs was left to "statesmen" while intellectuals regrouped. Some moved into policy-related areas of government (such as the RAND Corporation and other "think tanks") while others moved simply into professionalism, specialization, and political quietism.

To be sure, the values debates and the consensual "end of ideology" occurred simultaneously with other shifts and developments in North American intellectual life. Most conspicuously, many intellectuals involved in these debates, especially those who did not establish or already have unquestionable anticommunist credentials, were at risk of being investigated by anticommunist investigators from different levels of government (as discussed in chapters 12 and 13). What the values debates arguably provided, it would appear, was an intellectual foundation and outlook with which to make this enhanced professionalism and political disengagement seem natural and necessary in intellectual life. The best remembered icon of this view is Daniel Bell's collection of essays, *The End of Ideology: On the Exhaustion of Political Ideas in the Fifties.*

Ideology had reached its end for several reasons, Bell explained, some of which reflected the antiscientistic views of Meyer, Schlesinger, and those at the CSPR. One was the rise of "antirationalism" and "the intellectual vogue of Freudianism and neo-orthodox theology (i.e. Reinhold Niebuhr and Tillich), with their antirational stoicism." Since human beings were irrational and complicated, few could believe any longer "the rationalistic claim that socialism, by eliminating the economic basis of exploitation, would solve all social questions" (Bell 1960, 297). This skepticism killed not only socialist theory, but all claims to understand the mainsprings of society and politics. Theories had "lost their 'truth' and their power to persuade" (ibid., p. 373).

As science would have to concede, that which is not understandable is not consciously plannable. The promise of social and economic planning, taken seriously by many liberal and leftist intellectuals of the 1930s, now seemed empty and naïve, Bell explained. "Few serious minds believe any longer that one can set down 'blueprints' and through 'social engineering' bring about a new utopia of social harmony" (Bell 1960, 373). Bell even suggested that intellectuals tended to "fear" society and the "masses" of humanity that embody these irrational, uncontrollable forces. Perhaps just as Hook and other former socialists came to feel enormous anger and disappointment toward Stalin, the Kremlin, and those aspects of human nature that ruined their youthful hopes for the bright, socialist future of humanity, neo-conservative intellectuals of the 1950s, Bell suggested, "have begun to fear 'the masses,' or any form of social action. This is the basis of neo-conservatism and the new empiricism" (ibid., p. 16).

Piece by piece, Bell's essays rejected those parts of the Unity of Science movement that formerly had appealed to the many leftist intellectuals who had welcomed it in the 1930s. Science and a scientific outlook could not see the springs and forces of human nature; there was no point in cultivating the sciences for the benefit of social and economic planning (since such were futile); and the Enlightenment idea of intellectual activity helping to inform and educate "the masses" was now derailed by emotional resentment or "fear."

In ideology's wake, Bell explained, arose a new kind of intellectual and critic, one who accepts, even strives for, "a detachment which guards one against being submerged in any cause, or accepting any particular embodiment of community as final." Here, "the claims of doubt are prior to the claims of faith. One's commitment is to one's vocation."[7] In broad outline, this was the path of logical empiricism through the Cold War as it became detached from the leftist agendas for enlightenment and social reform embodied in the Unity of Science movement. Logical empiricism, under the leadership of Reichenbach, Feigl, and others, would argue for a well-defined set of problems and techniques for philosophy of science that cleanly excluded such engagements as planning, social philosophy, and ethics.

---

[7] Bell 1960, 16. As sociologist Seymour Lipset put it a year earlier, intellectuals in the 1950s were torn between postwar "prosperity with its concomitant improvement of the position of the workers and the intellectuals" and "the uneasy feeling that they are betraying their obligation as intellectuals to attack and criticize." "Their solution to this dilemma is . . . to vote the Democratic ticket and basically to withdraw from active involvement or interest in politics and to concentrate on their work, whether it be poetry or scholarship" (Lipset 1959, 477).

9

Horace Kallen's Attack on the Unity of Science

The most direct political attack on the Unity of Science movement began early, in 1939, as its leading members commenced the Fifth International Congress for the Unity of Science at Harvard University. Those convening were happy to be reunited, but the occasion was not joyous. Most had probably heard the news while traveling to the United States or to Cambridge: Hitler had invaded Poland, and the situation looked grim. On the eve of the Congress, Sunday, September 3, they gathered around a radio to hear President Roosevelt's weekly radio address and learned that Hitler had not backed down from England's and France's ultimatums demanding Nazi withdrawal from Poland (Neurath 1946, 78).

This gloom affected the movement for several specific reasons. Its conferences, publications, and publicity (*Time* sent a reporter to this conference ("Unity at Cambridge" 1939)) were intended not only to inject empiricist reforms into philosophy, to eliminate spurious metaphysical thinking, and to popularize unified science. These reforms themselves would potentially improve communication and understanding among nations and thus facilitate international cooperation in social and economic planning. But this enlightenment agenda seemed to fall on deaf ears, for war was breaking out and the world was growing darker.

Another more proximate force against the movement appeared at this conference, as well. Horace Kallen was a New York philosopher who had embraced Neurath and logical empiricism both intellectually and socially. At the conference, however, he sounded his alarm that the movement was "totalitarian." To most in the audience, the charge must have seemed somewhat ridiculous. As Morris would say later in the published exchange addressing Kallen's charge, "it is historically clear now that the unity of science movement has no affiliations with dogma and dogmatism" (Morris 1946a, 509). Though dogmatism and totalitarianism are not identical, Morris's point was clear and simple to his readers. Neurath, his movement, and other Europeans committed to it were refugees *from* fascist totalitarianism, not promoters of it.

For Kallen, however, things were not so clear or simple. He first presented his charges one month after the Hitler-Stalin pact had become public, a time when many leftist intellectuals had decided that, at least in the case of the Soviet Union, progressive, socialistic promises were sometimes empty, if not cover for antiprogressive forces and agendas. Some such logic appears to have motivated Kallen, for there are many indications that he and Neurath, as they became friends in the late 1930s, were working productively toward a shared understanding of the progressive, democratic values of Neurath's conception of unified science.[1] Neurath indeed welcomed Kallen as a fellow devotee of pluralism and a fellow critic of absolutism and pseudorationalism. The stage was set, therefore, for Neurath to be surprised and, one suspects, angered about Kallen's relentless attack on alleged political and social mischief lurking in the Unity of Science movement.

As the war threw the movement into disarray, however, there were more immediate problems at hand than Kallen's claims about totalitarianism. Neurath returned to Holland after the Congress (against the advice of his friends and colleagues), while Kallen and Morris rallied their connections at the Rockefeller Foundation for $1,000 in emergency funds to move Neurath's institute and its several employees to the United States or to Norway.[2] That prospect became moot, however, when Neurath fled from Holland in May 1940 and, after his six-month internment, settled in England.

In the meantime, Kallen tried to publish the paper he read at the Cambridge conference. It finally appeared in the *Journal of Educational Administration* in 1940, where it lay without much fanfare until the close of the war. By that time, with new issues in science and scientific management presented by the development of atomic weaponry, Kallen believed the issues he had raised were even more urgent. In 1945, *Philosophy and Phenomenological Research* revived the debate by printing a long,

---

[1] Kallen to Neurath, April 29, 1939, JRMC.

[2] Morris to Weaver, October 14, 1939, USMP, box 1, folder 18.

seven-article exchange of letters among Kallen, Neurath, and, briefly, Morris. First, Kallen reiterated his charges from 1939 and argued that, even after the allied victories, totalitarianism's threat loomed larger than before (Kallen 1946b). Then Neurath and Morris responded and, as Morris dropped out of the picture, the volleying between Neurath and Kallen began. Since they addressed their letters to each other, the exchange has a personal quality that reached a climax especially when Kallen responded to the news of Neurath's sudden death. Kallen's last contribution is, in effect, an obituary in which he expresses his shock and praises Neurath as an important scholar and, oddly, a fellow libertarian pluralist (Kallen 1946c).

Kallen and His 1939 Paper

Kallen was born in Berenstadt, Germany, in 1882. At the age of five, he came to America with his family and father, an orthodox rabbi and Hebrew scholar. Kallen studied philosophy in England and France before returning to obtain a Ph.D. at Harvard, where he cemented his relationship to his main teacher, William James.[3] He taught at Clark University, the University of Wisconsin, and finally at the New School for Social Research, which he helped to found in 1919.

Though not as well remembered as his fellow New York philosophers Dewey, Hook, or Nagel, Kallen was well connected to their academic and political projects. He was active in labor causes, consumer co-ops, and in the American Civil Liberties Union (Kallen 1956, 104); he assisted Dewey and Hook in their Committee for the Defense of Leon Trotsky in 1936 and 1937 (Hook 1987, 225); he edited with Dewey the volume of essays about the Russell affair (Dewey and Kallen 1941); and he participated in Hook's anti-Thomist Conference on Methods in Science and Philosophy and the Conference on the Scientific Spirit Democratic Faith (Hook 1987, 347). Kallen was also a signatory for Hook's Committee for Cultural Freedom (along with Carnap) and joined his antitotalitarian protests against the Waldorf conference of 1949 (ibid., p. 384).

Kallen's career flexed easily among these various intellectual, social, and political causes. So too did his charges against Neurath's Unity of Science movement. In his talk at the Cambridge conference, he posed

---

[3] See Menand 2001, 388–408, for more on Kallen's connection to James as well as his formative personal and intellectual relationships with Dewey, Randolph Bourne, Alain Lock, W. E. B. DuBois, and other contemporary theorists of cultural pluralism.

the ordinarily technical and scientific question of what "unity of science" might mean. His answer, however, was strongly political:

To ask these questions and to keep on asking them may be all the more needful in view of the climate of opinion in which today's research after "the unity of science" develops.

For we are living in totalitarian times. . . . (Kallen 1940, 81–82)

Gesturing toward Germany, Stalin's Soviet Union, Mussolini's Italy, and Franco's Spain, Kallen explained that various notions of "unity" were official and entrenched in these regimes: "Not only are the lives and labors of the people 'unified,' their thoughts boilerplated; also the arts and sciences are 'unified' to the respective orthodoxies of the fascist, Nazi, Communist and clericalist dogmas" (ibid., p. 82). Was Neurath's "unity of science" very different?

By process of elimination, Kallen answered, "no." If unified science meant "the coordination of the different . . . sciences into a single system or order," he reasoned, "it cannot mean that kind of coordination, excluding alternatives." If it meant the adoption of a unifying scientific language, "it cannot be . . . composed of universal and invariant signs." Whatever unity of science may mean, it cannot carry within it any "kind of authority, enslaving thought, destroying the doubt which expresses at once the freedom and the creative power of intelligence" (p. 83).

The only acceptable concept of unity, Kallen argued, was one that specifically rejected such constraints on thought and intelligence:

Indeed, it might turn out that the first and last meaning of "unity of science" comes to nothing else than the congress of the plurality of the sciences for the unified defense of their singular freedoms against the common totalitarian foes. . . . [It] means, and need mean, no more than the mutual guarantee of . . . liberty by each science to each, collective security for the scientific spirit from dogmatic aggression. (ibid.)

On this view, "unity" meant not "a state but an activity," an ongoing vigilant patrol of the epistemic shoreline. Where Dewey had treated "unified science" as both a scientific project and a "social problem," Kallen acknowledged only the latter. Unity of science ought to consist, he believed, entirely in the sciences' opposition and resistance to human forces and agencies that sought to corrupt and control them.

For Kallen, this radical pluralism was the engine and strength of science. It was even proper, Kallen explained, that his nonunificationist conception of "unity of science" should seem strange or inscrutable to those in his Cambridge audience: "As one man's food is another man's poison, so one man's meaning may be another's nonsense." "Clearly, as there are meanings, not one meaning, so there are logics, not one logic" (pp. 84, 86). This is because the human mind, he explained, if it be "free and doubting," will ever "seek alternatives to the prevailing meaning[s]" (p. 85). For Kallen, semantic ambiguity, plurality, and historical change were the rule. The more strange his view of the unity of science might seem to others, the more likely correct he took himself to be.

Kallen versus Morris and Carnap

Kallen turned to Charles Morris's writings about semiotic as a "science of science" and quickly specified the totalitarian impulse he detected at its core. How, he asked, do we select practices or beliefs to count as scientific as opposed to those we reject? Criteria such as being systematic, mathematical, predictive, or strict in measurement invite noncontenders such as astrology or contract law into the fold along with physics and psychology. We cannot appeal to truth because, as the history of science suggests, all theories are destined to be proved false. From the beginning, Kallen argued, Morris's "science of science" would necessarily make selections and exclusions based on nothing but (Stalinesque) whim and caprice.

Kallen himself avoided this problem by rejecting demarcation altogether. Science *was* a clash of disagreeing, disunified ideas and persons, an "aggregation . . . of different men combating or confirming or both . . . one another's variances, innovations, or repetitions regarding this or that selected area of the multitudinous universe." The only unity, again, is their shared freedom: "an equal liberty to the claims of all, a free field" (p. 89).

When Kallen turned briefly to Carnap's work, it fared no better. The very idea of reconstructing theories in formal logical languages, he suggested, risked violating science's freedom by neglecting "alternatives in logical foundations to choose between." This species of pluralism would have been old news to Carnap and to anyone who had read his *Aufbau* (Carnap 1969) or was familiar with his Principle of Tolerance, the main thrust of which is to encourage freedom in one's choice of philosophical languages (Carnap 1937c, 52). Still, Kallen forged ahead. Taking little account of rational reconstruction and formalization as tasks self-consciously removed from concern with the practical use of languages (and possibly also from concern with meaning altogether, with Carnap's syntactic project), Kallen supposed that the purpose of logic and formal,

semantic analysis was to freeze and control meaning: "The very undertaking to keep [meaning] invariant changes it – in the language of the
logisticians, propositional functions, in spite of their best efforts to turn
them into propositions, persist in continuing as propositional functions"
(Kallen 1940, 90).

Kallen versus Neurath

Kallen's criticisms of Morris and Carnap held that their philosophical
work was doctrinal and legislative. The totalitarian threat they posed,
however, was much less than Neurath's. Neurath was not only the recognized leader of the movement. He also concerned himself with language
actually used in and outside science. The bulk of Kallen's attack, therefore, was directed against Neurath and his proposed "universal jargon" of
science. Kallen dubbed it "logpu" and, in the course of their exchange,
returned again and again to his main argument: Logpu would be one language among others, like Ido, Esperanto, English, or French competing
for dominance as a language of science. Each language has its supporters
and critics in society. For Logpu to be "the one unifying menstruum of
all the sciences," Kallen reasoned, the logical empiricists who promoted
it "would have to establish themselves in a position to impose it by *force
majeure*." Adding a hint of insult to injury, Kallen then suggested that
after bullying their way to dominance, the logical empiricists would subsequently function as a language-managing priesthood, as "consecrated
keepers" of "the sacred mystery," much as medieval priests maintained
scholastic Latin (ibid., p. 91).

Suppose this victory were achieved. Kallen then asked, Would this be
good for the progress of science? No, because the unification might
instead hinder scientific progress. Problematic areas of science would
function as brakes on the advancement of the rest. Instead of integrating sciences into one overarching structure or language, therefore, the
movement should promote merely the "orchestration" of the different
sciences in ways that preserve (and celebrate) their differences and contradictions (p. 92).

Morris and Kallen on Orchestration

Morris dismissed Kallen's worries that he and his colleagues aimed "to
regiment science or to impose 'by force' any limitations on scientific terminology or research." This part of Kallen's argument, he wrote, required

"no further discussion" (Morris 1946a, 509). Morris did wish to discuss Kallen's proposed metaphor of "orchestration," however, for he believed that Kallen had embedded an important misconception inside it.

Morris adjusted and altered Kallen's metaphor to make several defensive observations. If we take the different sciences to be the different kinds of instruments making up an orchestra, he suggested, then the "unity in the orchestra does not cancel differences, nor does scientific unity cancel the differences of the sciences" (ibid.). On this view, Neurath's proposal for a unified language of science was no more harmful or legislative than the common, unifying musical notation in which each instrument's part can be notated. Nor, adding Carnap's principle of tolerance to the metaphor, must we require that there is only one acceptable musical notation or one acceptable universal language of science. In this light, the goal of the movement was "to formulate a notation (not "the" notation) which will give a common language to the performers in the scientific orchestra" (pp. 509–10). This project was yet further removed from the totalitarianism Kallen feared, Morris argued, because one must distinguish a language from what is said using the language. Nothing about a universal notation would stifle musical freedom: "The scientific performers can introduce new instruments and compose new tunes as they will."

One of Morris's points was not so defensive, for he gently pointed to the inconsistency that lay at the heart of Kallen's pluralism. On what grounds could Kallen, champion of intellectual variation, creativity, and freedom, legitimately exclude any of logical empiricism's projects? It seemed plain to Morris that the movement should be free to choose its tools and "be allowed to try its hand at analysis and synthesis of the results of science" (p. 510).

Yet Morris did not recognize how extreme Kallen's pluralism really was. Kallen required any such scientific orchestra to "make place for different compositions on different themes in different keys" (Kallen 1946b, 496). He could accept Morris's version if the orchestra was committed to something like free jazz, played without a conductor and with any variety of instruments. Then, he agreed, the orchestra might have a "unity which would retain the feeling of the movement, diversity, novelty, and freedom which I believe intrinsic to living science" (Kallen 1946d, 524, 525).

If Kallen found Morris's orchestra somewhat palatable, he maintained his criticism of Morris's semiotic idea of a "science of science" or "metascience." If this "metascience" were just another player, there would be no point to its methodological ascent. It would give rise to a "meta-metascience" and there would be "no end to such compounding" (ibid., pp. 525, 526). The real point of any science of science, Kallen reasoned, would be to control the sciences. Such a program, that is, would introduce a conductor into the orchestra, a "hierarchical head of all the sciences with the place and authority of such headship" (p. 525).

Kallen's Worries about Postwar Science

When Kallen reissued his original arguments in 1945, his worries were sharpened by the war's effect on science and, more important, American science policy. The military science behind the atom bomb, he charged, was the epitome of epistemological totalitarianism:

> Some 65,000 human beings, in three different plants, their foremen, production managers, expert physicists, chemists, biologists, engineers, worked at this task, knowing only that it was important war work, very secret, or otherwise worked blindly at jobs whose nature was blank to them and whose purpose was unshared.... The fabrication of this bomb provides us with a supremely momentous instance of one mode of the unity of science – the military or totalitarian mode.

Of course these workers operated freely to solve the problems they were assigned, but that was no consolation for Kallen:

> The over-all liberty, the choices and the decision are the prerogatives of the command.... Postulate, plan, and purpose are laid like a harness upon the men of science and their helpers.... The insurance which their alternative ideas, the measure of their different knowledges and methods might, if allowed free play, bring to the enterprise, is cut off by the totalitarian organization of the enterprise. (p. 516)

From where Kallen stood in 1945, the immediate threat to science was not logical empiricism and the Unity of Science movement but rather calls for the peacetime management of science along wartime modes of command and organization. That summer, Kallen followed the debate raging around MIT president Vannevar Bush's *Science: The Endless Frontier* (1960), which called for a national research foundation.[4] The 1945 meeting of the Conference on the Scientific Spirit and Democratic

---

[4] Bush's *Science: The Endless Frontier* presented Bush's argument, based on his management of the Office of Scientific Research and Development (OSRD) during the war, for the establishment of a national research foundation for the support of postwar science (the National Science Foundation was the result). Kallen's personal papers (JRMC) include reviews of Bush's book as well as papers and letters to the editor commenting on it.

Faith raised the question, "Does private industry threaten freedom of scientific research?"[5] Kallen felt sure that it did and argued to Neurath and Morris that contracts among government, industry, and universities would "convert scientific knowledge, which depends on the open and free cooperation and competition of many, into a private monopoly of a closed corporation" (p. 517). Kallen argued that a even hint of military or national organization in science would kill scientists' freedom: "Experience shows ... that in practical situations such provisos [safeguarding the liberty of the scientist] do not in any way deter administration and direction from conforming the individual to the organization." Good scientists, able to pursue "the personal, the contingent, the accidental," could not be so-called organization men (p. 519).

Neurath's First Response

Neurath tried to find some common ground with Kallen. He adopted Kallen's phrase "orchestration of the sciences" and used it to express his own anti-absolutistic and anti-architectonic notions of unified science. The metaphor, he believed, complemented his notion of "encyclopedism," according to which the sciences are brought together and juxtaposed in ways that facilitate debate about how to build bridges among them. Neurath packed all of this camaraderie and agreement into the title of his first response to Kallen: "The Orchestration of the Sciences by the Encyclopedism of Logical Empiricism."

Aside from accepting this notion of "orchestration," however, Neurath rejected Kallen's main claim as a more-or-less complete misunderstanding:

I do not think, however, that [Kallen] deals properly with our "logical empiricism," the main features of which are of an antitotalitarian character. . . . We intentionally rejected the plan of forming anything like a programme, and we stressed the point that actual cooperation in fruitful discussion would demonstrate how much unity of action can result, without any kind of authoritative integration. (Neurath 1946b, 496)

To rectify Kallen's misunderstandings, Neurath offered an autobiographical sketch of how and why he developed his project. Understanding its origin, he supposed, would make plain its antitotalitarian credentials.

5 Nathanson, Jerome to Members of the Conference Planning Committee, March 23, 1945, JRMC.

Logical empiricism began with Mach, Poincaré, and Duhem, Neurath explained. They in turn inspired Schlick and Carnap to identify "meaningless" sentences of metaphysics or pseudoscience – something which "became a kind of game" during meetings of the Vienna Circle. But Neurath "very soon felt uneasy" about this game, for he doubted whether any sentences could be trusted to be entirely free from metaphysics. Perhaps even the word "philosophy" would have to be scrutinized – "it would force us one day to invent a metaphysic to weed that [word] out." Suspicious of an endless eliminative project, Neurath opted for a positive, constructive project: "Thus I came to suggest as our object the collection of material which we could accept within the framework of the scientific [nonmetaphysical] language" (ibid., p. 497). That collection of sentences is what Neurath called "unified science" and the ordinary, nonmetaphysical language in which those sentences lived was the "universal jargon" of science or, as Kallen put it, "logpu." Here, simple statements about physical objects such as chairs, instrument dials, or people coexisted with abstract scientific terms and statements that were nonetheless "connectible" to more basic, physicalist statements. Metaphysical statements, on the other hand, could not be fastened to this language. In the parlance of unified science, "metaphysical" meant isolated from the larger body of scientific language.

So understood, unified science was more like an encyclopedia than an axiomatic formal system. It is riddled with "gaps," "gulfs" (p. 497), and inconsistencies among its parts:

Very often scientists know perfectly well that certain principles applied to a certain area are very fruitful, while contradictory principles applied to a different area also appear to be fruitful. It would, of course, be nice to harmonize the demonstrations in both areas, but, in the meantime, scientific research progresses successfully. (p. 498)

Science's success, Neurath thus emphasized, had never issued from a comprehensive, systematic, or (potentially) totalitarian unity of science. If Kallen granted that Neurath's movement aimed to advance science, it could not follow, therefore, that the movement aimed to achieve a comprehensive, systematic unified science.[6]

---

[6] Here Neurath was able to go on record again as a critic of Popperian falsificationism. Neurath's view accepts that science as a whole, and its local parts, are much more impervious to falsification than Popper believed. In an encyclopedic body of scientific statements riddled with gaps and contradictions, a falsification is nothing more than one more contradiction between a theoretical and observational statement. There are no "isolated

Neurath Defends Physicalism

Toward Kallen's interpretation of "physicalism" as an ontological thesis, Neurath explained that it was really a manner of speaking about objects and events experienced in physical space and time (not necessarily in the technical language of physics). The physicalist "universal jargon" worked "from the bottom to the top," facilitating communication "when we talk of cows and calves" as well as abstract theoretical entities or processes (p. 499). Importantly, Neurath emphasized that his proposal for physicalist language was largely descriptive: Ordinary communication among scientists and human beings is possible and effective, he believed, because something like a physicalist jargon is already in use. With respect to a mature unified science of the future, the physicalist jargon is a *start*. "My thesis is that this start is common to human beings, past and present, all over the world." Kallen's argument that logpu could become popular only by being *imposed* therefore had no target.

We are not presenting [people] with some new unity; not at all, we only want to say that wherever people speak to one another, for example, marooned men on an island coming from different parts of the world, about fishes and trees, drink and sleep, pain and pleasure, they will have no particular difficulties in communicating.

Instead, "difficulties will usually appear," Neurath continued, when physicalism is abandoned, "when they *want* to tell each other of their different magical expressions, theological sentences, or metaphysical formulations" (p. 500).

Neurath's Antifoundationalism

Neurath additionally tried to deflate Kallen's charges by underscoring his antifoundationalism. He had earlier rejected the term "positivism," he explained, in favor of "empiricism" (in "logical empiricism") precisely because it was "connected with what Kallen would call an imperialist attitude" (p. 501). The word suggested the image of a philosophical and ethical system "not based on the consensus of mankind, but on the deductions brought forward by the positivists." So understood, Neurath's defense joined Morris's. There was "nothing definite" or doctrinal in the universal jargon by itself. It was just a language in which propositions

single definitely negative instances which could destroy any general empiricist assumption" (Neurath 1946b, 498).

and sentences could be introduced, refined, discarded, and so forth according to "the consensus of mankind." The jargon was not a body of
doctrine, but rather a tool whose effectiveness would be judged only in
experience: After all, Neurath reminded Kallen, "one cannot test the
future usefulness of a scientific technique beforehand" (p. 502).

In general, Neurath could see nothing but democratic, pluralistic
values in his project, and he was optimistic that Kallen might come to
agree:

> I think that this gives a picture of the democratic attitude of the Unity of Science
> movement, which acknowledges from the start a multiplicity of possibilities. It is
> the problem of any democracy, which any actual scientific research organization
> has to solve: on the one hand, the nonconformists must have sufficient support;
> on the other hand, scientific research needs some cooperation.

The nonconformists and the traditionalists must cooperate yet maintain
their separate beliefs and convictions. They must, that is, be "orchestrated."

Neurath's Defense in Context

It could not have been lost on Neurath that his engagement with Kallen
directly connected – often ironically – to other debates and projects he
was involved in. First, Neurath's views about Nazism and his experiences
as a refugee had led him to start writing a book about persecution in
history. It must therefore have been striking for Neurath to be accused
of peddling totalitarianism as a time when he was working out his ideas
about close relations among science, physicalism, and democracy. Some
of his ideas, it would appear, came out in his debate with Kallen and his
defense of logpu: "If priests and rulers have a language of their own,
they become separated from the ruled masses, and it is just the unification of language that is a step forward to some democratic possibilities"
(p. 502). With a common language, that is, the ruled masses may come to
understand and criticize those who assume control over them. A healthy
skepticism and encyclopedic view of knowledge would further oppose
totalitarianism:

> Since the encyclopedism of logical empiricism challenges any intellectual author-
> ity which pretends to preach the truth . . . it is out of the question that it should not
> challenge any attempt to misuse any kind of distorted empiricism for creating a
> similar authority. . . . [T]he skeptic pluralism of our empiricism is, in itself, hardly
> a suitable tool for suppressors. (p. 504)

It was, rather, metaphysical idealism that duped ordinary people into submitting to totalitarian control and complicity in persecuting others. Metaphysics, not physicalist language, was the language of "terrible means to lofty ends, which very often reduces the preparedness of people to object to the mercilessness of totalitarianism" (p. 503).

Neurath had expressed similar thoughts to Carnap in their dispute over semantics.[7] Theories of *the* meaning of a word or proposition, theories of truth and stipulated relations between language and the world, Neurath argued, dangerously encouraged belief in absolute, transcendent truths – truths deemed worth dying for or worth persecuting others for. Similarly, he told Kallen, one who is full of "zeal"

> may spread, without knowing it, some danger, even if his propaganda deals with nice things; because it accustoms people to focus on one particular, overestimated, and emphasized point that may later be transformed into a super human being, *the* state, *the* leader, or something else. (p. 504)

Neurath's defense also echoed his recent review of Hayek's *Road to Serfdom* (1944; Neurath 1945b). Much as Hayek had argued that a little socialism will surely put us on the road to serfdom, Kallen argued that attempts to plan or organize science will probably, if not necessarily, descend into dictatorship. But that could not be the outcome for the Unity of Science movement, he urged Kallen, because scientists – as empiricists and pluralists – know that their job is "to discover as many alternatives as possible" (Neurath 1946b, 505). Their role is not, that is, to choose one of these alternatives and impose it on society.[8]

Neurath's Counter-Attack

Satisfied with his many-faceted defense, Neurath then argued that Kallen himself had fallen prey to some metaphysical (and totalitarian) dangers. Could Kallen's devotion to antitotalitarian ideals itself be some expression or remnant of metaphysical dogmatism?

> I think the "exaltation of diversity" is rather the concern of the metaphysicians, who are very proud of being different. Since they have no work in common, they can build their ivory towers *ad libitum*. For these metaphysicians, Kallen's "one man's meaning may be another man's nonsense" may be valid.

---

[7] Neurath's late debate with Carnap over semantics is treated in chapter 10 and in Reisch 1995.

[8] Neurath 1946b, 504. Here Neurath cited his paper "International Planning for Freedom" (Neurath 1942).

But in the daily lives of scientists and others, "we get on quite well, within limits" using logpu and steering clear of metaphysics (ibid.).

Neurath also questioned Kallen's reliance on popular writings by scientists to support his claim that communication was ordinarily (but properly) distorted and hindered by the pluralistic vagaries of language and meaning. Neurath found Kallen's examples of this to be unpersuasive:

> But the problems Kallen mentions occurred when famous physicists such as Jeans and Eddington . . . made up fairy tales of an evening, as Lewis Carroll did. . . . Unfortunately, many people take the fairy tales of physicists as seriously as others took Newton's interpretation of some biblical books.

Even worse, these "physical fairy tales have a totalitarian touch, in so far as they try to create the feeling of something miraculous around us." They conceal scientific argumentation and creativity, as if it really were controlled by some priestly caste. The movement opposed such hermeticism, Neurath wrote: "We logical empiricists want to show people that what physicists and astronomers do is only on a grand scale what Charles and Jane are doing every day in the garden and the kitchen" (p. 506).

Kallen's First Reply to Neurath (and Neurath's Turkey Dinner)

Kallen seemed surprised by Neurath's defense. Apparently, he noted in reply, Neurath had shifted his views in the last six years "from a monistic base to a pluralistic base." But that did not keep Kallen from trying to push through his argument. For despite the particulars of Neurath's views, the unity of science project was still, at least potentially, in the pocket of those seeking to regiment and control postwar science: "I cannot be sure that his logical empiricism might not serve as a ready-to-hand rationalization for the totalitarian program of research now being advocated as a political project" (Kallen 1946d, 520).

Kallen changed his target. Now the totalitarianism he attacked lay in the gaps and gulfs that Neurath acknowledged with his pluralism. These would focus scientific attention and, he seemed to fear, effectively coerce or force scientists to address specific problems at the expense of others and to devise particular solutions to them. Unified science itself, that is, would "cut a pattern that . . . requires the filling to conform to." "This pattern is an *Einheitswissenschaft* which Neurath Englishes into 'unified science' but which could more appropriately be translated as 'science of unity.' The unity seems to me prior" (ibid. 520). The project would also force scientists to submit to the totalitarian control of protocol statements:

With all the formal acknowledgments of the fact of change, of the reality of alternatives, [the encyclopedia] would still look *back* to protocol statements and make sure that statements fitting them are "connectible" – i.e., fitting. Its data would be derivative, its unity a pre-existence assumed and found, not made. (ibid.)

After six years, Kallen wrote, and despite this alleged shift from monism to pluralism, Neurath's views had not "inwardly changed" (p. 521). They remained essentially totalitarian.

With the totalitarian core of unified science again laid bare, Kallen proceeded to detail the dangers and confusions of Neurath's project. Did possession of a universal language really help to ensure peace or democracy, as Neurath suggested? No, Kallen answered, for mutual understanding among peoples might just as well lay foundation for attacks and wars (ibid.). Did use of a common language really help to ensure communication? No, Kallen insisted, as he took up Neurath's example of the expressive powers of logpu: a handful of persons from different cultures and nations meeting in New York City to share a lunch of hot turkey.[9] Neurath had argued that, despite all of their differences, the physicalist jargon would allow them to understand each other. Kallen disagreed and claimed that understanding would be difficult to achieve: words such as "'turkey,' 'cold,' 'hot,' 'happy,'" he wrote, "may have to start as incommensurable diversities and work their ways toward a consensus" (ibid.). The ideas, thoughts, and feelings of these guests may indeed remain incommensurable and beyond the reach of Neurath's scientific language:

What is edible and inedible to these guests will ... depend far less on the sensory data – with or without gestures – than on attitudes, feelings, and judgements which may be meaningless illusions and superstition vis-à-vis Neurath's *Protokolsaetze*. Inexpressible in Logpu, they would nevertheless be the dynamics of behavior around that luncheon table. (p. 522)

The only mutual agreement and understanding that these people having lunch could quickly achieve, Kallen wrote, was that Kallen was right and

---

9 This illustration or thought-experiment that Neurath and Kallen toss back and forth probably took root in an actual lunch they shared during one of Neurath's visits to New York.

Neurath was wrong. The guests would collectively understand that they are free *not* to understand each other, to "live and let live." Their unity would be "the collective security all brought each to lunch in freedom according to his own lights in his own way" (ibid.).

Kallen closed his reply by attacking Neurath's high-handed dismissal of metaphysics and popular writings about science:

Pharisee-like, [Neurath] speaks of Logpu as if it was superior [to other languages] in its nature and not by its consequences; serious where those others together with telepathy, astrology, numerology, and the like are frivolous. As if *serious* and *frivolous* were metaphysical – forgive me, physicalist – ultimacies and not personal judgments!

Neurath had, however, defended logpu precisely on the grounds of its future consequences, its effectiveness as a medium of communication in science, and a tool for further unification. And it was plain in Neurath's outline of his pluralism that science and a scientific approach to problems *required* "personal judgments" and free decisions.

Kallen's Misinterpretations and Neurath's Growing Frustration

Neurath seemed to know at this point that he was not going to win this debate. In part, this was because Kallen consistently misunderstood his project. In Kallen's hands, for example, the universal jargon of science was not something more or less already in use (among empirically inclined persons) but rather a proposed new competitor whose victory could come about only via force. And it was not just a language, but a language containing theories about the nature and structure of the world that effectively ruled out alternatives. Kallen wrote of Neurath's *Encyclopedia*, for example, as if it were a factual, positivistic survey of the world of objects and events – a compendium of controlling, protocol sentences to which future research must conform:

Behind the actual experiences of the daily life which for us involves what James calls the stream of consciousness and Bergson hypostatizes as *durée réele*, would be a static order called physicalism, ultimately reducible to an arrangement of *Protokolsaetze*. (ibid., p. 521)

For Kallen, these *Protokolsaetze* were metaphysical and ontological. Despite Neurath's attempt to correct his claim (at the 1939 conference) that physicalism "is usually called mechanistic" (Kallen 1940, 94), Kallen continued to compare physicalism to "mechanistic determinism and ineluctable causality" (Kallen 1946d, 520), to "mechanistic materialism or determinism" (Kallen 1946b, 494).

Kallen misunderstood not only Neurath but some basic features of science and scientific practice. For his claims that the Unity of Science Movement would mischievously restrict scientific thought and behavior apply just as much to ordinary scientific practice. In Kallen's eyes, for example, a physicist seeking to understand poorly understood boundaries between quantum mechanics and relativity theory or biology and chemistry would become a scientific serf, beholden to two despots of science that together define the gap to be filled and constrain available answers. His condemnation of protocol sentences, similarly, amounts to a rejection of the way scientists respect empirical reports about nature or experimental systems. As human beings interact with and observe it, nature itself appears in Kallen's framework as a totalitarian dictator that controls scientific thought and deprives would-be innovators of their epistemic liberty. Only science *fiction* emerges from Kallen's critique as scientifically and ideologically respectable, for it is unencumbered by other areas of science or accepted empirical reports.

Yet these considerations probably do not exhaust Neurath's surprise and dismay about Kallen's attack. Before the Cambridge conference, Neurath was intrigued by Kallen's expertise in pragmatism and invited him to "write an article about [William] James and the ideas propagated by our Unity of Science movement, with the principle of tolerance and the multivalued solutions."[10] Neurath considered him a supporter and ally of the movement and appreciated his help in seeking Rockefeller funding to save the institute from Nazi advance.[11] Even after Kallen read his paper at the Harvard Congress, Neurath did not see any insurmountable disagreements between them. In December 1939, after returning to Holland, he wrote to Kallen:

I agree absolutely with you. My Slogan will be: MORE THAN ONE UNIFIED SCIENCE IS TO BE DISCUSSED. The unification does not mean any reduction of competition or pluralism, but means only a program for the INNER ARRANGEMENT OF THE SCIENCES. The Unification of physics (as realized during the last decades) does not mean any uniformalization. Therefore, PLURALISM always as a programmatic principle.[12]

---

[10] Neurath to Kallen, January 25, 1939, JRMC.
[11] Neurath to Kallen, July 7, 1939, JRMC.
[12] Neurath to Kallen, December 23, 1939, JRMC.

A month later he reminded Kallen,

I am very interested in the problem of "Pluralism" for years, better for decades – I think it would be very nice if we could publish an article of you on this subject matter in the JOURNAL OF UNITED SCIENCE, as an addition to the synopsis of your Cambridge-Lecture.

Neurath envisioned a published dialogue or discussion that would be particularly effective since "we could eliminate before[hand] all misunderstandings and publish our real variations (I do not think we have divergences) of a common idea").[13] When Kallen was having difficulty finding a place to publish his first critique (Kallen 1940), Neurath offered to help find a European journal that might publish it.[14]

Neurath's Final Response

By the time he wrote his final response, however, Neurath had realized that he and Kallen were at an impasse. Neurath gave up and, in his final response, signaled that he considered the debate finished. "I shall not attempt at this point to clear up some misrepresentations of my opinions" (Neurath 1946a, 526). Its title was "Just Annotations, Not a Reply." If Neurath could not get through to Kallen, that is, there were some items he would like to be on the record for the benefit of his readers.

First, he denied there was a "shift" in his views, from monism to pluralism, as Kallen had suggested. He quoted his early paper, "The Lost Wanderers of Descartes and the Auxiliary Motive" (1913), to show he had always endorsed pluralism. He also quoted other papers he had written about the "gaps" and "gulfs" that will remain in any unified science and in the new *Encyclopedia*, which, he had written, "does not ask the collaborators to subject themselves to a common program." "What Horace says about the fitting into a system and its postulates is just the contrary of what I am saying in the papers quoted above."[15]

Neurath also returned to the hot turkey lunch, which Neurath now found incomprehensible on Kallen's view. On the one hand, Kallen had

---

[13] Neurath to Kallen, January 9, 1940, JRMC. (This letter is dated "Jan. 9, 1939," by Neurath, but is clearly from 1940 judging from its content and Kallen's response, which is dated February 6, 1940.)

[14] Kallen to Neurath, November 2, 1939; Neurath to Kallen, January 9, 1940 ("1939"), JRMC.

[15] Neurath 1946a, 527, 528. Neurath quoted himself from "Radical Physicalism and the Real World" (Neurath 1934) and "Individual Sciences, Unified Science, Pseudorationalism" (Neurath 1936c).

written that this gathering would involve "good fellowship" and "a free exchange of some diets, dialects and table manners" (as well as "an agreement to disagree"). On the other hand, Kallen insisted that many of the basic feelings and beliefs of each guest would remain inexpressible to the others. Neurath's sometimes shaky English grammar got turned around as he marveled that Kallen could imagine people communicating without using some kind of shared language:

I can hardly believe, that he thinks people could argue with other people without using phrases, which are not "connectible" [to each other within a shared jargon].

Communication is based on this connectibility which is all Carnap, Frank, [Richard von] Mises, myself and others are asking for. That this is full of totalitarian danger surely cannot be inferred, as Horace thinks, from the wording of my papers. (Neurath 1946a, 528)

Neurath could not have elaborated or further debated these points if he had wanted to, for he died of a sudden stroke in December of 1945, probably days after sending his "annotation" to Kallen.

Kallen's Annotation

Kallen received Neurath's "annotation" and then submitted his own – "An Annotation to the Annotation." He pursued Neurath, almost as if he were not willing to end the debate, and goaded him into defending himself. For example, he wordsmithed an apparent concession – "I owe it to [Neurath] to withdraw the observation that he has shifted from a monistic to a pluralistic base. I will regretfully agree that he has not" – into another repetition of the original claim:

I will not agree, however, that he has not on the ground that he has always been more of a pluralist than I knew [but] on the ground that he has always been more of a monist than he knew. His annotation underscores this monism of his.

Neurath merely talked about pluralism, Kallen suggested. He did not, in fact, embrace it. "The diversities which Otto champions are clearly not radical diversities, diversities *ab initio*, [and] qualitative." Using terms and references that could not have been more inflammatory in 1945, Kallen blustered that Neurath's pluralism

would excommunicate non-physicalist statements altogether from the fellowship of meaningful propositions: that they would be to such a fellowship as Jews are to Aryans in the Nazi anthropology. Hence the editors of Otto's encyclopedia of unified science could hardly find acceptable as contributors a neo-Kantian, an

Hegelian, a Berkeleyan, a phenomenologist, or an existentialist as such. None of these could be coordinated as Otto thinks . . . scientific workers should be coordinated. (Kallen 1946a, 528–29)

If Neurath had seen these words of Kallen's before he died, they must at least have raised his blood pressure if they did not precipitate the stroke that killed him.

Kallen's Obituary and Some Final Ironies

When he learned that Neurath had died, Kallen suddenly praised him as a fellow crusader for epistemic and social freedom. Neurath was, he wrote,

a veteran comrade of the struggle for freedom on whose sacrificial courage even the most unwilling were wont to count. Of course he had his intolerances and rejections – who has not? – but the doctrines and disciplines which he excommunicated were those which experience had led him to hold as enemies of free men in a free society, as superstitions employed by malicious power to degrade and starve the soul and body of mankind. (Kallen 1946a, 529–30)

If Neurath's crusade was suddenly so acceptable and admirable to Kallen, why had he so vigorously attacked it? One clue is that Kallen distinguished Neurath's philosophy of science and its (allegedly) totalitarian aspects from other, different features of Neurath's character and personality:

Witnesses of my own differences with Otto knew that I attacked these [intolerant] propensities of his, but that in attacking I could not fail to testify to the humane and prophetic urge for righteousness which sustained them. Neurath's whole career attests this disposition. (ibid., p. 530)

Neurath's unassailable "righteousness," Kallen believed, led him to be intolerant (of metaphysics, for example) in ways that Kallen believed could only backfire and become, themselves, totalitarian and freedom-destroying. True to form, Kallen believed that Neurath himself would have agreed, and he described the last letter he received from Neurath as proof. Neurath signed off, as he often did, with his signature self-portrait – a drawing of an elephant. In this case, the elephant

holds a peacock feather aloft in his trunk, and he is backing up against a cactus almost his own size. The implications involve an irony the most Socratic of Socrateses could not match. (p. 533)

Neurath knew, Kallen suggested, that his arguments led him ironically into the thorns of totalitarianism.

Kallen and the American Idea

The real irony of the Kallen-Neurath debate, of course, is quite different and much more severe. As Neurath and Morris detected, Kallen's own extremism and dogmatism about freedom and intellectual pluralism points to Kallen, not Neurath, as the agent of totalitarian control over science. For Kallen insisted that science did – and philosophy of science, in turn, must – adopt, implement, and promote the pluralistic doctrines Kallen upheld.

These pluralisms sustained nothing less than the "American Idea" that Kallen championed in his other writings. As opposed to the popular metaphor of the American "melting pot," Kallen's "American Idea" was emphatically pluralistic and suggestive of NewYork City's mosaic of ethnic neighborhoods:

[A]n orchestration of diverse utterances of diversities – regional, local, religious, ethnic, esthetic, industrial, sporting and political – each developing freely and characteristically in its own enclave, and somehow so intertwined with the others, as to suggest, even to symbolize, the dynamic of the whole. Each is a cultural reservoir whence flows its own singularity of expression to unite in the concrete intercultural total which is the culture of America. (Kallen 1956, 98)

Just as the American Idea was an "orchestration" of radical differences, so unified science, Kallen argued, should be a parallel orchestration of different scientific projects. To understand Kallen's objections to Neurath's program, however, the crucial point is that this American Idea must be regularly and loudly proclaimed and renewed from all quarters:

As a plan of human relations in the human enterprise, [the Idea] receives like repristination from the utterances of poets, statesmen, dramatists, men of letters. A sequence of such renewals makes up the Bible of the people of Israel. ... In the same way, a selected sequence may make up a Bible of America. Its book of Genesis would of course be the Declaration of Independence, which is also the simplest, clearest, most comprehensive, yet briefest telling of the American Idea. (ibid., p. 87)

Kallen required every institution to participate in this trumpeting of the America Idea. That is why he found Neurath's Unity of Science movement so threatening and offensive. He attacked so relentlessly because, to him, the ideals and goals of the movement threatened to subvert those that he attached to "America" itself.

So understood, Kallen's attack stands as a harbinger of what was to come, not only for the Unity of Science movement but for leftist academics who were investigated or prosecuted during the Cold War. Official members of the Communist Party, unofficial fellow-travelers or indepen-dents who promoted socialist ideas were variously prosecuted or pro-fessionally and socially shunned because they allegedly promoted ideas or actions that were subversive to the American political system and its cultural values.

The exchange among Kallen, Neurath, and Morris appeared in print during the same year that trend-setting anticommunist investigations be-gan at the University of Washington. With anticommunist investigations and loyalty-oath controversies sweeping across the American landscape of higher education during the late 1940s and early 1950s, and with "totalitarianism" typically equated with both communism and fascism in the public mind, it is difficult to imagine that Kallen's attack did not at least suggest to those philosophers, scientists, and graduate students who followed the debate that the Unity of Science movement was ide-ologically defunct. The irony Kallen described in Neurath's picture of an elephant backing into a cactus is thus again dwarfed by the irony with which Morris, some eleven years after their exchange, responded to Kallen's inquiries about anthologizing Neurath's writings. As if Kallen did not know or had nothing to do with it, Morris reported to him that Neurath and his movement were seen as "communistic" (see chap-ter 13). It was Kallen, however, who had so politicized criticism of the movement years before as he obscured real questions about how the Unity of Science movement's ideals would be implemented in practice with alarmist, hyperbolic parallels between (for instance) physicalism and "Nazi anthropology."

Kallen's Denunciation

Several years later, in 1953 at the height of McCarthyism, Kallen tasted something of what Neurath may have felt during their exchange. The anticommunist magazine *American Mercury* published one of many ar-ticles it featured describing how communists had infiltrated American institutions such as government and churches. This time, the article was "Communism and the Colleges" by leftist-turned-informant J. B. Matthews. He warned that "in [their] endeavor to corrupt the teachers of youth, agents of the Kremlin have been remarkably successful, espe-cially among the professors in our colleges and universities" (Matthews 1953, 111).

In a long, rambling article, Matthews rehearsed accusations about the pink and "morally disintegrating" credentials of prominent academics.

He attacked John Dewey's progressive education program (because its moral vacuum is "communism's opportunity"), professors who wrote to Roosevelt on behalf of Earl Browder (then jailed in Atlanta on perjury charges concerning inquiry into his Russian-born wife's naturalization), the "top ten" professors (out of "close to thirty-five hundred") known to have collaborated with Communist Front organizations, college presidents who sponsored the Waldorf conference, academics who supported the American Committee to Protect the Foreign Born and other academic or intellectual front organizations (such as the American Association of Scientific Workers), scientists ("much less familiar" than Klaus Fuchs and Harry Gold) who probably sent atomic secrets to the Soviets, and, finally, University of Chicago President Robert Maynard Hutchins, who (suspiciously) criticized anticommunist investigations in higher education. All this was preliminary to issuing an indictment of advocates of "academic freedom" who, Matthews claimed, were the same crypto-Communists running the National Council of Arts, Sciences and Professions and the American Civil Liberties Union (ACLU).

Whenever he could think of them, Matthews named names. Harlow Shapley, Kirtley Mather, and Dirk Struik appeared several times (here, and in other articles Matthews wrote). Coming closer to the Unity of Science movement, Matthews named William Malisoff (as a contributor to communist textbooks used in colleges), Abraham Edel and Rudolf Carnap (billed as "top collaborationists"), and E. C. Tolman and Ralph Gerard (whom Morris once proposed to write for the *Encyclopedia*).

As Cornell professor of labor relations Milton Konvitz thumbed through the long and rambling article, he was surprised to see that Matthews had named members of the ACLU's Academic Freedom Committee and its advisory council. Konvitz was on that committee along with his good friend Kallen. The two had known each other for years, probably since Konvitz studied philosophy at NYU. In Matthews's article, Kallen received top billing: He was one of "eight professors [who were] among the top hundred academic collaborators with the Communist-front apparatus in the United States" (Matthews 1953, 141).

Alarmed, Konvitz wrote to his friend:

Have you seen the April AMER. MERCURY? It has an article on Communism and the Colleges by J. B. Matthews. He names many professors as fellow travelers, singles out Cornell for special rough treatment but doesn't mention me; but he does mention you. He devotes some space to the ACLU committee on academic freedom and its advisory council, of which both of us are members.... [You] are, he says, "among the top hundred academic collaborators with the Communist-front apparatus in the U.S."[16]

Slapdash as Matthews's charges were, Konvitz took them very seriously and carefully relayed them to Kallen. He reasonably feared the worst if Kallen's name moved from the *American Mercury* to the House Committee on Un-American Activities, for example. Konvitz therefore suggested that Kallen enlist Sidney Hook (whose anticommunist reputation would carry weight with Matthews) to request some kind of retraction:

Sidney has just dedicated his newest book to you. Wouldn't it be in order for Sidney to write to [Matthews] about this matter? I am going to suggest this to Sidney. ...I think it would seem quite the natural thing for him to do.

I hope this will not upset you – not as much as it has upset me. I am writing to you about it because you will find out anyhow, and one can't afford to bury one's head in the sand in respect to such a matter.[17]

Kallen was as angry as Konvitz, and so was Hook when he was informed of it.

About a year before Matthews had fingered him as a communist, Kallen described to Konvitz his views about McCarthyite Red-baiting and black-listing. When dealing with anticommunists such as "McCarran and that ilk," he explained, one must "hit hard and clean and never stop punching."[18] Kallen indeed hit hard (if not clean) and never stopped punching Neurath and his Unity of Science movement, but when it came to his own indictment from J. B. Matthews, he and Konvitz turned to Sidney Hook not to argue the case but to call on Matthews to issue a retraction and, as much as possible, erase the charge. Given the logic and dynamics of Red-baiting, one should never take the bait and be drawn into defending one's legitimacy. Kallen seemed to know that. Neurath did not.

[16] Konvitz to Kallen, April 19, 1953, JRMC.
[17] Ibid.
[18] Kallen to Konvitz, January 2, 1952, JRMC.

10

Creeping Totalitarianism, Creeping Scholasticism

Neurath, Frank, and the Trouble with Semantics

As Neurath fended off Kallen's attacks in the last year of his life, he was also immersed in a long and frustrating debate with Carnap. It began in 1942 with a renewed skirmish over the viability of semantics and Tarski's theory of truth. By 1944, it had become exacerbated by a dispute over Neurath's encyclopedia monograph, *Foundations of Social Science* (Neurath 1944). As we see below in this chapter, Neurath charged Carnap and semantics with metaphysical mischief that had potentially severe political consequences. In some ways, Neurath's complaints against semantics paralleled Kallen's complaints about the Unity of Science movement, and they tended to the same overall historical effect: They helped to widen and sustain a rift within the movement that would later help facilitate logical empiricism's subsequent break with the Unity of Science movement. The debate arrayed Neurath, Frank, and Morris, on the one side, against most other logical empiricists whom Neurath and Frank believed were veering into formal, logical modes of philosophical inquiry that, were they to become dominant, would reduce the practical utility and relevance of philosophy of science.

Carnap and Neurath

Histories of the Vienna Circle usually adopt the view that the circle was intellectually and politically divided into a more radical left, led by Neurath, Carnap, and Hahn (who together wrote the circle's manifesto), and a more conservative right, led by Schlick and Friedrich Waismann. The split reflected different views and philosophical styles: political neutrality versus partisanship of philosophy, some personal animosity between

Neurath and Schlick, and the divisive effect on the circle of Wittgenstein, who, for example, captivated Schlick and Waismann but eventually refused to meet with Carnap (Carnap 1963a, 27).

After Hahn's death in 1934, it was this left wing of the circle – mainly Carnap and Neurath – that led the Unity of Science movement and brought it to America. But these two approached scientific philosophy with different skills and values. These differences sometimes sparked creative innovations (as with Neurath's reaction to Carnap's *Aufbau* and the protocol-sentence debates (Uebel 1992)) and sometimes only stubborn, exasperating and, unfortunately, personal disputes. This clash of intellects and personalities deserves a book of its own. But for purposes of understanding the Unity of Science movement in America, the relevant starting point involves Carnap's pursuit of semantics and Tarski's theory of truth in the mid-1930s. Neurath greatly admired Carnap's earlier syntactical program for its power to identify metaphysical language and to help "establish harmony between the statements of unified science" (Neurath 1932/33, 99). But semantics, he believed, committed several grave sins.

During the movement's most successful years, these disagreements remained submerged, though still detectable. In 1938, for example, Neurath corresponded extensively with the English logician L. Susan Stebbing, whom he cajoled into organizing the Fourth International Congress at Cambridge, England. Although Stebbing was short on time and energy, Neurath explained that she had a natural and important place in the movement and its three, main areas. Those were: logic, represented by Carnap and Stebbing; empirical procedure and probability and induction, represented by Joergen Joergensen, Reichenbach, and Louis Rougier; and the *Encyclopedia* and unified science, represented by Frank, Morris, and Neurath.[1] In Neurath's mind, that is, he and Carnap led different aspects or parts of their movement. At the time, moreover, Neurath was not much bothered by Carnap's and Tarski's semantics. In this letter to Stebbing, he asked her to invite Tarski to the congress.

The Trouble with Semantics

In late 1942, most likely, Carnap sent Neurath a copy of his latest book, *Introduction to Semantics* (Carnap 1942), and reignited their dormant

---

[1] Neurath to Stebbing, May 25, 1938, ONN.

disagreement about the scientific respectability of formal semantics. Neurath read the book and sent Carnap his assessment: "I am really depressed to see here all the Aristotelian metaphysics in full glint and glamour, bewitching my dear friend Carnap through and through."[2] In light of the central themes of Neurath's philosophy of science – semantic holism, pluralism, his critiques of "systematization" and pseudo-rationalism – Neurath saw Carnap's semantics as a kind of metaphysical atavism with potentially destructive consequences for their movement and, specifically, the projects Neurath anticipated on the postwar horizon. In their correspondence, Neurath's criticisms often came rapid-fire. Taken together, they suggest at least three arguments against semantics: one philosophical, one scientific, and one political and cultural.

The first holds that Carnap's semantics (and specifically the theory of truth and confirmation theory) presupposed or, at least, practically encouraged an absolutistic, nonpluralistic conception of scientific language. Scientifically, Neurath objected that semantics would be of little use for the practical scientific work involved in further unifying and connecting the sciences. Last, Neurath suggested that semantics would encourage enemies of the movement and support metaphysical modes of thought that had been historically conducive to totalitarianism and persecution.

Semantics as Absolutistic and Ontological

As a longtime friend and collaborator, Neurath was of course well versed in Carnap's work and its style, and despite his criticisms of semantics he understood its intellectual, even aesthetic appeal. He wrote to Carnap,

Sometimes I look in mathematical and physical books and enjoy the clearness and exactness of this field. I understand very well, why so many thinkers try to anticipate such a clearness in all sciences.

Still, this attraction to "clearness and exactness" was fraught with danger because the actual language of science was pluralistic and "aggregational."[3] Neurath would not be seduced:

I guess that the pleasant systems of logic and mathematics seduce people like you, Hempel, Tarski to desire systems of a comprehensive kind in empiricism,

---

[2] Neurath to Carnap, January 15, 1943, ASP RC 102-55-02
[3] Neurath to Carnap, July 17, 1942, ASP RC 102-56-04.

where . . . there are only islands of systematization . . . and, in principle, some unpredictability.[4]

Semantics, as Neurath saw it, overlooked the pluralism of scientific language and falsely assumed a precision and systematicity within it.

Carnap responded that his conception of semantic truth had nothing to do with an Aristotelian metaphysics of things and properties:

> That the semantical concept of truth is not metaphysical can very easily be shown by the following translation: "the sentence 'this tree is green' is true" means not more and not less than "this tree is green." (If the latter sentence does not occur in your strangely restricted language you may take instead any other sentence which you regard as meaningful).[5]

Neurath did not accept Carnap's disclaimer, however. He seemed to think that, regardless of how Carnap interpreted the semantic conception of truth, it contained implications that would mislead others (if not also Carnap) and that had to be vigorously opposed.

One implication was a realistic metaphysics. Recalling his position in the protocol sentences debates, Neurath replied, "I think the subject becomes so difficult, because you start from 'this tree is green,' whereas I start from 'we are using the sentence "this tree is green"' and do not overstep this threshold."[6] Similarly, Neurath's model of protocol sentences included a proper name of the protocolist and contextualized the facts reported to that individual and their experiences and motivations. "Otto's protocol at 3:17 o'clock: [At 3:16 o'clock Otto said to himself: (at 3:15 o'clock there was a table in the room perceived by Otto)]" (Neurath 1932/33, 93). His protocol sentences did not attempt to go beyond this "threshold" between language and the world, as if it were possible to describe the entire situation *without* language, as if it were "absolute" and exempt from pluralism.

Carnap's mistake, Neurath once suggested, was to misconstrue the distinction between a language and a semantical metalanguage – or between "levels" *within* a language (in case the two are specified in the same language) – as this threshold between language and the world. "Of course, the many levels principle I praise everyday," Neurath wrote to both Carnap and Morris in the midst of the dispute. But Carnap, he

---

[4] Neurath to Carnap, September 25, 1943, ASP RC 102-55-03.
[5] Carnap to Neurath, February 4, 1944, ASP RC 102-55-04.
[6] Neurath to Carnap, April 1, 1944, ASP RC 102-55-08.

charged, seemed to believe that "that the lowest level is an 'ontological' one."[7]

Semantics Obscures Empiricism

Carnap's semantics surely did not commit him to ontological realism. Nonetheless, in the context of Neurath's other worries (and with some interpretive charity), his point was not entirely empty. Though Carnap's semantics did not presuppose or argue for metaphysical realism, it did not explicitly reject it in the ways, and to the degree, Neurath believed necessary. Semantics, that is, was not emphatically pluralistic and "empiricist" in Neurath's preferred antimetaphysical understanding of the term. For Neurath, the aim of their movement was not to build ideal models of theories or scientific language, but to help unify the sciences as they existed and were practiced. Systematization, to the extent it would be achieved, would be a bottom-up (and not a top-down process) through which scientists would revise their methods, terminologies, and theories to create bridges and connections among the sciences. The problems involved belonged to "empiricism," as Neurath understood it, and had little to do with "systematic" models of language or scientific procedures that philosophers might devise. "You see," he urged Carnap, "I do not see how all these Semantics problems fit into the discussions of scientific practice with which I am highly concerned at the moment." "I ask how [semantics] affects empiricism."[8]

Neurath did not make his position clearer or more persuasive when he occasionally wrote approvingly of formal, systematic projects such as semantics and probability theory. He understood that scientific practice and progress often do involve abstract, highly articulated theories. As long as the messy, pluralistic character of most scientific practice was kept in mind, ideal philosophical models and systems (or "calculi") might be useful. Thus he urged Carnap to "take care" and always mind "the distinction between CALCULUS and EXPERIENTIAL [i.e., actual scientific] statements":

Probability seems to be a pure calculus term and the degree of confirmation seem[s] perhaps only possible within the framework of a MODEL, not a concrete theory. But, of course, the results may allow us some applications.[9]

---

[7] Neurath to Morris and Carnap, November 18, 1944, CMP.

[8] Neurath to Carnap, January 15, 1943, ASP RC 102-55-02; Neurath to Carnap, September 25, 1943, ASP RC 102-55-03.

[9] Neurath to Carnap, April 1, 1944, ASP RC 102-55-08.

As long as one did "not give the impression that the systematization of a construed language is in any extent transferable AS SUCH to the experiential scientific language," as long as one steers clear of pseudorationalism and the idea "that in principle the Laplace's Demon is the picture of the scientist in action" and the idea that the sciences will approach the one, true world ("the LIMIT of knowledge more and more"[10]) – then, Neurath told Carnap, semantics may be useful in the progress of science.

"But I do not see that you, Popper etc. stress this point," Neurath complained. By focusing only on the systematic and formal aspects of science, Neurath believed, Carnap and others threatened to create the impression that science was *entirely* systematic and formal, as if pseudorationalists and absolutists were correct in their intuitions that science was comprehensive, infallible, and required no personal decisions, conventions, or active interpretation for its advancement or use. "Just at the moment," he continued, "it is important to see, that decisions will be made based on some scientifically reached results, but that these decisions will be based on unsufficient material, in principle."[11] Pseudorationalism and its partner, absolutism, were seductive and had to be fought continually, Neurath reminded Carnap: "People sometimes cannot bear, that we start with many divergent statements, and remain with divergent statements FOREVER, as it were. There HAS TO BE SOMETHING ONE."[12] Neurath believed that the Unity of Science movement could help people to avoid the temptations of absolutism – but not, or at least not so easily, if the projects ongoing within it (such as semantics) were not emphatically pluralistic and empiricist.

At other times, however, Neurath denied that anything but metaphysical mischief could come from semantics, regardless of how it was presented, packaged, or qualified. Speaking specifically about inductive logic and confirmation theory, he told Carnap, "I am very doubtful about your system of inductive logic. . . . You see the degrees of confirmation etc. seem to anticipate that you are sure of ONE LIMIT OF THE SERIES." The most confirmed statement, Neurath argued, was merely a deceptive replacement for old Cartesian ideals of epistemic certainties anchored to metaphysical realities:

I do not know any empiricist material, which permits us to apply all this stuff on confirmation. . . . [H]ow should a group of protocol statements look . . . that

[10] Ibid.
[11] Ibid.
[12] Neurath to Carnap, September 25, 1943, ASP RC 102-55-03.

you can speak of a "degree of confirmation"? Can you give me any example? Of course you can always make an utopian structure on which you can demonstrate that, but then – you will see – you always will speak of THE REALITY (in some or another way, more or less concealed) and not of the possibilities (pluralism).[13]

At this point, Neurath believed that Carnap's semantics necessarily led him back to ontological metaphysics of the sort that logical empiricism always strived to identify and avoid.

Semantics, Persecution, and Hopes for a World Commonwealth

The last of Neurath's objections involved both world politics and philosophical politics. On one occasion, he needled Carnap with the idea that his *Introduction to Semantics* capitulated to the scholasticism of logical empiricism's neo-Thomist enemies. A few months after Hook, Nagel, and Dewey attacked Adler and other neo-Thomists in *Partisan Review*'s symposium on "The New Failure of Nerve," Neurath told Carnap, "I shall expect that some Neo-Thomists will find a way how to use your and Tarski's semantics."[14]

Yet Neurath was less concerned with the North American science wars of the early 1940s than with the recent past and immediate future of Europe. In one letter to Carnap, Neurath illustrated how he looked at the Unity of Science movement, the world situation, and their dispute over semantics through historical and political lenses:

I often ask me, to what extent we are responsible, too, for all that happened, by doing something or by failing to do something. . . . [B]ut I want to imagine a little, what kinds of streams lead to the Nazidom. Direct ones . . . but also indirect ones, e.g. the supporting of totalitarian habits as such, and so on. A difficult problem, really.

These "totalitarian habits," he suggested, included the absolutistic, antipluralistic tendencies he saw in Carnap's semantics. Through his research in the history of persecution, Neurath explained, he had

---

[13] Neurath to Carnap, November 18, 1944, ASP RC 102-55-06. On another occasion, referring in part to Carnap 1936/37, Neurath wrote that Carnap was being "often more systematic, than empiricism allows us to be" (Neurath to Carnap, September 25, 1943, ASP RC 102-55-03). On another occasion, he wrote, "your 'degrees of confirmation' do not fit into the pluralistic and non-predictability scheme, etc." (Neurath to Carnap, June 16, 1945, USMP, box 2, folder 14).

[14] Neurath to Carnap, September 25, 1943, ASP RC 102-55-03.

discovered a correlation: the "merciless habit" of persecution

is connected with absolutism in metaphysics and faith. If one thinks there are many possibilities in arguing then one cannot be very hard with argumentative conviction, only indirectly or by heart, but not in argument: Of all the possible world systems one is the best in coherence, or the "relative" best (that does not alter the habit) one moral way of living is "the best" given by the "categoric imperative," but "the decision of an authority" etc. may be combined with merciless destruction of other people. Whereas a sceptic habit as such does not give a "reason" for aggression and merciless action as deserved by THE ONLY BEST SOLUTION.

If pseudorationalism is kept at bay, aggression and persecution cannot so easily take root in pseudorationalistic authority. In part, Neurath simply reported to Carnap one result of his latest research. But he also nudged Carnap to see that his objections to semantics were vitally connected to these humanitarian issues. Nodding to Carnap's view that Neurath was being intolerant of semantics, Neurath wrote, "perhaps you will not think too strong of my 'intolerance' toward metaphysics when you think of this possible correlation."[15]

When Neurath looked to the future of the Unity of Science movement, these concerns remained paramount. He hoped the movement could grow and assist in the task of reconstructing Europe by promoting an appropriately scientific, antimetaphysical, and pluralistic outlook for modern life. When Neurath told Morris that "this nazified Germany and Europe will need good dishes[;] we shall present them,"[16] his menu included educational reforms. Neurath and Joseph Lauwerys argued that Plato's writings were absolutistic, metaphysical, and therefore ripe for elimination from curricula, especially in Germany (Neurath 1945a; Neurath and Lauwerys 1945). Another item was the movement's new *Journal of Unified Science* that would continue *Erkenntnis*. Against the wishes of Morris and Carnap, who both insisted that the *Journal* be published by the University of Chicago Press (alongside the *Encyclopedia* and the *Library of Unified Science*), Neurath contracted with Basil-Blackwell in order to keep the movement planted closer to European soil. In part, he opposed the view that war had ruined England and Europe – "I shall not share this 'point of view' no, no, no." But his main goal was to maintain "our British-American branch" of the movement so that it may "play its

---

[15] Ibid.
[16] Neurath to Morris, December 28, 1942, CMP.

role in the building up the intellectual atmosphere of future Europe."[17] Similarly, Neurath refused requests and suggestions from his colleagues that he emigrate to the United States. "You know I feel we should work here, indirectly against Hitler by promoting an attitude with is really antinazi."[18]

This "attitude" was the most important "dish" Neurath wanted to serve to postwar Europe. It was his attitude opposing metaphysics and pseudo-rationalism not only in Nazi ideology but in education, scientific philosophy, and ordinary daily life. After settling in England, Neurath beamed with delight as he described to Carnap, Morris, and others the scientific, antimetaphysical, and pluralistic attitude and habits he observed in his new English neighbors. In the same letters accusing Carnap's semantics of being absolutistic and unempiricist, he marveled at how his new friends avoided metaphysical pitfalls quite naturally, at least more readily than most Europeans (including, of course, the Nazis):

It is impressive to listen to plain people here, how they avoid boasting and over-statements in daily matters. I collect "expressions," e.g. fire guard leaders explaining how people should get a feeling to be sheltered by the neighbours etc. and then explaining what is needed to act "quickly," to be "calm," and to have the "usual commonsense." I like this type of habit much more than the continental one, with "highest duty," "national community," "self sacrifice," "obedience," "subordination," etc. "eternal ideals" wherever you give a chance to open the mouth.

The British way of living is nice, the compromise habit, the not believing in too many arguments, usual common sense, instead of skyhigh principles from which one tries to deduce concrete details – in vain, of course.[19]

These were people after Neurath's heart: They prized happiness as a leading goal in life, they preferred common sense to formal arguments and principles, and they avoided unscientific, metaphysical language.

Once settled in England, Neurath joined the movement to promote a world commonwealth. He joined the editorial board of the British *World Commonwealth Quarterly* and published in that journal a paper in which this pluralistic attitude again took center stage. In "International Planning for Freedom" Neurath argued that social and economic planning need not eventuate in serfdom, as Hayek (1944) would soon argue. The task for postwar planners and social engineers, Neurath explained, was to

---

[17] Neurath to Morris, December 1, 1941, USMP, box 2, folder 14.
[18] Neurath to Morris, January 7, 1942, USMP, box 2, folder 14.
[19] Neurath to Carnap, September 25, 1943, ASP RC 102-55-03.

formulate plans that maximize human happiness (Neurath 1942, 423). This can occur, however, only in a cultural atmosphere where happiness is valued and not sacrificed in the pursuit of abstract or metaphysical doctrines or principles. To that end, pluralism must reign. Arguments and clashing points of view must be encouraged. These arguments would help to specify the many possibilities the future held, so that Europe could plan its "plurifuture." "Cautious" scientists and experts would come to understand that "very different patterns of organization may exist side by side" (ibid., p. 439). Even individuals, Neurath wrote, could come to see a kind of pluralism within themselves:

> In such an atmosphere the mode of being tolerant towards a multiplicity of attitudes within one's own behavior sometimes develops. "I am not a wittily constructed work of fiction; I am a human being and full of contradiction."[20]

Surrounded by his English neighbors given to "common sense," Neurath described "international planning for freedom" as a process driven not by systematic theories but rather by "common sense arguments" (p. 426) about social and economic possibilities supplied by science and social planners.

This was the task that Neurath had in mind for the Unity of Science movement. Not unlike the lunch of hot turkey shared by representatives of different cultures that Neurath discussed with Kallen, Neurath envisioned an international dialogue about reconstructing Europe that could only benefit from pluralistic sensibilities and scientific language. "Whatever the future COMMONWEALTH OF NATIONS WILL BE," he wrote to Morris,

> to have a language in common for expressing comprehensive ideas will be of importance. How important will it be, to have expressions which enable a better understanding between thinkers from Europe, America, and the Far East, etc.[21]

This, in its largest aspect, is the main reason why Neurath was so suspicious of semantics and, more generally, formal, abstract philosophy of science. To contribute intellectual tools to this international task, they had to be sufficiently accessible, popular, and "humanistic." However telegraphically, this was the point Neurath made when he defended his decision to

---

[20] Neurath 1942, 429. Neurath quotes Conrad Ferdinand Meyer.
[21] Neurath to Morris, December 1, 1941, USMP, box 2, folder 14.

publish the new *journal* in England:

> Most important the future evolution of Europe. Logical Empiricism as an element of really scientific attitude, democratization, etc. We could do something. . . . Popularizing and humanizing Logical Empiricism seems to be important.[22]

Resuming the International Congresses for the Unity of Science after the war was important to Neurath for the same reason. Suggesting that the first postwar congress be held in Holland, he wrote, "Imagine these people hungry for news, hungry for civilization. If only a few people from the USA came over, it could be a symbol of a new scientific world commonwealth."[23]

Carnap's Response

Carnap earnestly tried to make sense of Neurath's objections to semantics. He denied that Tarski's theory of truth invoked any epistemological certainty or metaphysical realism and argued, besides, that any theory deserved a chance to prove its mettle in the commerce of ideas. Most frustrating for Carnap, it seemed, was the way that Neurath's arguments cut across the distinctions among syntax, semantics, and pragmatics that he upheld as the skeleton of any complete theory of language. To Neurath's complaints that semantics had little to do with "empiricism," for example, Carnap responded, "I am in complete agreement with your description of the *scientific procedure*. I should classify this as belonging to the methodology of science. I do not see what it has to do with semantics."[24] As Carnap saw it, the two were merely emphasizing different aspects of science, each of which was legitimate and deserved study. Their dispute, Carnap once suggested, concerned not "a difference of opinion" but a difference "of emphasis": Those whom Neurath had diagnosed with the "systematization disease"[25]

> emphasize the importance of the task of systematization in science; you, on the other hand, emphasize the fact that the statements accepted by scientists at a certain time do not form a well connected system and you point to the dangers involved in overlooking this fact. I think you are right in both points.

[22] Neurath to Morris, January 7, 1942, USMP, box 2, folder 14.
[23] Neurath to Carnap and Morris, November 18, 1944, CMP.
[24] Carnap to Neurath, February 4, 1944, ASP RC 102-55-04.
[25] Neurath to Carnap, April 1, 1944, ASP RC 102-55-08.

All the situation required was more tolerance:

the people on the one side who see [in science] the more turbulent whirl of material in all its colorfulness and vagueness and those on the other side who love nice structural schemata [should] not polemize [*sic*] against each other but rather realize that the work of both is necessary for science.[26]

Having so aggressively maintained his own tolerance about different kinds of intellectual approaches, Carnap strained to remain patient with Neurath's intolerance and his "absurd" rhetorical posture:

Seriously, I think it would be better if you would at long last abandon your habit of calling people who are empiricists and antimetaphysicians metaphysicians if you do not share their opinions. First, it is absurd; and second, it does not help for a successful discussion.[27]

Indeed, the discussion between Carnap and Neurath was becoming less successful in light of Neurath's anger and hurt feelings regarding his monograph, *Foundations of the Social Sciences* (Neurath 1944).

Neurath's Monograph

Though Carnap accommodated Neurath's main points about the nature of science, he never accepted Neurath's claims that semantics could have reactionary effects that would oppose the movement's social and cultural aspirations. Since that was the main thrust of Neurath's objections, and the aspect of them that he cared about most, Carnap's unyielding posture fed Neurath's growing fear that he was losing influence and prestige among his (now) American colleagues. That fear, in turn, came to the fore when Neurath's and Carnap's dispute about semantics became a dispute about Neurath's monograph.

As they argued over semantics in 1943, pressure mounted for Neurath to deliver a completed *Encyclopedia* monograph to the University of Chicago Press. To persuade the press from suspending the project some months before, Neurath had promised that monographs would be soon forthcoming. Though he encouraged others, such as Philipp Frank, to finish their monographs quickly, the only monograph Neurath could possibly deliver was his own. Having left behind his preliminary manuscript when he fled Holland, Neurath hastily rewrote it and sent it to Morris

[26] Carnap to Neurath, February 4, 1944, ASP RC 102-55-04.
[27] Ibid.

in Chicago. Morris rushed it into print and Carnap, then in Santa Fe, became angry as he learned – too late to influence the outcome – that he would have no opportunity to edit it. Even worse, Carnap found the monograph poorly written and unorganized. He therefore relinquished his role as editor for that particular monograph by asking the printer to add a small, factual disclaimer on the inside front cover: "Due to special circumstances, Rudolf Carnap does not share the editorial responsibility for this monograph."

When Neurath learned of this, he became furious. He saw it as a personal and intellectual insult that was not unrelated to their earlier argument over semantics. The disclaimer, he believed, took their in-house dispute public and announced to the readers of the *Encyclopedia* that Otto Neurath and his ideas did not merit the editorial attention of the eminent philosopher Rudolf Carnap. Carnap denied that the disclaimer was anything but a responsible, professional response to the situation, but Neurath remained unconvinced. Their correspondence thus began to sink into a morass of accusations and bad feelings that, in light of Neurath's sudden death, dominated the end of their life-long friendship.

These last, frustrating letters of Neurath and Carnap's correspondence especially reveal Neurath's fears and insecurities concerning his role and reputation in philosophy of science. On one occasion, Neurath recounted how both Schlick and Feigl had refused to acknowledge him "as a member of their community." Anticipating his later reputation as primarily the "indefatigable organizer" (Joergensen 1951, 43) of logical empiricism, and not one of its luminaries, he commented, "My friends and potential friends do not regard me as much important within their circle – except as far as I act as a kind of manager.... OK. Let them."[28] "I think that I, to a certain extent, am one of the pillars of this movement, not only its 'promoter' as people sometimes like to treat me."[29]

Neurath felt that, intellectually, he was being ignored. For example, Hempel's Deductive Nomological model of explanation (in which explananda are logically deduced from general laws) had deterministic and mechanistic overtones that misleadingly suggested a kind of

---

[28] Neurath to Carnap, September 22, 1945, ONN. Neurath did not send this letter to Carnap (marking "not sent" across it), though he rephrased much and mentioned Schlick's personal treatment of him in Neurath to Carnap, September 24, 1945, ONN.
[29] Neurath to Carnap, September 14, 1945, ONN.

"old-fashioned" Laplacian predictability about historical events: "Take Hempel," Neurath explained to Carnap,

he is writing on laws of history etc., he quotes a through and through metaphysician as his choice in discussing problems. Then he continues in explaining the old-fashioned approach. But he does not even mention the fact that there does exist another opinion, too, about unpredictability IN PRINCIPLE.[30]

Hempel knew Neurath's views about pluralism and unpredictability, for they had discussed it personally. He did not accuse Hempel of deliberately ignoring his contributions to the debate. Rather,

what I guess is, that Hempel READING my papers LISTENING to my arguments, did not REMEMBER afterwards these points, BECAUSE THEY ARE SO FOREIGN TO HIM.[31]

This was one reason why Neurath emphasized the theme of unpredictability in the monograph that Carnap found so objectionable. As he told Morris just after sending him his manuscript to put in the pipeline for publication, "many of our 'young men' like Hempel, Zilsel, etc. do not sufficiently realize the importance of NON PREDICTABLE ITEMS."[32]

Neurath's Reputation

For the story of the Unity of Science movement after Neurath's death, it is important to see that Carnap was not the only fellow logical empiricist who was offended by Neurath's manners. One part of Neurath's campaign to keep the movement on its antimetaphysical track was to control the titles of several monographs that had been planned for the *Encyclopedia*. In early 1943, as he argued with Carnap over semantics, he also announced that he could not accept certain unempirical and "dangerous" terms appearing in the titles of forthcoming monographs. Meyer Schapiro's "Interpretation and Judgment in Art,"[33] Hempel's "Foundations of Systematic Empiricism," and Herbert Feigl's "Principles of Scientific Explanation" utilized terms and phrases ("interpretation" and "judgment," "systematic empiricism," and "explanation") that Neurath thought committed

---

[30] Neurath to Carnap, June 16, 1945, ASP RC 102-55-11. This "metaphysician" is probably Maurice Mandelbaum, whom Hempel cites in note 1 of Hempel 1942.

[31] Neurath to Carnap, June 16, 1945, ASP RC 102-55-11.

[32] Neurath to Morris, November 6, 1943, CMP. For more on Neurath's unpredictability arguments and their relations to Hempel's philosophy of social science, see Reisch 2001a.

[33] Neurath to Morris, February 26, 1943, CMP.

the sins of absolutism and metaphysics. They obscured the "backbone of empiricist discussion."[34]

In Feigl's case, Neurath wanted a different title partly because he would possibly not have an opportunity to edit the monograph itself.[35] He suggested the title "Empiricist Argument" because "explanation" did not adequately reflect the pluralism and ambiguity of scientific language and procedure. There was no singular schema for explanation, Neurath believed, and he did not want the title of the monograph to create the impression that there was. After informing Feigl of this and other objections, Neurath emphasized that Feigl was free to choose his title and write his monograph as he wished:

I think, as usual, we shall find a compromise, (democracy is compromise). Please, discuss the matter with Morris. . . . I hope you will send me a short "inventory" of your monograph, what you have in your stores for us. I shall make a few remarks only today. It is your business whether you want to use them or not, you are free in your writing, as you know. The only point is, that we discuss manuscripts before they appear.[36]

If Feigl was free to write what he liked, however, Neurath remained free to disagree with him. After tangling for roughly one year over this matter, Feigl decided to cancel his obligation to the *Encyclopedia*. There were other reasons involved, he explained to Morris, but

frankly speaking, Neurath's senile termino-phobic objections (against the term "explanation" etc.) and his complete misunderstanding of the aim of my contribution did not help in inspiring me to greater speed in finishing the piece.[37]

Ultimately, Feigl did contribute, with Morris, the final monograph of the *Encyclopedia* to appear (Feigl and Morris 1970). By abandoning his monograph on explanation, however, he effectively distanced himself from Neurath's programmatic agenda, as he and others would continue to do in the 1950 and '60s.

When Neurath's monograph appeared, it encouraged perceptions of Neurath that fed into his fears about becoming marginalized. As Carnap had noted, the monograph is disorganized and difficult to understand.

---

34 Neurath to Feigl, February 26, 1943, CMP.
35 Morris sensed late in 1942 that the press was nearing its decision to suspend the project, and he promised them that Feigl's manuscript would soon be ready for the printer. He wrote to Neurath and explained, "Carnap, Hempel and I will read it. I am afraid there will not be time to get it to you" (Morris to Neurath, January 31, 1943, CMP).
36 Neurath to Feigl, February 26, 1943, CMP.
37 Feigl to Morris, August 16, 1944, CMP.

And as Ernest Nagel pointed out to Morris, its poor quality fed a growing reputation of Neurath as a somewhat dictatorial leader of the movement. After reviewing the monograph's faults, Nagel wrote,

> I have finally been stimulated to wonder what the Advisory Committee [of the *Encyclopedia*] is good for. Is its function primarily that of serving as window-dressing? Or is the editor-in-chief so high and mighty a personage that he will not submit to any advice or correction?

Nagel was disappointed not only with Neurath's suspiciously authoritarian methods, but also with the *Encyclopedia*'s content. Recent monographs besides Neurath's, he implied, were substandard and he was not impressed with the political partisanship of the latest author whom Neurath had added to the roster:

> And this leads me to ask one final question: Why has [Lancelot] Hogben been selected to do the pamphlet on biology? I admit the Encyclopedia could have made a worse selection – but the real point at issue is whether it couldn't have made a much better one. For my part, I don't enjoy the prospect of having the foundations of biology class-angled.[38]

Hogben was a British Marxist biologist, foreign editor (with J. D. Bernal and others) of *Science & Society*, and inventor of the artificial language Interglossa. Neurath met Hogben in 1942 and immediately asked him to write a monograph for the *Encyclopedia*. At first, Neurath wanted him to write about international languages; later, the topic was switched to biology. From Nagel's point of view, such politically partisan authors were as inappropriate for the *Encyclopedia* as were Neurath's "high and mighty" methods.

Neurath managed to offend others besides Feigl, Carnap, and Nagel. He had earlier showdowns with Hans Reichenbach and Louis Rougier, both of whom pulled out of the *Encyclopedia* project.[39] Most important, however, Neurath's manners and methods – along with Kallen's charge that the Unity of Science movement was "totalitarian" – survived him into a period in American history that upheld the values of freedom and individuality high above all things collective, planned, organized, or authoritarian. After Neurath's death, for example, the Rockefeller Foundation's decision to fund Frank's new institute was based in part on the contrast the foundation's officers saw between Frank and what they had heard

---

[38] Nagel to Morris, November 16, 1944, CMP. Hogben never wrote this monograph and the topic of biology was passed to Felix Mainx (Reisch 1995).

[39] For more on Neurath's run-ins with Reichenbach and Rougier, see Reisch 1995.

and learned about Neurath. He ran his movement and the *Encyclopedia* in "a very individualistic and indeed almost dictatorial way."[40] Though Neurath's pluralistic and antitotalitarian values were firmly embedded in his philosophy of science, he succeeded in creating the opposite impression.

[40] Interview: WW, December 13, 1946, RAC RF 1.100, box 35, folder 281 (WW stands for Warren Weaver of the Rockefeller Foundation, whom Morris, Neurath, and Frank consulted about possibilities for funding the Unity of Science movement).

11

Frank's Neurathian Crusade

Science, Enlightenment, and Values

At the end of 1943, Neurath was managing several problems in the Unity of Science movement. In the midst of war, Kallen's charge had been made (and awaited debate in the pages of *Philosophy and Phenomenological Research*), Neurath had just resumed his dispute with Carnap about semantics, and the University of Chicago Press tapped its fingers waiting for Neurath and his editors to deliver long-overdue monographs for the *Encyclopedia*. Earlier in the year, the press had threatened to suspend publication until after the war, but Neurath promised them he would soon produce a monograph. Though it was his own (Neurath 1944) that kept the promise, he hoped that his solution to this problem might be his old friend Philipp Frank, who was slated to write a monograph about philosophy of physics. Frank and Neurath were old friends, but they had corresponded little during the early years of the war. When they reconnected, they were quickly reminded how similarly they felt about scientific philosophy and its future.

The Neurath-Frank Alliance

Like Neurath, Frank especially hoped that the movement would prosper after the war. He kept the faith that it would still be in tune with the times – "I think that the longing for a unified scientific view exists everywhere"; "I think that the prospect for Unified Science in the English speaking countries is not bad."[1] Frank also shared Neurath's worries about the intellectual direction that logical empiricism was taking. Not unlike what Neurath

---

[1] Frank to Neurath, December 19 and 15, 1945, ONN.

said about Carnap's semantics, Frank said he detected the beginnings of a "scholasticism" and abstractness that could prevent the movement from growing and gaining influence after the war. Many logical empiricists were moving "more and more into pure formalism which means almost into a new scholasticism" – a scholasticism that was scarcely able "to influence the real world."[2] As if anticipating the publication problems that he would encounter several years later as head of the Institute for the Unity of Science, Frank observed that "the overemphasis on formal logic discourages scientific people [from] participat[ing] actively and passively."[3] If this trend continued, Frank feared, scientists would lose interest in the movement, and one of the main causes Frank championed – regular collaboration and conversation between scientists and philosophers of science – could be lost.

Neurath agreed. Smarting from his run-in with Carnap over his monograph, he told Frank that he blamed semantics for at least some of this scholasticism:

You see I am reading now carefully what Kaplan, Feigl, Hempel and others are writing and I cannot help [but] get the impression that something goes wrong. The Tarski-Carnap coalition becomes more and more unempiricist. . . . That would not be too bad, but I see how they all together discuss more and more scholastically problems without touching the empiricist consequences.[4]

The new scholasticism meant that philosophers were selecting and tailoring scientific problems for their own disciplinary and professional purposes (such as building logical models or "systems" of science) without concern for whether those philosophical projects engaged or assisted the progress of science as it was practiced in empiricism's trenches: "I have a feeling that the Viennese Circle people become formalists and less and less interested in empiricism as a living thing."[5] A week later, Neurath wrote again to Frank and articulated the charge: "Our friends do not try to analyze scientific arguments" – instead, they "evolve some complicated logical devices just connected with scientific argument by some occasional problems arising from scientific reflection."[6] This concentration on singular, technical problems, Neurath felt, threatened the movement's connections to other disciplines, the public, and a scientifically

---

[2] Frank to Neurath, December 19, 1943, ONN.
[3] Frank to Neurath, December 15, 1945, ONN.
[4] Neurath to Frank, November 18, 1944, ONN.
[5] Neurath to Frank, June 16, 1945, ONN.
[6] Neurath to Frank, June 21, 1944, ONN.

informed approach to "life": "I think we [i.e., the movement] should be a little more interested in social and personal life from our point of view," Neurath remarked to Frank – "not only in physics and biology, etc."[7]

With his alliance with Frank restored, Neurath hoped he could solve not only this problem of creeping scholasticism, but also the problem of delivering a monograph to the University of Chicago Press. Neurath pulled some of his favorite notions and slogans from his debate with Carnap – "partial systematization only," "common sense" – and urged Frank to speak powerfully for pluralistic empiricism and the nonscholastic legacy of the Vienna Circle:

JUST YOUR MONOGRAPH IS SO IMPORTANT. . . . It is so important, that EMPIRICISM speaks a clear word. I always fear that systematization eats up the cake, whereas in empiricist research work we always have to do with partial systematization only, the pluralist point of view is important. . . . You see no systematization is needed only plain commonsense.[8]

Oh dear, think of the Vienna Circle and be faithful in writing.[9]

Frank was happy to write a monograph with these emphases, for he was already planning other, similar publications. The scholasticism arising within scientific philosophy in North America, he explained to Neurath, was perhaps more widespread than Neurath could see from his English vantage. Frank had diagnosed Aristotelian scholasticism in his students:

Now I give courses in "Basic Physics" for the army students at Harvard. . . . One thing is striking: students who have had very little scientific training are unconsciously influenced by a sort of vulgarized scholastic philosophy. They may have picked [it] up from the church or from "philosophical" introductions to science textbooks or what not. Before the science teaching got a certain grip on these boys they were genuine Aristoteleans. I have been planning recently a work by which I hope to revive interest in the empiristic and unifying job we used to do. I shall try to write a book on "Logic of Science" for the future science teacher. [I want to] make science teaching really an element in the progressive evolution of the mind. Science itself without this logical background can even have a reactionary effect as we have seen recently in many cases.[10]

Just as Neurath debated the goals and curricular structure of education in postwar Europe, Frank aimed to contribute to science education tools of logical analysis that would counteract this Aristotelianism in both his

---

[7] Neurath to Frank, July 1942, ONN (exact date unknown).
[8] Neurath to Frank, September 26, 1943, ONN.
[9] Neurath to Frank, April 26, 1943, ONN.
[10] Frank to Neurath, December 10, 1943, ONN.

students and, as we see below, his colleagues at the Conference on Science, Philosophy and Religion. Given the practical sensibilities for which America was famous (or infamous) in the eyes of many Europeans, Frank suggested that working in these educational venues might be the surest way to revive the movement after the war:

I think that a general interest in our work can only . . . arise if we are able to show that it is a "practical" help in some "job." [Since] in this country education has been regarded as a very important and serious job, it seems to me very helpful to make use of this interest as a starting point . . . as the thin end of a wedge. The article in the Ency[clopedia] will be written in a spirit to serve later as an element of this future work. I think that it is now time to try a "revival." I shall try to cooperate also with Bridgman in the matter. I gave already speeches about these problems at several colleges.[11]

Neurath had months before approved of this strategy. He agreed, "I think it will be of some use also for logical empiricism that you are now more interested in educational courses."[12]

As much as Neurath and Frank agreed with each other about the problems and what was to be done, one problem remained: Frank had to stop planning and start writing. At this point in his life, Frank was notoriously slow and forgetful when it came to fulfilling his promises. An ocean away, Neurath did what he could by locking his typewriter's shift-key:

PLEASE WRITE YOUR PAPER ON PHYSICS, PLEASE WRITE YOUR PAPER ON PHYSICS, PLEASE WRITE YOUR PAPER ON PHYSICS.[13]

PLEASE START WRITING, PLEASE START WRITING.[14]

Though Neurath probably wished to send Morris Frank's monograph instead of his own to the University of Chicago Press in late 1943, his did at least move Frank along. Frank finished his monograph in 1945, but Neurath died before it was published a year later.

Frank's Program for Philosophy of Science

It was not mere coincidence that when Frank and Neurath resumed their correspondence in the early 1940s they found that they shared

[11] Ibid.
[12] Neurath to Frank, July 1942, ONN (exact date unknown).
[13] Neurath to Frank, July 11, 1943, ONN.
[14] Neurath to Frank, September 26, 1943, ONN.

objections to the directions logical empiricism was taking. Roughly thirty years before, in what is now recognized as the *first* Vienna Circle (Haller 1991), they helped to set the parameters for logical empiricism as it developed around Schlick's famous Vienna Circle. At the time, Frank's main interests were conventionalism and the unity of the sciences. The first led to his book *The Law of Causality and Its Limits* (Frank 1998), in which he argued that our notions of scientific causality are conventions or "disguised definitions" (Frank 1949b, 57). Given a causal law according to which a system regularly evolves from state A to state B, and given an empirical observation that violates it, Frank argued, scientists typically will not revise their theoretical beliefs or consider the law falsified. Instead, they will conclude that the initial state of the system was not actually what it was believed to be. The causal law, in other words, functions as a definition for what "state A" means, not as a factual proposition. Frank aimed to show that scientific theory was neither built up entirely out of experiences (as some Machists held) nor structured and constrained by a priori concepts and categories (as Kantians held). Rather, scientists freely used logic and their imaginations to build conceptual frameworks that, in turn, "experience serves to fill in" (ibid., p. 58).

Frank's picture of scientific theories was also an explication (indeed, a celebration) of Einstein's theory of gravitation. In his contribution to the volume *Albert Einstein: Philosopher-Scientist*, Frank cited Carnap's description of theories as freely created logical structures that could utilize "abstract" terms that need not even be defined or interpreted in any observation language. As a whole, however, the terms of a theory could still be tested and confirmed "if by logical conclusions statements about observations can be derived which can be confirmed by actual experience" (pp. 275–76). As far as Einstein's work in general relativity went, Frank explained, this was an accurate and informative picture of scientific theory:

> This conception of logical empiricism seems to be fairly in accordance with the way Einstein anchored his theory of gravitation in the solid grounds of observable facts by deriving phenomena like the redshift of spectral lines, etc. (p. 276)

> I do not see in the question of the origin of the fundamental concepts of science any essential divergence between Einstein and twentieth century logical empiricism. (p. 282)

Here, as well as in his biography of Einstein (Frank 1947), Frank portrayed Einstein as a scientist and philosopher who understood that science defied characterization along inductivist, deductivist, or Kantian lines. It was instead a complex interplay of free imagination and rigorous empirical testing.

Frank's Neurathian Philosophy of Science

Several parallels and points of intersection between Frank's and Neurath's conceptions help to substantiate their alliance and shared hopes for the Unity of Science movement.[15] Frank, for example, envisioned philosophy of science as a discipline closely involved with actual scientific practice. Indeed, Frank's conventionalist thesis crucially takes into account how scientists behave and respond to experimental and observational events, and his anti-Kantian and anti-inductivist views highlight the creative activity of scientists. Where Carnap, against Neurath's advisement, partitioned philosophy of science into pragmatics, semantics, and logical syntax, Frank's philosophical target was broader and included scientific discovery and justification as well as pragmatics and theory.

Frank was equally committed to the unity of science. He approached the topic via Mach, as he sought to rescue Mach from the idealist interpretation (popularized by Lenin (1908)) holding that the world we experience is somehow "built up of perceptions" (Frank 1949b, 83). This metaphysical reading of Mach threatened to obscure what Frank saw as the real motive behind Mach's interest in sensations and scientific phenomena: the unity of science. Mach used sensationalism, Frank wrote, "to take, in physics, a standpoint which does not have to be abandoned immediately when we look over into the field of another science. For all the sciences ultimately form a whole" (Ibid., p. 81). Unity of science, moreover, meant for Mach an economical unity, free of spurious metaphysical concepts that essentially support no empirical interpretation:

Therefore, if we demand of science an economical representation of our experiences, that is, a representation by a unified system of concepts, we must admit only propositions that are reducible to propositions containing only perception terms as predicates.

This is the real meaning of Mach's doctrine that all propositions of science deal with perceptions.... The unification of science is not possible except by the

---

[15] For a compelling account of the unity behind the "left wing" of the Vienna Circle (namely, Neurath, Frank, Carnap, and Hahn), see Uebel 2003b; and for an account of programmatic ties shared by Frank and Neurath, see Uebel 2004. For more on Frank and his cultural engagements with the CSPR, see Nemeth 2003 and Uebel 1998, 2003a.

elimination of metaphysical propositions. Then only propositions of a homogeneous type remain. Hence we can form from them a coherent logical system. (p. 84)

Properly understood, Mach taught us that unity of science and economy in science were conjoined goals. To the extent that Mach stands at the heart of logical empiricism and its historical origins – as Neurath, Carnap, and Hahn suggested in their manifesto (Neurath et al. 1929, 302) – Frank thus joined Neurath in defending their Unity of Science movement as an expression of a distinctly Austrian (as opposed to German) tradition in philosophy of science (Uebel 2003b, 154–55).

Frank's views also accorded with some of the more subtle and technical features of Neurath's program, including its emphasis on historicity, theoretical holism, and the underdetermination of theories by evidence. Theoretical meaning, Frank once noted, belongs only to "a system of statements or principles. . . . An isolated word or even an isolated statement has meaning only indirectly. We call it 'meaningful' if it is fit to be a part of a meaningful system or doctrine" (Frank 1950, 31). It would be foolish, Frank agreed with Neurath, to regard a scientific proposition in isolation as simply "true," if only because the meaning of the claim resides partly in the larger body of accepted statements. In addition, that larger body has a historical life of its own and continually changes, sometimes abruptly. Such truths, therefore, would hardly count as timeless and permanent.

This holism allowed Frank to join Neurath in rejecting metaphysics as "isolated" statements that lacked connections with other accepted scientific statements. Frank emphasized a historical aspect to this argument that owes much to Duhem:

In metaphysics a statement or a system . . . is regarded as "true" if our common sense understands the validity of the principles immediately without having to draw long chains of conclusions from these principles and without checking some of these conclusions against our observations. (Frank 1949a, 296)

Metaphysics, that is, typically appeals to our "common sense" without ever facing any tribunal of contemporary science. As did Duhem and James, Frank regarded common sense as a historical product (see Duhem 1954, 259–68; James 1981, Lecture V). Traditional philosophy and popular beliefs tend to lag behind the development and progress of science, Frank observed, with the result that "established philosophic principles are mostly petrifactions of physical theories that are no longer appropriate to embrace the facts of our actual physical experience" (Frank 1949b, 215). Those who take refuge from science in seemingly incontrovertible metaphysics are actually taking refuge in *old science*. Critics of Einstein, for instance, who found ideas and consequences of relativity theory fantastic and in violation of "common sense" did not understand that the "established philosophical principles" of Newtonianism to which they appealed "are nothing else than physical hypotheses in a state of petrifaction" (ibid., p. 214). Only because these hypotheses have become isolated from living, dynamic science, they can take on an illusory quality of being eternal and naturally immune to revision.

Finally, though he did not share Neurath's expertise and training, Frank was also committed to sociology of science. Both acknowledged interactions between science and the social and economic contexts in which scientists work and urged that science and scientific philosophy study these interactions. Frank especially leaned on the underdetermination thesis to justify his call for cultivating sociology of science within his Institute for the Unity of Science. Since "it is certainly possible that several scientific theories may account for the same facts" (Frank 1951b, 19), there was opportunity to study how social and cultural forces affected scientists' choices of theories and their practical understandings of their meanings. Equally important, there were resources within the Unity of Science movement to explore these issues. Frank connected them to Carnap's interest in "pragmatic" and "external" questions concerning choices among linguistic frameworks, to Hempel's writings on meaning and verifiability, and to Quine's criticisms of the distinction between analytic and synthetic propositions (Frank 1951a, 7–8). Indeed, Frank championed even more than Morris the idea that the evolution of logical empiricism, rightly understood, was a vector pointing directly to North American pragmatism.[16]

Philosophy of Science, History of Science, and Politics

Given Frank's interest in sociology of science, his program for the Unity of Science movement also pointed directly to politics. Since "the

---

[16] In his polemics against absolutism about values and semantics, Frank regularly lumped together logical empiricism, Bridgman's operationalism, and pragmatism. See, e.g. Frank 1950, 4. Frank saw American pragmatism, in particular, as providing one of the first effective critiques of Kantianism (Frank 1949b, 47–48).

'logico-empirical' criteria are compatible with two or more different theories," Frank explained,

state or church or public opinion can select from among these doctrines the one which is most useful for the training of "good citizens".... Many people have claimed that there cannot be any influence of state or church on scientific doctrines because no authority can "change the observed facts." This is certainly true. But an "authority" can require the abandonment or acceptance of scientific *theories* without requiring a "change of the facts." (Frank 1951b, 19)

In his books and essays, Frank examined several historical intersections of philosophy of science, history of science, and politics. These included debates between mechanism and vitalism, Newtonianism and its critics, the condemnation of Mach by Lenin and the alleged cultural and "spiritual" meanings of quantum mechanics and general relativity (Frank 1949b, 186–97). One of Frank's favorite episodes was the Copernican controversy and heliocentrism's eventual victory as "a symbol of liberation from church authority" (Frank 1951b, 20). He viewed scientific theories as a composite of a formal, logical structure, on the one hand, and observations or "facts" that theories were built to explain or illuminate, on the other. But this dichotomy was not strict, for Frank knew that observations were "theory-laden" (to use the term that later came into vogue) and that it was incorrect to conceive of facts as wholly extralinguistic. Not facts, but descriptions of facts, Frank emphasized, were relevant to understanding scientific theory.[17]

While these complexities of scientific practice and language underpinned Frank's call for cultivating sociology of science, they also introduce the political tone of his program and connect it to the Marxist philosophical left of the 1930s and '40s. One task of sociology of science, Frank noted, would be to "examine how the social conditions of the scientist have affected or colored his work" (Frank 1950, 86). Frank's call can be seen as a response, that is, to criticisms from Feuer, Schlauch, Cornforth, and others that logical empiricism promoted a false consciousness about science's social and ideological functions in society. Frank's interest in sociology of science was not motivated by any allegiance to dialectical materialism or Communist Party policy, however. Primarily for scientific reasons, he believed, science educators, especially, should better understand the many unexplored connections between science and the world in which it operates and evolves. That

---

[17] "The jobs of finding and interpreting facts are indivisible" (Frank 1950, 71).

understanding, in turn, would promote beneficial social, cultural, and political effects.

Science Education

Part of Frank's crusade was directed against popular science writing that ignored, if not obscured, the intellectual, social, and political aspects of contemporary science. In Frank's eyes, even famous scientists who should know better were guilty. One antidote, therefore, was to promote logical empiricism in science education. It was the only philosophy of science both faithful to the character of modern science (especially physics) and promising (via the underdetermination thesis) potential understanding of where and how science interacts with political and cultural forces. Not unlike his experience at the Conference on the Epistemology of the Exact Sciences, however, where scientists sat speechless through talks by Frank, Carnap, Reichenbach, and others, Frank observed that North American scientists were not equipped to explore (much less popularize) a logical empiricist view of science. Consequently, students in science and nonscience students taking Harvard's general education courses had little opportunity to break out of their Aristotelian preconceptions (as Frank described them to Neurath) and take advantage of logical empiricism's offerings.

After the war, these issues became especially urgent. College students in the 1940s and '50s would soon inherit problems (such as nuclear technology and the Cold War) that were at once scientific, ethical, and political. Because science education failed to teach a view of science that was conceptually sophisticated – "90 percent of the textbooks of physics on the college level present the law of inertia in such a way that its meaning is obscure" (Frank 1949b, 235) – and failed to put science's historical, philosophical, and cultural realities in plain view, it

> misses an opportunity to teach the student a reasonable and scientific approach to all problems of human interest. For in all these fields the central problem is the relationship between sensory experience (often called fact finding), and the logical conclusions that can be drawn from it. The failure to grasp exactly the nature of this relationship accounts for the confused attitude of many people toward the complex problems by which they are faced in private and public life. (ibid., pp. 234–35)

What students of science fail to learn, of course, will soon be reflected in what science teachers fail to teach. The cycle was already under way,

Frank noted, given the unreceptive attitude of most science teachers toward "philosophy": "If the question is whether a science student can find some information about the foundations of his own field in philosophy classes, a great many science teachers have only words of disdain or even sneers" (Frank 1950, 62–63). Relationships between facts and their "logical conclusions" should be emphasized in mathematics classes, too, for

> the role of mathematics and physics in the understanding of geometry is perhaps the simplest example by which the student can learn how to discern the role of facts and logical conclusions in the involved problems of human relations. (Frank 1949b, 235)

Yet because of excessive professionalization and specialization dividing mathematics from science, and the sciences from philosophy and the humanities, there was little hope that this situation could be easily righted.

Frank's Activism

Frank had few reservations about making his criticisms public and direct. He criticized science educators by taking his case to the *American Journal of Physics*, in which he urged that "elementary science courses themselves should contain a good deal of the philosophy of science" (Frank 1949b, 249). Turning to scholars in the humanities reading *Etc: A Review of General Semantics*, he suggested that humanists study "*the values which are intrinsic in science itself.... [I]nterest in humanities is the natural result of a thorough interest in science*" (ibid., p. 261, emphasis in original).

Frank also turned to his colleagues in philosophy and suggested that they were best equipped to build bridges between science and the humanities. But they had little interest in doing so as they separated themselves from science:

> Even the departments of philosophy have kept to a policy of isolationism. Instead of working toward a synthesis of human knowledge, they have proposed a kind of truce between science and philosophy. (p. 262)

> Philosophy (as taught in most departments of philosophy) has become a special science itself which is more separated from mathematics, physics, or biology, than these branches are separated among themselves. (p. 267)

Frank most likely had analytic philosophers in mind, possibly his colleagues Quine and Nelson Goodman, when he criticized "professional philosophy" for accepting the epistemic power of modern science, yet deliberately avoiding substantive professional contact with it:

> Among philosophers the apology is current that it is just impossible for them to have an exact insight into a scientific issue because the sciences have become, in our time, so highly specialized that only the specialist can have a thorough understanding.... In this situation a great many philosophers have chosen to establish as their redoubt a special field of philosophy outside the field of science. To master this field one supposedly needs only an acquaintance with the prescientific knowledge that is familiar to the man in the street.... On this level we make free use of words like "time," "space," "existence of external objects" in the sense in which the man in the street uses these expressions. The special sciences like mathematics, physics, biology, as isolated branches of knowledge, are taken for granted and the policy of nonintervention is upheld. (pp. 268–69)

The philosopher does not inquire into the origin or nature of these sciences and does not want them "to intrude into his 'living space' which is located between and above and below the domain of these isolated special sciences" (p. 269). Like the neo-Thomists who turned away from science, but for different reasons, Frank believed that, by turning away from the goals and values of logical empiricism and the Unity of Science movement, "professional philosophy" was giving up on "the chief goal of liberal education" (p. 271).

Frank, Values, and the CSPR

Those to whom Frank most consistently appealed on behalf of logical empiricism were audiences at the Conference on Science, Philosophy and Religion in Their Relation to the Democratic Way of Life (CSPR). Following Adler's infamous talk "God and the Professors" (Adler 1941) at the first meeting in 1940 (see chapter 3), the tone was set for the conference to celebrate the study of social, cultural, and ethical values that, most of these congregants believed, had nothing to do with science.

For Frank and some of his colleagues, this view was entirely mistaken. Besides Hook's, Dewey's, and Nagel's rebuttals to Adler in *Partisan Review*, the early 1940s saw several efforts by logical empiricists to defend pragmatic and instrumental approaches to values in science (see, e.g., Feigl 1943, 404; Carnap 1944). While Morris, Hempel, and Carnap participated briefly in the CSPR, Frank attended these conferences yearly and tried to understand the dominant critiques of science on their own terms in order to diagnose and correct them most effectively.

Frank's book *Relativity: A Richer Truth* (1950) presents his many presentations to CSPR audiences from 1940 to 1949. It responds to the accusation (filed by Adler, Hutchins, and others) that because science is "value-free," it is responsible at least in part for the century's social and political upheavals. More specifically, the claim went, science proudly waved this value-free banner in the form of Einstein's celebrated theories of relativity. Those theories, many believed, underwrote ethical relativity in the form of "a disbelief in absolute values" (Frank 1950, xiii). That disbelief in absolute values, in turn, underwrote social and cultural disasters, such as the rise of totalitarianism.

There was much here for Frank to disentangle and correct. Frank began by explaining that relativity theory did indeed illuminate a kind of relativism that led to progress in science throughout its history. Yet this relativism did not mean that science was value-free or inimical to liberal democracy:

I have attempted to show that so-called "relativism" of science has not the slightest thing to do with agnosticism or skepticism, that *it is in no way hostile to the belief in ethical or democratic values,* that it is accompanying every advance in science and is nothing but a significant representation of the enrichment of human expression which is inseparably connected with our gradually increasing experience. (ibid., p. xv)

Properly understood, Frank believed, the relativizing tendencies of science and scientific progress were powerfully antitotalitarian. By attacking science, therefore, Adler and his sympathizers obscured these important issues and mistook friend for foe:

I think that the fight for the values of our Western civilization has been compromised frequently by attempts to tie it up with a fight against the scientific attitude, an attitude which so far, has been instrumental in human progress. These anti-scientific efforts ignore the obvious fact that not only modern science but also liberal Christianity and reformed Judaism are offsprings of the doctrine of the "relativity of truth." They ignore the equally obvious fact that this "relativism" has been for centuries the only effective weapon in the struggle against any brand of totalitarianism. (pp. 4–5)

The "relativity of truth" that science's critics so tightly associated with Einsteinian relativity theory was in fact, Frank explained, a general feature of progressive intellectual and cultural change. Developments in physics no less than these in theology exhibited a typical pattern, whereby propositions once taken to be "absolute" become, through intellectual and or social reforms and struggles, relativized to specified contexts.

To anchor his argument, Frank introduced a simple, paradigm case of relativization:

For ages there was no truth which had been as clear and evident as the statement "my head is above my feet and my feet are below my head." And "if we proceed in the direction from our head to our feet the objects we meet are located the lower the farther we proceed." Then the existence of the antipodes was discovered. People lived on the opposite side of the globe. According to our well-established "truth" each of these people had his head below his feet. (p. 7)

This situation was ripe for relativization, if only because "men of equally good sense could make statements which were contradictory."[18] Progress came when these people "learned that the direction of gravity is different at different points of the earth" (8) and then further specified their claims with reference to their different locations:

I say: "My head is above my feet relative to the field of gravity at my place," and my antipode says that his head is above his feet relative to the field of gravity at his place. Both are right and they are furthermore in agreement with each other and with all other correct statements. (pp. 8–9)

Frank presented similar stories to explain aspects of relativity theory, such as the relativization of mass or time intervals to different inertial frames of reference. He emphasized throughout that nothing in this process of relativization brought objectivity into doubt or suggested that scientific knowledge somehow depended on "subjective whim" (p. 9). Instead, this kind of relativization and qualification of our propositions was an inevitable aspect of our increased knowledge about the world and ourselves in it.

On this basis, Frank outlined a bridge connecting science and the humanities: a similar kind of relativization was in play when we meaning-fully and usefully talked about cultural and social values. For Frank, state-ments about values are not theoretical, scientific propositions. Rather, they express "preferences" such as those implicit in the claim that we would be happier living with, as opposed to without, the political, artistic, ethical, or social values we endorse and the institutions that implement them. But when we speak of values themselves, he proposed, such as "lib-erty," "charity," or "democracy," we risk conceiving them as abstract and independent of the institutions, behaviors, and experiences that sustain

[18] However clear, Frank's example is mistaken. Even in antiquity, the roundness of the earth was plainly known. On the widespread legend of flat-earth theory, see Russell 1991.

them. They become "absolute," unqualified, and in that form generally useless verbalisms (pp. 40, 41).

Unfortunately, Frank wrote, this was the starting point for most discourse about values. Those who tried to defend and protect their chosen values from any dilution or "relativization of truth" by regarding them as "absolute" got things backward and worked against themselves. For the more abstracted and "absolute" our values become, Frank explained, the less meaning they have and the less practical help they can provide:

> It is trivial to know that one must not kill an old lady to take a couple of dollars from her handbag. The cases in which we need the guidance of religious and ethical principles are the rather doubtful ones – the cases of self-defense, of foreign war, of civil war, of revolution, of sabotage, etc. Precisely in these cases we notice almost regularly that even the most orthodox and consistent ethical and religious creeds somehow avoid giving definite advice. . . . If and when they must finally decide an issue their decision is rather a "relativized" one. The actual argument by which they confirm their decision in a specific case makes little use of the doctrine of "absolute unqualified truth" [and this is] . . . quite natural and, as a matter of fact, practically unavoidable. For this so-called doctrine of the "relativity of truth" is nothing more and nothing less than the admission that a complex state of affairs cannot be described in oversimplified language. (p. 45)

Relativity theory was no nemesis of values or objectivity. It could and should be seen, Frank urged his audiences, as a parable of enlightenment and hard-won lessons about the complexity of the world. When it comes to values, "the most ardent advocates of 'absolute truth' . . . are in the situation of the physicist who has to avail himself practically of the Theory of Relativity" (p. 46).

Absolutism, Persecution, and Totalitarianism

Frank's crusade for relativism (properly understood) further helps to substantiate his alliance with Neurath in the early 1940s. Like Neurath, Frank diagnosed "absolutism" as a common, multifaceted mistake. First, it mistakenly held that concepts (be they the values defended at the CSPR or, in Neurath's debate with Carnap, the concept of truth in semantics) required *less* interpretation than was the case. For Neurath, Tarski's theory of truth only begged questions about truth, since two disputants might accept Tarski's theory (that, for instance, "the statement 'snow is white' is true if snow is white") and still disagree in any particular case as to whether some patch of snow they both observed was white. For a pluralist such as Neurath, interpretation of competing claims in an observational language was necessary for empiricism, but never exhaustive and complete; there could be no interpretation enough to justify claims as "absolute." For Frank, defenders of "absolute" values made a similar, but more extreme, mistake: They believed that their cherished absolute values required *no* interpretation at all. The main goal of his book was to show that values, far from being deflated and dismissed, are actually "richer" for requiring interpretation and relativization.

A second flaw in absolutism that Frank attacked was its ignorance of historical change. What Adler, Hutchins, and other neo-Thomists found so objectionable in "progressive" thinking about science was its historicity and plasticity:

A sort of general opinion has taken shape among the progressive groups. In science as well as in its application to human conduct these groups have refused to take anything for granted. They have ventured the idea of an unlimited progress in thought and life based upon the methods of science. Essentially the new attitude does not believe in statements the truth of which, once established, can be trusted forever. This attitude has been called positivism, pragmatism, relativism, operationalism, and so on. (Frank 1950, 4)

On the one hand, the history of science shows the futility of belief in unchanging, absolute values and propositions. But that does not mean that the history of science and broader intellectual history are unintelligible. Rather, Frank explained, there is a cyclic structure that helps to illuminate the institutional and intellectual features of the contemporary landscape. Given the continual advance of science – often consisting in revisions and relativization of its key concepts – those who stay abreast of developments must, like these "progressive groups," remain anti-absolutistic. Absolutists, on the other hand, cling to scientific beliefs that become "cornerstones of ethical and religious systems" but which, unbeknownst to them, are destined to be abandoned:

Again and again in history scientists led by experience, logic, and imagination had set up a system of principles from which the actually known facts could be derived. Again and again when these principles turned out to be a success the human mind cherished the illusion that these principles [such as geocentrism's "dignity of man"] could be derived from pure reason. Again and again new experiences demonstrated that not all the results of these principles were in agreement with the observed facts. Again and again principles that seemed to be demonstrated not only by abstract reasoning, but by the common sense of all people, had to be abandoned. (ibid., p. 87 [insert from 88])

Frank thus urged his colleagues at the CSPR not only to accept relativity theory as a general epistemological touchstone, but also to accept history and the fact of historical change. Then they could perhaps see their own place in the contemporary phase of this cyclical pattern, embrace science instead of misunderstanding it, and then use science's conceptual and methodological resources to more responsibly address the conference's concerns about threats of totalitarianism.

For as things stood, the absolutists at the CSPR appeared to Frank not only to be fighting the wrong battle (against science or "positivism") but to be using the wrong tools. As Neurath had suggested in his disputes with Kallen and with Carnap, it was *absolutism* itself that seemed to help drive fascism and totalitarianism. "Totalitarian ideologies," Frank wrote, "always rest on the assertion that all good lies in a definite principle, and that this principle must be maintained in all cases regardless of what follows from it in real life" (p. 113). As a result, this principle typically becomes "a mere banner" without operational connection to the world of experience. Nazi claims for the ultimate supremacy of the "Aryan race," Frank explained, were upheld only verbally. In practice, Finns, Hungarians, and other groups who were not Aryan according to Nazi racial theory were in many cases "regarded as friendly to the German Reich" (p. 115) and officially tolerated as not "non-Aryan." This latter concept was defined operationally (on the basis of Jewish religious ancestry), but it had no connection to the original anthropological definition of Aryanism and led even to gross violations of logic since "a man like a Hungarian may be neither an Aryan nor a non-Aryan" (p. 116). This and other displays of pretzel logic, Frank argued, gave the Nazi authorities a power over masses of Germans. They were manipulated by the emotional power of abstract, absolute doctrines and slogans about cultural and military supremacy and blinded to the fact that these slogans had no integrity and no coherent interpretation in concrete experience.

Frank especially worried, therefore, that his colleagues so resentful of relativism were merely fighting one absolutistic ideology with another. They upheld slogans and doctrines about "democracy" or "freedom" or philosophical systems such as Thomism as foils to different, but equally absolutistic, doctrines at the core of different totalitarian systems. What was required, Frank and Neurath agreed, was instead to steer our thinking and discourse consciously away from absolutism altogether. The task was perhaps most important in the classroom where science's critical and nonabsolutistic spirit could be adopted by democratic citizens: "The pragmatic spirit of science is a force for democracy and against totalitarianism. We must accordingly always cultivate this spirit in scientific training" (pp. 117–18).

Frank's Radical Educational Proposals

From this antitotalitarian point of view, the unity of science assumed special importance for Frank, as well. For even students of science, Frank explained, became susceptible to propaganda when their scientific education failed to provide them with any appreciation for the continuity of science with other areas of life and thought. Like everyone else, Frank noted, students desire some comprehensive world picture:

The longing for the integration of knowledge is very deeply rooted in the human mind. If it is not cultivated by the science teacher, it will look for other outlets. The thirsty student takes his spiritual drink where it is offered to him ... and he may become a victim of people who interpret recent physics in the service of some pet ideology which has been, in quite a few cases, an antiscientific ideology. (Frank 1949b, 230)

If therefore a student conceives of science as a body of "purely technical" knowledge disconnected with other areas, she or he may become "extremely gullible when he is faced with pseudophilosophic and pseudoreligious interpretations that fill somehow the gap left by his science courses" (ibid., p. 231). Here Frank joined Morris in seeing unified science as a comprehensive world picture that, however ridden with gaps, would function as a substitute for theological and metaphysical beliefs. And here again Frank effectively responded to philosophical critics farther to the left (such as Cornforth and the editors of *Philosophy for the Future*) who charged that philosophers had a duty not only to tear down regressive and metaphysical belief systems but to help erect in their place ideas and a world-view accessible to ordinary persons.

As Frank saw it, however, the task was less to supply students with a world-view or ideology than to supply them with critical tools for their individual and collective evaluation of *competing* ideologies. In his article in the *American Journal of Physics*, Frank must have surprised some readers with his call for training science teachers in philosophy of science, semantic theory, and the history of science. He surely surprised them all by urging that science classes offer comparative analysis of ideologies:

It has become almost a commonplace that the communist and other left-wing creeds have their philosophic basis in materialism, while the right-wing groups look for their foundation mostly to some kind of organismic metaphysics, for example, to Thomism. It is, therefore, very important that the student have a good training in these philosophic interpretations which have become the bases of political creeds. The combination of philosophic and political creeds is often referred to as "ideology." The student of science does not need to be ignorant of this important field. He can take science and its interpretations as his door of entrance.... [T]he future scientist should be taught to take advantage of these ties and get a real insight into historical and contemporary ideologies. (p. 259)

Frank described how one of his own courses enters ideological territory:

Dialectical materialism is discussed in the course in its application to the foundations of science. As in Kantian idealism, we find in dialectical materialism all varieties from almost pure positivism to almost pure Hegelian metaphysics.

These discussions of the philosophic interpretations of science are of great importance to the general education of future scientists... for they are the link connecting science with the humanities. (pp. 257–58)

To audiences of scientists, humanists, and the many theologians at the CSPR, Frank promoted integration of science and the humanities not only for the sake of better understanding science, but to make students better citizens who are able to exercise democratic freedoms, to evaluate critically their leaders, policy, and the claims they make, and to avoid being manipulated by unscientific appeals to "absolute" values.

Frank on Dialectical Materialism

Frank urged an open, critical attitude toward ideology and followed closely the career of dialectical materialism in the American intellectual scene. His classroom analyses of dialectical materialism, Thomism, and other global philosophical outlooks that he described in 1947 were nothing new. In 1935, at the first International Congress for the Unity of Science, he discussed dialectical materialism to show that, despite some distinct differences, there was hope that the Unity of Science movement and the many Soviet and European devotees of dialectical materialism could accept each other as allies in a shared critical project. He was nonetheless aware of widespread criticism, following Lenin, of logical empiricism and Machism. He cited a Soviet textbook criticizing Machists, who

regard science as a game with empty symbols and thus make it incapable of embracing the colorful fullness of the real multiform world. Idealism, mechanism, and logicism are only three ways of leading people to a fictitious supersensual world and of retraining them from occupation with the practical questions of the real world.... Philosophers who teach idealism, mechanism and logicism are in the service of the bourgeoisie, just like the clergy, and make their disciples unfit to work for the social reorganization of the world. (pp. 199–200)

But this critique was excessive, Frank explained, as he excused and apologized for it. On the one hand, like Lenin's critique of Mach, it was "polemical and tactical."[19] In fact, dialectical materialism ("diamat," as Frank abbreviated it here) and logical empiricism were not so dissimilar. Historically, they both developed "in the struggle against the idealistic metaphysics of the school philosophy, which with its odd mixture of faded theology and obsolete science has fulfilled a very definite social function" (p. 200). They also overlapped in three substantive areas, two of which Frank described in detail.

The first was the belief that "science should be 'materialistic' but not 'mechanistic.'" The problem with mechanism, Frank explained, was that it viewed nature as "analogous to a machine" (ibid.). Yet both programs agreed that "the properties of matter reveal themselves to us only in the course of the development of science" (p. 201) and that we should be suspicious of metaphors (such as that of a machine) that tend to anticipate science's results. Diamat's materialism was also an objectivity condition that was opposed to idealism and solipsism. Logical empiricists fully endorsed this condition but went further than diamat by finding these conditions in scientific language: "[A]ll scientific propositions are to contain only terms that occur in statements about observable facts." The "materialism" in diamat, Frank explained, was Neurath's and Carnap's "physicalism" (pp. 200–201).

The second area in common was diamat's doctrine of "concrete truth":

> The doctrine of concrete truth, if it is formulated conceptually, and wherever it is applied exactly, is nothing else than the view that the truth of a proposition can only be judged if the methods of testing it are given. If somebody states a proposition and fails to state the conditions, observable in practice, under which he would be ready to accept it as true, then it . . . is not scientifically applicable – it is meaningless for science. With the doctrine of concrete truth, diamat is therefore defending a standpoint which is very closely related to that of positivism and pragmatism. (pp. 204–5)

Dialectical materialism and logical empiricism found themselves rubbing elbows at the semantic core of American pragmatism.

The third commonality (which Frank only mentioned) is the fact that for both diamat and logical empiricism,

> the propositions of science are to be understood not only from their logical connection with the propositions of the previous stages of science, but also

---

[19] Frank 1949b, 200. "In reality, Lenin took issue with Machism because it is in many respects related to diamat, and he considered it especially suitable to bring out his own teachings very sharply by means of a polemic against it" (Frank 1949b, 199).

from the causal connection of scientific pursuits with other social processes. The investigation of this causal connection is carried on by a special factual science, the sociology of science. (p. 200)

This observation is striking because it shows that Frank could not have failed to know that his promoting sociology of science as the president of the Institute for the Unity of Science (almost twenty years later) could probably be perceived as politically improper. As we see in chapter 14, when Frank's colleagues in his institute were tending to become less political and more professional, Frank conversely called for them to initiate a new project whose affinities to Soviet philosophy he himself had outlined.

By no means, however, does Frank's sympathetic reading of dialectical materialism render him a devotee. As he called for cooperation between these two programs, he criticized the so-called laws of dialectical materialism ("the unity of opposites, the transition from quantity to quality, and the negation of the negation") as plainly unscientific. They "still wear their idealistic eggshells," he explained, and distracted diamat "from the path of establishing the laws of matter through the methods of exact research" (p. 202). Still, Frank was sanguine about cooperation between logical empiricists and dialectical materialists, especially if the latter would apply their doctrine of concrete truth more conscientiously when it came to evaluating other philosophical projects. If they read pragmatism and logical empiricism more accurately, they would see that "it is no longer appropriate to embrace the new empirical and positivistic groups with the idealistic school philosophy in *one* concept, 'the bourgeois conception of science.'" The Vienna Circle, for example, hardly sees science as "a formalistic game which avoids having to do with reality" because it "uses logistics only as an aid to a radical empiricism and positivism." If dialectical materialism would be more careful, it could better discern its friends and enemies and discover "very fundamental ties between diamat and logical empiricism" (p. 205).

Frank wrote and presented this essay long before the Cold War had made it socially and professionally unwise for intellectuals to appear to be sympathetic with Soviet life and thought. When it was first published, his audience may well have connected his call for philosophical cooperation between Western and Soviet philosophy to the politics of the popular front under which Moscow and the Communist parties reached out to other political persuasions jointly to oppose fascism. However, when Frank reprinted the essay in English in 1950, and when he published another remarkably sympathetic essay about dialectical materialism in Carnap's volume of *The Library of Living Philosophers* (Frank 1963), he did so at a time when many others consciously avoided even the appearance of sympathy with Soviet communism. Frank was less foolhardy than honest, it would seem. He seemed to have no desire to modify or repackage his views in order to avoid possible trouble, even though, as we see in chapter 13, he was by 1953 under investigation by the FBI as a potential subversive.

Philipp Frank, Thomas Kuhn, and Cold War Science Studies

While Frank's projects, their political character, and his political fearlessness will help to support claims in later chapters that Frank lost influence in the profession for political reasons, they also tell us something about the career of Thomas Kuhn. For in the late 1950s, as Frank's star was fading, Kuhn's was about to rise. Kuhn would become one of the most influential figures not only in history and philosophy of science but in twentieth-century intellectual history.

The contrast between Frank's and Kuhn's careers presents some intriguing historical questions. While Kuhn's opus *The Structure of Scientific Revolutions* (1962) was published as part of Neurath's *Encyclopedia*, a more telling coincidence is that Kuhn wrote his monograph and intellectually matured at Harvard where Frank taught physics and philosophy of science. Both, moreover, worked under (in one sense) Harvard's president James B. Conant, who shared their interest in science, and in teaching science, as a historical enterprise.[20]

Possibly because of these connections, there are several substantive features in common between Kuhn's *Structure* and Frank's writings on science. Both criticized excessively formal, abstract models of theory that had little resemblance to either contemporary or historical science; both looked to history of science (and often the Copernican controversy) for criteria of adequacy for philosophical accounts. When *Structure* appeared in 1962, Kuhn famously proclaimed, "History, if conceived as a repository for more than anecdote or chronology, could produce a decisive transformation in the image of science by which we are now possessed" (Kuhn 1962, 1). Fifteen years before, Frank told science teachers that "logico-empirical analysis" must be supplemented by "historical analysis"

---

[20] What follows owes much to but departs from Fuller 2000.

of scientific theories: "the history of science is the workshop of the philosophy of science" (Frank 1949b, 278).

Both Frank and Kuhn produced models of scientific change in which accepted beliefs are strained by the accumulation of contradictory evidence and, finally, overthrown by new theoretical innovations. Then, the vanguard becomes the old guard, and eventually, the cycle continues. Most features of Kuhn's well-known model had precedent in Frank's. The notion of paradigm, for example, was for Frank the notion of analogy (or pattern) that guided scientific thinking:

> Medieval science derived all observable phenomena from the principle that they are somehow analogous to the well-known phenomena in a living organism. Seventeenth- and eighteenth-century science, in turn, preferred the analogy to simple mechanisms which are familiar to us from our everyday life experiences. (ibid., p. 288)

These patterns would eventually be abandoned, however, as their successes (or rate of successes) diminished: "Toward the end of the nineteenth century the physicists came more and more to recognize that there are phenomena which can be fitted into this 'mechanistic' pattern only very artificially and incompletely" (p. 302). The result was a scientific "revolution." This is how Frank understood the "disintegration of organismic physics around 1600" and the decline of Newtonian, mechanistic physics around 1900: "In both cases, before the actual revolution took place under the impact of new discoveries of facts, the belief in the certainty of the ruling principles was shaken from within by logically minded critics" (p. 252).

Both Frank and Kuhn found science's discontinuities and periods of rapid change to be most illustrative and revealing about the nature of science. They were science's progressive leaps and, for Frank, they highlighted the progressive, anti-absolutistic quality of science.

> The progress of science takes place in eternal circles. The creative forces must of necessity create perishable buds. They are destroyed in the human consciousness by forces which are themselves marked for destruction. And yet, it is this restless spirit of enlightenment that keeps science from petrifying into a new scholasticism. (p. 78)

Indeed, when Frank told his colleagues at CSPR that science students should study the history of science, he emphasized the importance of studying revolutions – "the great turning points in the great evolution of science" (Frank 1950, 86–87). It was especially important to illuminate these dynamic periods because totalitarian ideologists often seize various "revolutionary" developments in science as significant for their political causes:

> The student who has been through the training in logico-empirical and historical analysis will assess the attempts that have been made to exploit the new physical theories for the benefit of particular religious and political ideologies. He will see through the argument by which the "overthrow of eighteenth- and nineteenth-century deterministic physics" has been used in the fight against liberalism and tolerance.... He will understand that the breakdown of mechanistic physics did not actually imply a return to organismic [medieval] physics, which was, historically, connected with the political and religious doctrines of the Middle Ages. He will understand why twentieth-century Fascism has gladly interpreted the "crisis of physics" as a return to organismic physics which could provide a "scientific" support for a return to some political ideas of feudalism. (Frank 1949b, 282)

History of science, for Frank, was not only the workshop of philosophy of science, but a workshop in democracy and enlightened skepticism about political rhetoric and manipulation.

What distinguishes Frank's program from Kuhn's is not mainly Frank's political interests. Kuhn's *Structure* utilizes a concept of "revolution" that comes packaged with political implications, and Kuhn himself pointed out the similarity between his notion of scientific incommensurability and Cold War politics.[21] Rather, Kuhn's primary interest was scientific change fueled by professional, sociological, and psychological dynamics. While it is true that Frank saw in these dynamics important political and cultural forces that Kuhn did not emphasize, the main difference is that Kuhn took his historical picture of science to recommend professionalization and specialization in science studies. For Frank, it recommended just the opposite.

Recall that according to *Structure*, scientists pay more attention to the philosophical assumptions guiding their different paradigms during times of revolution and epistemic crisis:[22] "It is, I think, particularly in periods of acknowledged crisis that scientists have turned to philosophical

---

[21] "Like the choice between competing political institutions, that between competing paradigms proves to be a choice between incompatible modes of community life" (Kuhn 1962, 94).

[22] "The transition from a paradigm in crisis to a new one from which a new tradition of normal science can emerge [involves] ... a reconstruction of the field from new fundamentals, a reconstruction that changes some of the field's most elementary theoretical generalizations" (Kuhn 1962, 84–85).

analysis as a device for unlocking the riddle of their field" (Kuhn 1962, 87). Frank made the same point:

> Practically, the separation between science and philosophy can be kept up strictly only during a period in which no essential changes in the principles of science take place. In a period of revolutionary changes the walls of separation break down. (Frank 1949b, 266)

But Kuhn and Frank part company over the question of whether this separation between science and philosophy is proper and desirable. As we have seen here, Frank consistently argued that history of science and philosophy of science ought to be intertwined and interconnected. In the laboratory, the classroom, and in democratic life, he believed, this connection would have intellectual, cultural, and political payoffs.

Kuhn, however, was more isolationist:

> Scientists have not generally needed or wanted to be philosophers. Indeed normal science usually holds creative philosophy at arm's length, and probably for good reasons. To the extent that normal research work can be conducted by using the paradigm as a model, rules and assumptions need not be made explicit. (Kuhn 1962, 87–88)

For Kuhn, the disciplinary isolation of science from philosophy was for the progress of science a desirable state of affairs.

In part, this divergence reflects two different backgrounds. Frank matured intellectually in revolutionary times: As a young man he observed the Russian revolution and the birth of the U.S.S.R. and as a young intellectual he studied controversies in physics over relativity and quantum theory. Borders between countries and borders between disciplines were unsettled and unclear. Kuhn matured in postrevolutionary times, in the wake of World War II, and well after relativity theory and quantum mechanics had become (in his model of scientific change) something like "normal" science. Despite widespread fears over the nuclear weapons that physics had facilitated in the Cold War landscape, the sciences were revered for their military and economic implications, and they were soon to enjoy unprecedented levels of federal financial support (via the National Science Foundation, for example) in universities and national laboratories. This raised pressing questions and debates about how to oversee distribution of funds and how to evaluate the resulting science. One result of these developments was the increased specialization, professionalization, and bureaucratization reflected in the growth of the postwar universities, the proliferation of specialized university departments, journals, and professional societies.

Structure championed these results and argued that discrete disciplinary paradigms in the hands of their respective professional communities was good for science. "In its normal state, then, a scientific community is an immensely efficient instrument for solving the problems or puzzles that its paradigms define" (ibid., p. 166). Kuhn also believed that science crucially depended for its continued success on the disciplinary and intellectual autonomy that paradigms and their adherents came to enjoy: "The very existence of science," he wrote, "depends upon vesting the power to choose between paradigms in the members of a special kind of community" (p. 167). From this point of view, the gap between Kuhn and Frank widens. The immense success and influence of Kuhn's book helped to promote and normalize a view of the sciences as isolated from each other (in their respective paradigms) and from philosophy of science and a view of scientists and experts as properly isolated from public life.[23]

As the subsequent history of Frank and his project emphasizes, there is most likely a political component to the different legacies of Frank and Kuhn, as well. In light of the realities of McCarthyism, philosophers and historians of science would most likely have agreed with Kuhn, and not Frank, that this specialization and independence was desirable. Much more than Frank's books and essays, that is, Kuhn's *Structure* spoke persuasively to intellectuals and scientists because professionalization tended to come with not only epistemic legitimacy but job security. Though anticommunist attacks on academics had declined by the end of the 1950s, they were not over, and many who survived the anticommunist investigations in academia had done so by making sure that their intellectual work avoided the appearance of political partisanship or controversy. At the end of ideology, as Daniel Bell put it, "one's commitment is to one's vocation" (Bell 1960, 16). Kuhn's *Structure*, from this angle, had perhaps an irresistible, two-pronged appeal: It suggested that the path to intellectual success and integrity in any discipline and the path to job security and freedom from political attack were one and the same.

---

[23] Questions about whether Kuhn personally understood or intended his book to have political and social effects are taken up in Fuller 2000.

12

"A Very Fertile Field for Investigation"

Anticollectivism and Anticommunism in Popular and Academic Culture

After 1945, in the wake of the war and Neurath's death, the Unity of Science movement and its leaders were subject to pressures that hastened the movement's demise and helped to fashion professional philosophy of science as it flourished in the 1950s and '60s. These pressures can be sorted into three kinds according to their generality and diffusion through Cold War society. The most general concerns intellectual fashions. The beliefs and values that became popular and influential in American and British academic and popular culture after the war specifically opposed some of the Unity of Science movement's basic ideals and methods. One such ideal and method was collectivism. It was attacked in two of the era's most influential books, Friedrich Hayek's *The Road to Serfdom* (1944) and William F. Buckley's *God and Man at Yale* (1951). These books attacked social collectivism and praised individualism, and did so to wide audiences both intellectual and popular. Buckley's witty disdain for Ivy League intellectual culture entertained popular readers (many of whom had themselves never attended a college or university), while Hayek's *Serfdom* was serialized in the popular *Reader's Digest*.

This mood of anticollectivism illustrated by Hayek and Buckley helped to support the second kind of pressure in play, one more localized to institutions of government, popular media, and education. For when Hoover's FBI, the House Un-American Activities Committee, the American Legion, or local politicians sought out communist professors in American colleges and universities, Hayek's and especially Buckley's books enlightened a largely approving public about the different kinds of un-American ideological perversity rampant inside them. In different ways, as the next chapters show, Morris, Carnap, and Frank directly experienced these institutional pressures as they sought to usher Neurath's movement into the Cold War.

The third kind of pressure was local and internal to the philosophical profession. It was exerted not by intellectual trends, anticommunist politicians, or administrators, but rather by specific philosophers, colleagues, and friends of Neurath, Carnap, Morris, and Frank. They variously attempted to persuade them or, in some cases, bully them to toe certain political lines and ignored their calls for the profession to continue the prewar work of the movement. With the exception of Carnap, these pressures tended to diminish the stature, influence, and – in Neurath's case – legacy of these former leaders in the eyes of the larger, growing profession. The kind of logical empiricism they represented and promoted, one that was connected to the agendas of the Unity of Science movement, was not the kind of logical empiricism that survived the anticommunist purges of academia and the advent of the Cold War.

This chapter examines these first two kinds of pressure, especially the extent to which the so-called hysteria over the "Red menace" in American culture helped to legitimate anticommunist persecution of suspicious intellectuals. Together with the next, this chapter documents that anticommunism on most campuses was well known and that ideas, activities, and individuals either involved in or close to the Unity of Science movement were among its targets.

Hayek, Planning, and Serfdom

Friedrich Hayek's *The Road to Serfdom* (1944) is revered today as a seminal argument against totalitarianism and liberal statism. As H. Stuart Hughes put it, Hayek's book was "a major event in the intellectual history of the United States" and "marked the beginning of that slow reorientation of sentiment – both in academic circles and among the general public – toward a more positive evaluation of the capitalist system."[1] The reorientation was indeed slow. From 1950 to 1962, Hayek taught at the University of Chicago, where some sixty copies of *Serfdom* sit in the library's stacks. But those copies were probably not initially read by students of economics, for Hayek was not well received by the university's economists. He was hired by the university's Committee on Social Thought (Ebenstein 2001, 173). Only later, after Hayek's views joined with those of Ludwig von Mises, Milton Friedman, and other anti-Keynesian economists at the

---

[1] This quote appears on the internet site www.hayek.org. Its source is unidentified.

annual meetings of Hayek's Mont Pelerin society (ibid., pp. 142–43), did Hayek win the Nobel Prize (in 1974) and become one of the most influential economists of the century.

One reason for Hayek's appeal was his persistence. Margaret Thatcher, for instance, said that *Serfdom* provided moral and intellectual sustenance to her and England's Tory Party long before her brand of conservatism ascended in the 1980s (quoted in Frowen 1997, xii). Another was the broad exposure the book received in the media. The book was condensed and featured by *Reader's Digest* in April 1945 and featured in *Look* magazine in the form of cartoons.[2] *Serfdom* also answered to both popular and intellectual anticommunism. On the one hand, Hayek's project was broadly political. The manifesto of his Mont Pelerin Society, for example, exalted individual social freedom against the menace of unnamed (but obviously Soviet Communist) ideology:

> The central values of civilization are in danger.... The position of the individual and of the voluntary group are progressively undermined by extensions of arbitrary power. Even that most precious possession of Western man, freedom of thought and expression, is threatened by the spread of creeds which, claiming the privilege of tolerance when in the position of a minority, seek only to establish a position of power in which they can suppress and obliterate all views but their own. (quoted in Ebenstein 2001, 145)

Hayek's preliminary drafts for this manifesto more closely track the omnipresent postwar concerns with ultimate values, especially those of "individual freedom" and "complete intellectual freedom," which – as Neurath's critic Kallen might have added in his denouncement of Neurath's project – "no consideration of social expediency must ever be allowed to impair" (in Ebenstein 2001, 368 n. 21). On the other hand, Hayek himself confessed autobiographically in *Serfdom* that he too had once been a woolly minded socialist who, like so many others, was unpleasantly disillusioned before embracing individualism and free markets in his intellectual maturity. While "disillusionment" about communism and socialism dominated leftist thinking, that is, *Serfdom* addressed the psychological dynamics of that illusion and the hidden evils that he and other former leftists had so easily overlooked.

Hayek's arguments against socialism and collectivism join the story of Neurath's movement in four ways. First, the argument helped to eliminate any socialist middle ground between capitalism and party-controlled communism (or "totalitarianism") as the Cold War took shape. What many

---

[2] The *Reader's Digest* and *Look* versions of *Serfdom* are published in Hayek 1999.

took for granted, Hayek justified and explained: One must be partisan *either* to Western capitalism and free markets *or* to totalitarianism in its many forms. Socialism supported no middle ground because, according to Hayek, socialism was embryonic totalitarianism: "Few are ready to recognize that the rise of fascism and Nazism was not a reaction against the socialist trends of the preceding period [in Germany] but a necessary outcome of those tendencies" (Hayek 1944, 3–4). Socialists and liberal sympathizers, usually unbeknownst to themselves, Hayek believed, were on "the road to serfdom."

The argument behind Hayek's claim that socialism is merely embryonic totalitarianism crucially involves social and economic planning. Totalitarian serfdom, he claimed, would be the outcome of the kind of collective, large-scale planning that socialists usually championed. Attempts to coordinate people in the interests of any global, coherent plan – economic, social, military, and so on – will be crippled by dissatisfaction. No plan will please everyone involved:

> The effect of peoples' agreeing that there must be central planning, without agreeing on the ends, will be rather as if a group of people were to commit themselves to take a journey together without agreeing where they want to go. . . . That planning creates a situation in which it is necessary for us to agree on a much larger number of topics than we have been used to, and that in a planned system we cannot confine collective action to the tasks on which we can agree but are forced to produce agreement on everything in order that any action can be taken at all, is one of the features which contributes more than most to determining the character of a planned system. (ibid., p. 62)

If planning is to go forward, therefore, someone or some unified minority must take control: "It will often be necessary that the will of a small minority be imposed upon the people"; "In the end somebody's views will have to decide whose interests are more important" (pp. 69, 74). If planning is to succeed, that is, it must degenerate into dictatorship: "Dictatorship is the most effective instrument of coercion and the enforcement of ideals and, as such, essential if central planning on a large scale is to be possible" (p. 70).

This argument was popular in intellectual as well as popular and political culture. Winston Churchill's infamous "Gestapo" speech in the election of 1945 was widely recognized as drawing on *Serfdom* to proclaim, "No socialist system can be established without a political police. They would have to fall back on some form of Gestapo, no doubt very humanely directed in the first instance" (in Ebenstein 2001, 138). Clement Attlee defeated Churchill because, Hayek's biographer suggests, voters

took offense at Churchill's identification of the wartime Labour government with the worst and most extreme aspects of Nazism. Attlee himself
took offense at Churchill's "second-hand version of the academic views of
an Austrian professor, Friedrich August von Hayek" (in ibid.). In France,
Hayek's outlook was promoted and popularized by the French political
philosopher Raymond Aron. In 1955, Aron directed his book *The Opium of
the Intellectuals* to the same audience to which Hayek directed *Serfdom* – not
official communists, but rather the broad population of liberals friendly
to the left and "whose sympathies are with the Soviet world" (Aron 2001,
xvii). One of the several opiates Aron deconstructed was a faith in social and economic planning, which he dissected in a now familiar way:
"Every impulse towards global planning is doomed to end in tyranny"
(ibid., p. 199).

By conceiving of social and economic planning as a road map to totalitarianism, Hayek's and similar arguments worked against the Unity
of Science movement in at least two ways. The movement not only was
friendly to collective planning in society and economy (the second point
of contact), it also sought to incorporate the ideals and methods of planning into science itself. The movement can be usefully understood, that
is, as an effort to plan the future of science (Reisch 1994). The new
Encyclopedia and the International Congresses, for example, would bring
together specialists of different fields to formulate and attack problems
that overlapped disciplines. Thus, Neurath proposed a universal jargon
of science to facilitate this collective project and not – as Kallen insisted –
to impose some body of theory on science from without. Always a pluralist, Neurath regularly emphasized that many different future unities of
science were possible. The movement's goal was to institute awareness of
that openness, not to constrain it with a priori theories or preconceptions
about what the future of science would look like in its details. That is why
"the whole business," as Neurath put it in his second contribution to the
Encyclopedia, "will go on in a way we cannot even anticipate today."[3]

Last, there is a personal connection, it would appear, between Hayek's
Serfdom and the Unity of Science movement. As if to confirm Hayek's hypothesis that planning must degenerate into dictatorship, and Kallen's
charges that such a transformation was almost certain to occur, Neurath
himself sometimes came to be seen as an overbearing, uncompromising
dictator. Neurath's personality was strong, and most found him either very

---

[3] Neurath 1944, 47. For a different, "postmodern" view of Neurath's unity-program, see
Cat, Chang, and Cartwright 1991 and, for criticism of that view, Reisch 1997b.

appealing or very unpleasant. Perhaps because of their common backgrounds as Austrian economists, Hayek seemed to have little respect for Neurath. In Vienna's overlapping intellectual circles, Hayek knew about Neurath and the Vienna Circle and gained enough of an impression to report later that it was Neurath and "science specialists in the tradition of Otto Neurath" that first led him to reject "positivism" in economics and philosophy of science.[4] Hayek was also personally and intellectually allied with his countryman Karl Popper, who even more aggressively targeted logical empiricism and Neurath, in particular.[5]

Darwin noted that competition is fiercest between rivals with much in common. Hayek's attack on planning is all the more striking because of several programmatic similarities between Hayek and Neurath. They both championed an Epicurean utilitarianism that took individual happiness as a basic value; both criticized naïve scientism holding that science supplies a complete and true world picture; and both opposed fascism and found it lurking in heretofore unsuspected places. While Hayek was writing *Serfdom*, for example, Neurath was debating Joad and other English classicists over the mischievous metaphysics and fascism he saw in Plato.

Perhaps suspecting that Hayek's arguments against planning might come to be very popular, Neurath wished to debate Hayek publicly.[6] But he succeeded only in publishing a review of *Serfdom* that began on this shared note of antifascism: "Let us be grateful to authors who show up concealed Fascism." But Neurath's politeness lasted about one sentence. He continued, "we cannot go all the way with Hayek in his relegation of all planning to this category [of totalitarianism]." Indeed, Neurath found the Achilles' heel of Hayek's view: Hayek simply ruled out, a priori and without justification, the possibility that collective interests might win out over individual interests, that societies or nations might succeed in "planning as a co-operative effort, based on compromise" (Neurath 1945b, 121). Neurath had faith that collectivism and compromise could possibly prevail over individualism: "World planning based on co-operation would perhaps give rise to a world-wide feeling of responsibility for other people's happiness" (ibid., p. 122). Hayek and the many influenced by him

---

[4] Ebenstein 2001, 157. Hayek's connections to the Vienna Circle are described in Stadler 2001, 143–61, 149.

[5] On the Popper-Neurath debate, see Cat 1995. For an argument that Neurath, as a vocal defender of both scientism and scientific planning, was among the unnamed, sinister targets of both Hayek's "Scientism and the Study of Society" and Popper's *The Poverty of Historicism*, see Uebel 2000, esp. 161.

[6] Neurath to Hayek, July 12, 1945, ONN.

simply did not share that faith, a faith that belonged to the progressive 1930s, not the distrustful, anticommunist 1940s and '50s.

William F. Buckley

While Hayek provided intellectual firepower to conservatism's rejection of collectivism and planning, William F. Buckley took up the cause with the wit and rhetorical skill that he made famous as one of the leading conservative writers, publishers, and talk-show hosts in the postwar United States.[7] Buckley's *God and Man at Yale*, published in 1951 shortly after his graduation from that university, is a founding document of American postwar conservatism. Written largely while Buckley was an undergraduate, its pages vividly document Buckley's growing outrage as he realized that state-of-the-art scholarship in history, philosophy, and economics had little, if anything, to do with the sermons he had earlier imbibed in his church and prep school chapel. Instead of seeking to create some kind of synthesis of science and religion, or simply abandoning theology as anachronistic metaphysics in an age of science, Buckley defended theology against the atheistic collectivism that he felt was rampant on ivy-league campuses.

To some extent, Buckley helped to bring into America's living rooms the antiscientific and anti-intellectual views promoted in the Conference on Science, Philosophy and Religion and in the anti-Stalinist press of the late 1940s. He rejected the secularism of his professors – "there are no inherent contradictions between sociology and religion" (Buckley 1951, 206) – and joined the values debate by insisting that America's free and healthy civilization rested essentially, but precariously, on certain extra-scientific values (associated with individualism and theism) that must at all costs be protected and sustained. Testifying to his ideological kinship with Hayek, Buckley praised *Serfdom* and once wrote that all institutions must encourage "cultivation of values only a painstaking care for which can guarantee the survival of free men" (Buckley 1976, 102).

Buckley shared a podium with Sidney Hook as an intellectual who defended anticommunist investigations at colleges and universities in the early 1950s. Those who argued that professors and teachers were immune to anticommunist scrutiny on the grounds of academic freedom he called "academic freedomites" (and thus winked at the idea that more

---

[7] Buckley founded the magazine *National Review* in 1955 and hosted the interview program *Firing Line* on American public television from 1966 to 1999.

than their politics might be deviant and un-American).[8] However entertaining, Buckley was no mere wordsmith. He and his approving readers genuinely believed that higher causes mandated campus anticommunism. For those who understood that "the duel between Christianity and atheism [was] the most important in the world" and who identified that duel with "the struggle between individualism and collectivism" (Buckley 1951, xvi–xvii), the "approaching fate of the communist teacher" (ibid., p. 189) was clear:

> I shall not say, then, what specific professor should be discharged, but I will say some ought to be discharged. I shall not indicate what I consider to be the dividing line that separates the collectivist from the individualist, but I will say that such dividing line ought, thoughtfully and flexibly, to be drawn. (p. 197)

For Buckley, the "communist teacher" and the "collectivist" professor were the same. Neither ought to take his or her professional security for granted.

Buckley nourished not only academic but also more general anticommunism through his social and professional connections with those who promoted anticommunism as one of the Cold War's main events. These included McCarthy, Richard Nixon, McCarthy's deputy Roy Cohn, and the former-communist and poet Whittaker Chambers. Chambers's testimony before J. Parnell Thomas's House Un-American Activities Committee helped to convince many that the Red menace was real. He claimed that Alger Hiss, a state department official and co-organizer of the United Nations, was a Communist whom he knew from his own days in the communist underground. In highly publicized trials in 1948, Hiss was convicted of perjury and his government career was ended. He continued to make a living selling stationery while Chambers published a best-selling book, *Witness* (1952) and was championed by anticommunists such as Buckley, who shared his theological interests and also employed him to write for his magazine *National Review*. Buckley had earlier promoted the partnership of "God and man," while Chambers's bestseller claimed that

---

[8] Buckley 1951, 149. It was not completely unnoticed at the time that McCarthy, J. Edgar Hoover, and Roy Cohn, who targeted (alongside communists) homosexuals in government (usually on the grounds that their deviance rendered them either susceptible to communism or candidates for blackmail by communists) were themselves suspiciously unmarried. Critic Granville Hicks observed in this connection that "long before Freud most people were shrewd enough to be suspicious of a man who rants loudly against a particular vice" (Hicks 1950, 9). For more on the antihomosexual aspects of anticommunism, see Johnson 2004.

"the Communist vision is the vision of Man without God" (Chambers 1952, 9).

The Red Menace in Political and Popular Culture

At least because of the role played by nuclear weapons in World War II, American universities had always been one focus in the Cold War. They were the professional homes of the physicists and chemists who developed the atomic technologies that were widely believed to have ended the war in the Pacific. Against the views of many scientists (such as those publishing the *Bulletin of the Atomic Scientists*) who opted for civilian control, many politicians came to believe that these technologies had to be controlled as military secrets and, above all, withheld from the Soviets. As this debate unfolded in the years immediately after the war (the founding of the Atomic Energy Commission was the result), headlines were often dominated by scandals involving military secrecy and, in particular, the passing of atomic (so-called) secrets to the KGB. On the heels of the Hiss trial, Americans avidly followed the declining fortunes of Julius and Ethel Rosenberg. Julius was found guilty of passing to the Soviets information obtained by Ethel's brother, David Greenglass, who worked at Los Alamos. The two were executed in June 1953. In another highly visible case, the scientist leader of the Manhattan Project, Robert Oppenheimer, came under anticommunist scrutiny and lost his security clearances (Klingaman 1996, 288–89).

As the Cold War progressed, however, anticommunist investigators easily broadened their focus from scientists to intellectuals and professors in nearly all disciplines. To understand how and why anticommunists were so effective and encountered little resistance as they scrutinized higher education, it is necessary to consider some of the leading players and their methods. The main engine behind the anticommunist fervor of the Cold War was not Senator Joseph McCarthy, but rather the FBI of J. Edgar Hoover. Partly because of his skill at empire building and partly because there really were Soviet spies and informers to be pursued in the United States, Hoover propelled Cold War anticommunism more than any other individual. Though McCarthy became the public face and namesake of the anticommunist hysteria, most historians agree that, like Richard Nixon, he joined the crusade in its early days in order to boost his otherwise unremarkable political career and, in McCarthy's case, to minimize political damage from scandals involving his personal state income taxes (Klingaman 1996, 255). He became a crusader in

1950 and, after Republicans gained control of the Senate in 1952, he ascended to chairmanship of the Senate's Permanent Subcommittee on Investigations. He oversaw anticommunist investigations until his downfall and condemnation in the Senate in 1954. Three years later, he died from alcohol poisoning.

Hoover, on the other hand, remained in place before, during, and after McCarthy's brief celebrity. He maintained a library of files on individuals inside and outside government that linked them, actually or merely suspiciously, to the USSR, to the Communist Party, to "front" groups secretly sponsored by the party, or to other known or suspected communists. The library empowered Hoover in private and public sectors. In the private sector, he was known to prevent individuals from being hired by ordering files or papers to appear mysteriously on the desk or in the mailbox of a prospective employer. In cases where employers contacted the FBI to request information about the political rectitude of a specific individual, Hoover would sometimes have a letter hand-delivered in which he would officially deny the request for legal reasons. Along with the official letter of denial, however, the employer would nonetheless find the information requested. In the public sector, if Hoover believed evidence was sufficient, he would forward names of suspected subversives to Senate or House committees that would issue subpoenas and hold hearings. Countless government officials and spies were investigated and scrutinized in public, asked if they were, or ever had been, involved in the Communist Party, and asked to name names of others who may have been. In most cases, these appearances permanently damaged careers and livelihoods.

Hoover thrived on anticommunism and so did many others. Politicians and public officials almost always found that doing battle with the Red menace was good for public relations. For many, the tone was set by the presidential election of 1948. The election saw the resounding defeat of Henry Wallace, Roosevelt's vice president from 1941 to 1945 and candidate for the newly formed Progressive Party. Wallace was aggressively liberal, internationalist, and skeptical about the Red menace. He also defended indicted figures (such as the Hollywood Ten and Harry Dexter White, a State Department official accused of soviet espionage) and responded to Red-baiting during the campaign by calling it a "red herring" designed by his political enemies to obscure real social and economic issues. The public, however, was unimpressed. Wallace received only slightly more than one million votes that November. Taken together with the Hiss trial and the Soviet clampdown in Czechoslovakia, the election of 1948 functioned as something of a mandate for Cold War anticommunism. A year later, the founding of the German Democratic Republic would cement separate economic futures for the different parts of a divided Germany, while the entirety of China would be "lost" to the communism of the People's Republic of China. By 1950, the Korean War would be widely seen as the first phase of a coming third world war pitting the Soviet Union against the United States in an epic battle of societies devoted to opposing, irreconcilable values.

In this tense, worrisome climate, certain beliefs and stereotypes about the Soviet Union, the Communist Party, and their followers began to thrive in American popular culture. One popular belief held that Moscow, having taken over much of Eastern Europe, aimed next to take over the United States with a two-pronged, military and psychological, strategy. Others warned that the takeover may be entirely psychological. On this view, Moscow sought to recruit converts (who, in turn, would recruit their friends and neighbors) until a sufficient number of voters would be able simply to elect a communist-friendly regime. As a typically alarmist essay in *American Mercury* put it in 1953, there were only "twenty-four steps to communism," and none involved firing bullets, exploding bombs, or even placing the word "communist" on voter ballots. Prior to elections, writer John T. Flynn breathlessly explained, communists would masquerade as socialists, infiltrate institutions, and promote their "collectivist slant" (Flynn 1953, 4) in areas such as entertainment, education, publishing, labor, and government. With this collectivist mood firmly secured, the stage would be set for socialism to move necessarily into communism, as Hayek's *Serfdom* argued it would. "Once we are penetrated by a heavy percentage of socialist operations," Flynn concluded, "a ruthless Communist dictatorship will take over" (ibid., p. 6).

This fear of a silent-but-total ideological takeover supported other features of Cold War anticommunism. Naturally, converts infiltrating American institutions needed to hide their true motives and their connections to other operatives. Since communists worked in a cell structure where the identities of others were often unknown, and cells were orchestrated externally by Moscow, inquisitorial methods seemed necessary. Only by naming names of others they had known, as Whittaker Chambers named Alger Hiss, could hidden networks be identified and uncovered. For this reason, Hoover, McCarthy, and other leading anticommunists opposed legislation that would outlaw the Communist Party. Such legislation, they claimed, would only drive communist activity further underground and make cells and networks even harder to discover and disable.

One psychological component of the Red menace was based on the view that Soviet scientists had perfected techniques in "mind control" and "brainwashing." After having been suitably conditioned in the Soviet Union, spies deployed in the United States, it was believed, functioned like puppets on strings under control of Moscow. The necessities of orchestrating from afar a vast cultural, economic, and political takeover, that is, played into the values debate by enforcing an image of communists as robotic, unthinking, unfree, and bereft of all Christian values. These notions came to life in movies such as *The Manchurian Candidate* (about an American soldier "brainwashed" by Chinese communists and then deployed into the American political system) and in the subtext of *Invasion of the Body Snatchers.* Its chilling effect lay in the way that characters in the movie cannot discern who has or who has not been taken over by the invaders. To the more knowing audience, however, those whose bodies were now controlled by an evil foreign power seemed slightly robotic, mechanical, and lacking in human traits.

America's Cold War interest in brainwashing and mind control was not limited to geopolitics, as suggested by the debate over the rise of consumer culture, a topic that Morris mentioned in his book (1946b, 240–41) and that made Vance Packard a household name in the late 1950s. Packard's *Hidden Persuaders* (1957) claimed to expose techniques of subliminal suggestion used by advertisers and marketers to control consumer behavior. But the topic was firmly connected to political and even intellectual debate, as suggested by the decision of the editors of the anti-Stalinist *New Leader* to advertise their magazine on the grounds that as early as 1950 they had scooped 1953's headlines about "brainwashing" and Cold War "brain warfare" (see Fig. 4).

Viewed with some fifty years of distance and perspective, writers and producers who were the most vocal and concerned about the Red menace seemed themselves taken over by fear, anxiety, and lack of balanced, critical judgment. The popular movie *I Was a Communist for the FBI* promoted the stereotype of the communist operative as a short, dark, middle-aged man with thick glasses and Nazi-esque accent who faithfully awaited instructions from Moscow. Inspired by the case of Gerhard Eisler, whom the FBI pursued for years, this highly dramatized film nonetheless won nomination from the prize-giving academy for best *documentary* in 1951 (Schrecker 1986, 122). The distinction between fiction and nonfiction became blurry in unlikely venues, as well. The magazine *House Beautiful,* under the editorship of Elizabeth Gordon, claimed to expose ideological mischief in the popularity of modern, European design. Both the

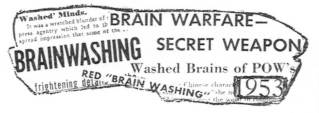

THREE

YEARS

AHEAD of Today's Page One News

SPECIAL
INTRODUCTORY OFFER

27 Weeks—$2 *(28 weeks if you enclose your remittance!)*

ACCIDENT? Luck? Shot in the dark? No, THE NEW LEADER'S world scoop on brain-washing was the result of none of these. The reason THE NEW LEADER knew about brain-washing three years ago was a man: Edward Hunter, our roving correspondent in East Asia for more than a decade. Hunter's *experience* covering the Asian battlefronts of World War II and the cold war enabled him to make the contacts novices don't dream of. Hunter's *honesty*, his uncompromising dislike for all despotisms from Tojo's to Mao's, compelled him to seek out and print a story that made other reporters uncomfortable. Hunter's *thoroughness* made him follow up this October 7, 1950 article on brain-washing with many others, detailing the entire system of Chinese Communist indoctrination which the rest of our press is only now beginning to note.

These qualities—experience, honesty, thoroughness — characterize all of the 16 correspondents who cover the globe for THE NEW LEADER and enable it to stay ahead of the headlines *consistently.* If you are not already a subscriber, why not fill out the adjoining coupon and receive *tomorrow's headlines today?*

FIGURE 4. Subscription advertisement in *New Leader* from 1953 boasting of a "world scoop on brain-washing" three years ahead of other political affairs magazines and newspapers.

sensational tone and the argument in Gordon's editorial, "The Threat to the Next America" (1953), illustrates the emotional power and appeal of Cold War anticommunism: "What I want to tell you about has never been put into print by us or any other publication, to my knowledge. Your first reactions will be amazement, disbelief, and shock. You will say 'It can't happen here!'" Few people enjoy the look of modern design, Gordon argued, because the modernist aesthetic is

unscientific, irrational, and uneconomical – illogical things like whole walls of unshaded glass on the west, which cause you to fry in the summer.... Or tricks like putting heavy buildings up on thin, delicate stilts – even though they cost more and instinctively worry the eye. Or cantilevering things that don't need to be cantilevered, making them cost more, too. A strong taint of anti-reason runs through all of their houses and furnishings.

No wonder you feel uneasy and repelled!

The explanation was simple: European design was one component of the communist psychological strategy:

If we can be sold on accepting dictators in matters of taste and how our homes are to be ordered, our minds are certainly well prepared to accept dictators in other departments of life. The undermining of people's confidence is the beginning of the end.... Break people's confidence in reason and their own common sense and they are on their way to attaching themselves to a leader, a mass movement, or any sort of authority.... So, you see, this well-developed movement has social implications ... [and presents] a social threat of regimentation and total control.

For Gordon, the psychological and cultural significance of Mies van der Rohe's famous slogan "less is more" had little to do with the fact that Mies, like many other European intellectuals, found refuge from totalitarianism in North America. In fact, it was a "mystical" brain-washing mantra designed by foreign powers to help lull North Americans away from their attachment to private possessions and into a collectivist regime.[9]

The Problem of Communist Professors

The logic and popularity of these beliefs about communists and communism helped to launch and sustain campaigns to purge schools of communist teachers and professors. Students, after all, are the political future of any society. They were seen as ideal targets for communist ideologues who could use their authority in the classroom to make the perversions of

[9] My thanks to Richard Creath for pointing out to me Gordon's editorial.

communism seem normal, admirable, and perhaps – as Marx predicted – historically inevitable. Universities were additionally revered as bastions and symbols of free thinking. They needed to be protected, therefore, from communist professors who, beholden to a foreign power and party, had betrayed their professional dedication to free inquiry. Patriotic regents, trustees, administrators, parents, and students sometimes panicked as they realized that battles in the epic war between communism and freedom were, as Buckley had revealed about Yale, unfolding in their very midst.

Though controversies over communist professors and teachers had erupted earlier in North America, this amalgam of logic and suspicion created in the late 1940s an unprecedented "climate of fear" on most university campuses. Many professors had good reason to fear anticommunist scrutiny because they had been members of the Communist Party or fellow-travelers in the 1930s. Federal loyalty-review programs were so strict as to specify even past association with communism as grounds for dismissal (Klingaman 1996, 367). The climate was also sustained by administrations, especially at public universities, that struggled to remain in the good graces of state and federal officials who both controlled state and federal funding and used anticommunism to advance their careers or power. Typically, once Hoover or McCarthy or an aspiring politician announced suspicions or reported rumors about a particular campus, administration officials would form committees and pursue their own internal investigations. Since none wished to be perceived as "soft on communism," many campuses endured two sets of investigations or hearings about suspicious professors.

The University of Washington

The case of the University of Washington set the pattern for most of the investigations of the 1950s. It began in 1946 when state senator Albert Canwell, at the prodding of the Foreign Legion, investigated communist infiltration among organizations for the elderly. Soon, Canwell had targeted the University of Washington where, an aid of Canwell's reported during preliminary hearings, "there isn't a student who has attended this university who has not been taught subversive activities." Reports put the number of communist professors on campus "as high as thirty." Others believed the number was as high as 150 (Schrecker 1986, 94).

Hearings began with testimony from a handful of influential former communists who came to make a living out of testifying against their former comrades. As Schrecker tells it, they "trooped out to Seattle to tell Canwell and his fellow legislators how the CP was planning to overthrow the American Government by force, violence, and subterfuge." Some spoke about the conspiracy on a national scale, perpetuating rumors that, for example, Eisenhower, then president of Columbia University, was protecting communists on that faculty. Others told stories about the political left in Seattle, and they named names (ibid., p. 96).

Professors under suspicion were then called to testify about their present and past affiliations. If they admitted to being active in the party, they were then asked to name others. There were three general ways to respond: One admitted past membership, named names, and declared oneself a *former* member who now disowned communism; one admitted past membership but refused to name names that would incriminate others (a strategy known as the "diminished Fifth"); or one remained uncooperative and refused to answer questions at all, usually invoking the Fifth Amendment. The game was additionally complicated because one could easily be caught lying or evading a question on a trivial matter unrelated to international espionage or devotion to Marxist theory. Since communists, qua conspirators, were prone to lying and evasive language, white lies or exaggerations would often feed investigators' suspicions.

In the Canwell hearings, eleven professors were called to testify. After the hearings, and after the university had conducted its own, separate investigations, three of these eleven were fired in January 1949: the philosopher Herbert Phillips, the medievalist Joseph Butterworth, and the psychologist Ralph Gundlach. The first two were active members of the Communist Party; the latter probably was not. In each case, however, their lives and careers were upended. None was able to secure an academic job elsewhere (ibid., p. 104).

By the middle to late 1950s, major investigations had taken place at the University of Washington, Harvard, City College of New York, the University of Buffalo, Wesleyan, the University of Minnesota, the University of Arkansas, the University of Michigan, the University of Chicago, the University of California, Reed College, Temple University, Ohio State, and Rutgers. Nearly all state university systems implemented loyalty oaths that required faculty to sign that they were not active in subversive groups and that their political loyalties lay exclusively with the United States. Those who refused to sign, on whatever grounds, almost invariably received added scrutiny from administrations or investigators. Because some dismissals were discrete and unpublicized, however, the precise extent of these purges remains unknown. Schrecker's research suggests that the groups most often targeted were physicists, English professors, and economists. Once the sizes of departments are normalized, however, as John McCumber tallies the numbers in Schrecker's research and in American Association of University Professors (AAUP) bulletins, it was philosophers who "were more likely to be attacked by witchhunters than were members of any other discipline"(McCumber 2001, 37). Extrapolating from the available record, McCumber estimates that roughly thirty-five professors of philosophy were in some way attacked by anticommunist administrations or investigators during the Cold War.

Communist Professors and Popular Anticommunism

In 1952, Justice William O. Douglas editorialized in the *New York Times* that the nation was rapidly devolving into "arrogance and intolerance" for dissenting and unpopular opinions and points of view. Fear of communism and fear of each other, Douglas explained, had begun to eat away at national dialogue that was previously often coherent, free, and relatively rational:

> Once we could safely explore the edges of a problem, challenge orthodoxy without qualms, and run the gamut of ideas in search of solutions to perplexing problems. Once we had confidence in each other.

In light of the Cold War, Soviet expansion, and the militarization of U.S. foreign policy, Douglas wrote, fear of a conspiracy leading to Soviet world domination changed everything. He continued,

> Now there is suspicion. Innocent acts become telltale marks of disloyalty. The coincidence that an idea parallels Soviet Russia's policy for a moment of time settles an aura of suspicion around a person.

That logic had already led to Horace Kallen's totalitarian suspicions about the Unity of Science movement when Douglas observed that this fear was spreading throughout the professions. "Fear of losing one's job, fear of being investigated, fear of being pilloried" had

> even entered universities, great citadels of our spiritual strength, and corrupted them. We have the spectacle of university officials lending themselves to one of the worst witch-hunts we have seen since early days. (quoted in Schrecker 2002, 273)

Two years later, Robert Hutchins offered similar observations to a popular audience in *Look* magazine. In an article titled, "Are Our Teachers Afraid to Teach?" he answered, "Yes, they are" (see Fig. 5). Standing firm against

a the University of Chicago, where Prof. James Cate was photographed at the time Hutchins was its head, a probe, the author recalls, left "paralyzing" effects.

ARE OUR TEACHERS Afraid to teach?

BY ROBERT M. HUTCHINS

Dr. Hutchins, former president of the University of Chicago, is one of America's most brilliant, independent and provocative educators. His many educational innovations have included the now famous Great Books course and the abolition of course credits. Today, a quarter century after he made his bow at Chicago at the age of 30, he is an associate director of the Ford Foundation.

EDUCATION is impossible in many parts of the United States today because free inquiry and free discussion are impossible. In these communities, the teacher of economics, history or political science cannot teach. Even the teacher of literature must be careful. Didn't a member of Indiana's Textbook Commission call Robin Hood subversive?

The National Education Association studied no less than 522 school systems, covering every section of the United States, and came to the conclusion that American teachers today are reluctant to consider "controversial issues." But what does that mean? An issue is a point on which the parties take different positions. A noncontroversial issue, therefore, is as impossible as a round square. All issues are controversial; if they were not, they would not be issues.

In Los Angeles, Houston and Pawtucket, a teacher would hesitate to mention UNESCO, the United Nations Educational, Scientific and Cultural Organization, because school authorities have made it plain that they are afraid of it. Since those who oppose UNESCO generally oppose the United Nations, the teacher should probably not refer to the U.N. either. Since those who oppose the U.N. believe that the United States should somehow isolate itself from world affairs, the teacher would be unwise to say very much about international relations. How, then, can teachers say anything worth saying about the world in which we live?

continued

Produced by LEO ROSTEN

FIGURE 5. University of Chicago President Robert Maynard Hutchins's article for *Look* magazine criticizing anticommunism for making college teachers afraid to teach or discuss "controversial issues" (Hutchins 1954).

Sidney Hook, who minimized civil-libertarian inconveniences by arguing that very few professors had actually lost their jobs in the course of campus investigations, Hutchins fumed, that "you do not need to fire many teachers to intimidate them all." And in the wake of such intimidation, Hutchins observed, curricula were changing. In their seminars and lectures, professors were now reluctant even to mention, much less discuss, topics as ordinary as the story of Robin Hood (in light of his economic antics) or the programs of UNESCO (the United Nations Educational, Scientific and Cultural Organization) on behalf of peace and humanitarianism (Hutchins 1954, 27). The mechanisms of professional and social ostracism in place, Hutchins observed, were keyed to a hair-trigger:

A person becomes controversial when a question is raised about him. If you want to get rid of a teacher, make loud charges against him – then demand that he be fired because the charges have been made. All that anybody will remember about him when he tries to get another job is that he was in some kind of trouble about being a Red. (ibid., p. 28)

Hutchins was especially agitated because he wrote his article just as investigations at Harvard University were set to begin:

The attack on Harvard will not be lost on other, weaker institutions. If this can happen at Harvard, think what can happen to them. And so professors everywhere will hesitate before they express opinions contrary to those of Senator McCarthy, or before they say anything that can be twisted – somehow, sometime, by someone – into an unpopular statement. (p. 29)

According to the writer Mary McCarthy, it was not always a matter of what a professor said, but sometimes what he or she did. In her fictional short story "The Groves of Academe" for the *New Yorker* in 1951, she described a professor of literature named Mulcahy who casually laughs as he reads a letter from his president informing him that he has been fired. At first, he thinks the letter is just another round in his ongoing tug of war with the president. Soon, however, he realizes that the letter is quite serious and that he has, in fact, been fired. He remembers attending a rally of "Partisans for Peace" and recalls that his president, Maynard Hoar, had "tipped him off" beforehand. At a holiday social weeks before the rally,

Maynard Hoar, in heavy ribbed sweater, genially assert[ed] that he could as soon, *in these times*, as president of a small struggling college, appear at a "peace" rally as be found playing strip poker on Sunday in a whore house.

Mulcahy "clapped his hand to his head. '*Dummkopf!*' he cried. '*Dummkopf!*' – awestruck at his own blindness. The clue had been in his hands long ago as the Christmas reception." But now it was too late. "In these times," conferences and rallies promoting "peace" were routinely suspected to be "front" events, sponsored by communists as part of their so-called peace offensive to recruit new party members. Thus, even though Mary McCarthy's Maynard Hoar was on the record as a defender of academic freedom (which was perhaps why she named him with a nod to Robert Maynard Hutchins), Mulcahy knew that these niceties were irrelevant. The president with whom he had wrestled through the years was now using him "as a scapegoat to satisfy the reactionary trustees and fundraisers" (McCarthy 1951, 31).

This coverage in the popular media of the anticommunist investigations and loyalty oaths could only have added to the pressures academics experienced. In part, the coverage probably emboldened investigators because it supported the postwar image of "intellectuals" as victims of Soviet propaganda in the 1930s who were therefore extremely susceptible to disinformation and, again, primary targets for Soviet control. Economist Carl Landauer, who observed the loyalty oath controversies at the University of California, noted in 1950 that his campus still contained "the usual quota of liberals who have been slow to rid themselves of illusions concerning Communism" (Landauer 1950, 13). Writings by University of Washington president Raymond Allen called out both to the popular image of communists and communist spies as unfree, unthinking robots and to the themes of intellectual freedom championed by Horace Kallen and Sidney Hook. In an essay defending the firings at his university, Allen referred to the "American Idea" (championed by Horace Kallen) and quoted Sidney Hook to argue that party members should be fired from teaching jobs because "their conclusions are not reached by a free inquiry into the evidence. To stay in the Communist Party they must believe and teach what the party line decrees" (Hook, in Allen 1949, end of §1). Intellectual radicalism was and should be permitted at any university, Allen took pains to emphasize. What was unacceptable, he explained, was the kind of intellectual radicalism later suggested, it would seem, by *Invasion of the Body Snatchers* or *The Manchurian Candidate* – the kind of radicalism in which one's philosophies are "subject to dictation from outside the mind of the holder" (Allen 1949, §2). Concluding his defense with a nod to Buckley's theological concerns, Allen found it nothing less than a "sacred duty" to fight communism and expunge from the academy "the false and robot prophets of Communism or any other doctrine of slavery that seeks to be in, but never of, our traditions of freedom." "Through education alone," he concluded, "can we combat the tenets of communism" (ibid., §4).

The Blacklist and the Transformation of Curricula

In light of Schrecker's research, Hutchins was right to hint that academia collectively created and maintained a blacklist. It was neither official nor public, Schrecker explains, but was nonetheless a

blacklist at least as comprehensive and far less well known than the one in the entertainment industry. . . . It lasted for years, beginning for some people in the 1940s and lasted for others throughout the 50s and often into the 60s. (Schrecker 1986, 265)

The logic of ostracism was tight and self-validating in light of popular fears of ideological invasion: on pain of being suspected and investigated as a potential member of communist networks, academics and administrators needed visibly and publicly to separate themselves from "Red" or "pink" colleagues.

This is perhaps one reason why academic professional organizations offered little effective resistance to these purges. The AAUP was caught off guard by the Washington investigations and this mistake snowballed. Because of the poor health and poor administrative skills of national leader Ralph Himstead, as well as its own administrative inertia, the AAUP took years to issue its report in defense of the dismissed professors. This delay held up all the other cases that the AAUP came to investigate, with the result that it was late and ineffective in defending dismissed professors throughout the 1950s. Its most vigorous actions occurred only after "the worst of McCarthyism was over" (Schrecker 1986, 315). Still, local departments and faculties often failed to organize on behalf of an accused colleague or to confront administrations or investigators on their behalf, precisely because defense of academic freedom was one of the AAUP's *raisons d'être* and that organization was thought to be more powerful than any handful of colleagues. The American Philosophical Association reasoned the same way. Though they issued statements in defense of academic freedom, one specifically in defense of William Philips at the University of Washington after his dismissal, the Eastern Division officially deferred to the AAUP in declining to investigate anticommunist attacks against philosophy professors (McCumber 1996, 39).

Higher education, like Hollywood and other American institutions, survived McCarthyism by largely succumbing to it. On the whole, few historians of the era find merit or courage in the behavior of the nation's academics or academic organizations. Many academics were sacrificed to anticommunist investigators or trustees, while those who retained their jobs avoided "controversial" topics. As Schrecker puts it,

> political reticence . . . blanketed the nation's colleges and universities. Marxism and its practitioners were marginalized, if not completely banished from the academy. Open criticism of the political status quo disappeared. . . . Teachers . . . played it . . . safe, pruning their syllabi and avoiding controversial topics. (Schrecker 1986, 339)

Even the AAUP's Ralph Himstead embraced the argument that drove the academic purges. Offering advice to his successor in 1955, he explained that he had been judicious and cautious in publishing reports in defense of controversial professors because "if the Association should get a left-wing or pro-communism tag, this would certainly end the effectiveness of the Association" (quoted in Schrecker 1986, 328). With leaders like Himstead, Schrecker could easily conclude, "the academy did not fight McCarthyism. It contributed to it" (ibid., p. 340).

The Case of Owen Lattimore

The case of Johns Hopkins sinologist Owen Lattimore illustrates how the prevailing anticommunist mood simultaneously implicated geopolitics, pedagogy, and espionage. (It is also worth examining because Lattimore's career was in many ways similar to Charles Morris's.) Lattimore's ordeal began to take shape with the rise of the Chinese Communist Party in 1949 after four years of civil war. When Mao Zedong rose to power and inaugurated the People's Republic of China, anticommunists in Washington fumed that China had been "lost" to communism, and they set out to find the guilty parties.

One target was the Institute of Pacific Relations (IPR), which, prior to the establishment of Eastern studies programs at American universities, functioned as a professional forum for scholars of Far East culture and politics. Trading on the left-leaning reputation of the institute and its journal, *Pacific Affairs*, Nevada Senator Pat McCarran seized some of the institute's records in February 1951. He used them to pursue a network of scholars and alleged communists who had, McCarran was convinced, colluded to cause the fall of Chiang Kai-shek and the rise of Mao. Such

charges had been circulated for over a year by McCarthy and his aide
Roy Cohn, but McCarran's investigation had more traction (Schrecker
1986, 162). Using standard techniques of questioning professors about
their political affiliation, relying on testimony from "professional" ex-
communists, and painting many suspects as Fifth Amendment communists, McCarran's IPR investigations helped to establish the charge that
the Truman administration had been duped by communist intellectuals
in its midst.

Perhaps because Lattimore had edited *Pacific Affairs* in the 1930s,
McCarthy decided to proclaim that he was "the top Russian espionage
agent" in the country (Klingaman 1996, 193). But it was not Lattimore's
communism (for he was not communist) but his pride and intellect that
drove him further into trouble. He chose to battle his accusers with logic
and legal maneuvers that only angered them and drove them on. He
was also accused by a former colleague, sinologist Karl Wittfogel, who
testified that "the absence of any trace of Marxism in Lattimore's writings
was a ruse" (Schrecker 1986, 165–66). In his senate testimony, Wittfogel claimed that Lattimore was among "those elements of the periphery
who are really closely coordinated and integrated into the movement,
but who try to promote the advantages of the movement without exposing themselves" (quoted in ibid., p. 166). Lattimore was therefore in the
impossible position of having to demonstrate the nonexistence of his alleged communist connections when, aside from that logical impossibility,
at least one of his accusers believed that such a nonexistence proof, were
it given, would have circumstantially incriminated him even more.

Lattimore was indicted twice. By 1955, both indictments had been
thrown out of court and he was technically victorious. Though he never
officially lost his tenured position, his career was nearly ruined. Johns
Hopkins abolished the school of international relations he directed; his
speaking engagements dwindled; his graduate students had trouble finding jobs; and Lattimore found it more difficult to publish. Passports and
security clearances also became difficult to obtain for him, his students,
and other China scholars named (if not investigated) during the years of
proceedings (ibid., p. 166). Eventually, Lattimore relocated to England
and spent the rest of his career at the University of Leeds.

A Climate of Fear, a Climate of Unreason

The Lattimore case, the general warnings from Raymond Allen, William
Buckley, and others, and the popular hysteria about the "Red menace"

sent a message to intellectuals that the merest appearance of sympathy with or even dispassionate interest in communism would not be tolerated in higher education. The message held that, especially at the federal level, academics would be powerless to defend themselves against the pretzel logic of conspiracy that usually structured investigations and interviews. The various standards of evidence, logic, and rules of rhetoric that an intellectual might command in a lecture hall or seminar room were visibly irrelevant to hearings conducted by House or Senate investigative committees who were just as much grandstanding for their electorate as they were pursuing real or imagined threats to American democracy. Few commentators made this clearer than editor of the *National Guardian*, Cedric Belfrage, who reported that McCarthy "talked almost pure nonsense at such roller-coaster speed that courtroom and parliamentary veterans gave up the struggle to bring any colloquy with him down to earth" (Belfrage 1973, 119).

Belfrage's views may not be objective because he himself, a British native, was deported for alleged communist activities and later named in the Venona files as a KGB confidant. Still, even when considered in light of his extreme antipathy for anticommunists, his observations and reports capture the often bizarre logic of anticommunism. Consider, for example, this editorial remark, cited by Belfrage, in Manchester, New Hampshire's *Union Leader:* "As is well known, this newspaper prints letters whether or not we agree with the letter-writers, so long as the letters do not libel anybody, are not obscene and are not written by Communists. This is what we believe freedom of the press means" (ibid., p. 262). Consider Belfrage's report of an exchange in 1952 between House Committee on Un-American Activities chair Harold Velde of Illinois and Tom O'Connor, managing editor of the magazine *Compass:*

VELDE: Are you a Communist now?
O'CONNOR: No.
VELDE: Were you a year ago?
O'CONNOR: No.
VELDE: Were you ten years ago?
O'CONNOR: No.
VELDE: I can draw only one inference, that you are not only a past member of the CP [Communist Party], but you continue to be a member and that you are an extreme danger to the country. (p. 178)

The same year, philosopher Barrows Dunham lost his job at Temple University, Lattimore remained mired in hearings, and Velde announced that he would organize probes of Johns Hopkins, Harvard, MIT, and the University of Chicago (p. 183). A year later, Velde told the *New York Times* that "the field of education" would be "a very fertile field for investigation" (quoted in Klingaman 1996, 379). So, too, was the field of philosophy of science. As we see in the next chapter, at least two leaders of the Unity of Science movement had already become focal points of interest for Hoover's FBI.

13

Anticommunist Investigations, Loyalty Oaths, and the Wrath of Sidney Hook

Education was indeed, as Velde put it, "a very fertile field" for anticommunist investigations in the 1950s, if only because so many academics, if they had not actually joined the party, had been sympathetic to communism in the 1930s. Because the University of Chicago and Harvard are private institutions, Charles Morris, Rudolf Carnap, and Philipp Frank were sheltered from some of the anticommunist pressures applied to their colleagues at public institutions. Still, all three experienced various kinds of anticommunist pressures in the late 1940s and early 1950s.

Hearings at the University of Chicago

Morris witnessed two investigations at the University of Chicago, the first in 1935 before Carnap had arrived at the university. That investigation was precipitated by radical students who had distributed revolutionary literature to the city's poor. At the same time, the university's May Day celebration of 1934 drew the press's attention to "the enormous gap that had opened between [the] University's students and the climate of opinion prevailing in the city" (McNeill 1991, 62–63). The university's reputation as a hotbed of radicalism was also fed by the early anticommunist book *The Red Network* (Dilling 1934). The issue ignited when Charles Walgreen, of drugstore fame, wrote to President Hutchins that he was withdrawing his niece from the college: "I am unwilling to have her absorb the Communist influences to which she is assiduously exposed" (in McNeill 1991, 63). William Randolph Hearst's paper, the *Herald Examiner*, printed the letter (even before Hutchins responded) and within weeks formal investigations and hearings about communism at the university were under way. The investigators were Illinois state legislators and their main targets were political scientist Frederick L. Schuman (who, Walgreen believed, was indoctrinating his niece), English professor Robert Morss Lovett, and economist Harry Gideonse (Schrecker 1986, 70). Gideonse, future president of Brooklyn College, had at least one connection to Morris. He edited the series *Public Policy Pamphlets* in which Morris had debuted his theories about democracy and pragmatism a year before (Morris 1934b).

The hearings went nowhere. Walgreen's niece testified that she had not in fact attended Schuman's classes, while another student's anticommunist testimony was impaired by his well-known "psychological instability" (McNeill 1991, 64). The author of *The Red Network*, Chicagoan Elizabeth Dilling, reduced credibility further by testifying that, besides the Chicago faculty, Supreme Court Justice Louis Brandeis and Senator William Borah were communists. On the other hand, testimony from Hutchins and others readily persuaded investigators and the public that free intellectual exploration, and not political subversion, was taking place at the university.

The second investigation took place in 1949, the same year in which the University of Washington's professors were fired. These Chicago hearings were also initiated by the American Legion, which in 1947 prompted the state to create a commission of lawmakers and citizens, the Broyles Commission, to investigate communist influences in education. The commission soon decided that the University of Chicago was sufficiently pink for investigation, especially after students there loudly protested the commission's proposed bills against the hiring of communist academics. The commission held hearings on campus and heard testimony from professional anticommunists (including J. B. Matthews, Benjamin Gitlow, and Howard Rushmore) who testified to alleged connections between university faculty and the Communist Party (Schrecker 1986, 113).

Like the Walgreen investigations, the hearings quickly lost momentum. Hutchins again forcefully stood up to the commission's investigators and turned the tables, describing their agenda as "the greatest menace to the United States since Hitler" (quoted in ibid.). Though he was at the end of a long and combative tenure as Chicago's president, Hutchins probably inspired his faculty to pull no punches as they testified and responded to the commission's charges. According to Schrecker, this victory was at least assisted by the fact that the Broyles Commission's final report was "a masterpiece of illiteracy" (ibid.).

Charles Morris, China, and the Open Society of Open Selves

Morris was not involved in these hearings, though he brushed up against circumstances that proved dangerous for others. In the late 1940s, for example, he traveled abroad to conduct international studies of personality types and value choices. Though his original plan to return to Russia (as part of a tour through China and India) did not materialize, he traveled to India and China, leaving China in late 1948 and narrowly avoiding revolutionary violence in Peiping, where he had been staying.[1] Morris was surely aware of the anticommunist controversy over China and apparently contributed to the Lattimore Defense Fund.[2]

During years when even an anonymous letter or phone call to the FBI could trigger an investigation, Morris's travels to China and his contacts there were risky. In the early 1950s, for example, his contact in China, Professor Li An-che, was foiled in his efforts to publish a book in America on Tibetan religion. Morris was told by the publisher that "it became rumored that Professor Li was holding a high position in the communist Chinese government." Whether or not this rumor was true, the manuscript became a hot potato and the foundation that was to oversee its publication, the Wenner-Gren Foundation for Anthropological Research, happily gave it to Morris. The foundation's director of research, moreover, made a point to cover himself in print: "Our sending this to you, you understand, could not be considered as consent on our part that it be published. It is being sent because of the interest which you have expressed and with hopes that it will be helpful to you."[3]

Morris also confronted the issue of loyalty oaths that swept through most public universities in the late 1940s and early 1950s. Though he

---

[1] Morris wrote to his friend Li An-che from Peiping on November 13, 1948, reporting that the city had displaced Prague as his favorite city. One month later, on December 17, Li An-che replied, "I trust that you left Peiping early enough to escape the besiege!" (Morris to Li, November 13, 1948; Li to Morris, December 17, 1948, CMP).

[2] George Boas to "the contributors of the Lattimore Defense Fund," September 2, 1955, CMP. Morris was also recipient of a mass mailing in support of sinologist John W. Powell, who described abusive treatment by FBI and customs officials on returning to America after a stay in China. Powell attributed this to the fact that he was on record for claiming that most people's lives were improved under the Communist regime. Unlike most other mass-mailed letters in Morris's files asking for financial or letter-writing support for an intellectual or humanitarian caught in a political or international snare, this one lacks annotations. Morris usually annotated these appeals to record what actions he took – in this case, apparently, none.

[3] Fejos to Morris, November 17, 1954, CMP.

was apparently willing to sign a mandatory statement of patriotic loyalty when he taught at the New School for Social Research in the 1946–47 academic year, some ten years later he refused a temporary position from the University of Illinois's Institute of Communications Research in protest of the university's oath requirement.[4] He wanted to accept the position, which had been arranged partly by his colleague Charles Osgood at the institute. But he decided in advance that were he offered the position, and were Illinois governor Stratton to sign into law the oath requirement (the latest recommendations from the state's Broyles Commission), he would decline the offer in protest.

Osgood encouraged Morris to take a public stand on behalf of academic freedom and constitutional rights:

> I would sign the appointment papers [and thus accept the position] and then, if you wish to go through as you told the Governor, I would later refuse to sign the special Loyalty Oath – this would automatically fire you! But . . . I would make sure that the Illinois newspapers are notified in advance of your plans and that you make a public statement. . . . [T]he whole point, as I see it, is to impress on the public mind the distinction between disloyalty and unwillingness to be pushed into essentially unconstitutional behavior by politicians.[5]

Morris disliked the plan. "[I]f I am not a state employee[,] I see no moral obligation to become one just in order to provoke trouble," he replied. Besides, he reasoned, "it would be at least something like a protest not to come to the University under the present circumstances." Morris seemed to recognize that this "something like a protest" would be a relatively weak, if not imperceptible, gesture. "I hope you don't feel that I am letting you down," he told his more defiant colleague.[6]

Osgood's defiance, however, was paper-thin. After the oath requirement became law, he wrote to Morris,

> I shall probably sign it – I'm afraid it would be simply playing into the hands of those who want to get rid of liberals to do otherwise, and further I shall be way off in Arizona on Sabbatical and thus in a poor place to fight. It hurts my conscience to do so, though.[7]

Osgood's about-face illuminates how at least one professor's resolve to stand up to loathsome and "essentially unconstitutional behavior by politicians" was ground down by the realities of circumstances and the hardball

4 Dean Clara Mayer, New School for Social Research, to Morris, June 13, 1946, CMP.
5 Osgood to Morris, July 11, 1955, CMP.
6 Morris to Osgood, July 14, 1955, CMP.
7 Osgood to Morris, July 19, 1955, CMP.

politics of anticommunism. Refusing to sign a loyalty oath only invited suspicions and, usually, a "fight" with investigators, administrators, or colleagues that could jeopardize one's professional and social standing.

With Lattimore and other persecuted academics as a model, Morris could legitimately have looked at himself and seen a potential target for anticommunists. Besides his visits to China and Russia (some fifteen years before), he was an editor for the journal *East-West*, which was devoted to synthesis of intellectual trends and ideas; was a member of the American Humanist Association, whose atheism and "naturalism" most anticommunists would find noxious; and belonged to or supported several groups promoting worldwide social and cultural unification. For example, Morris's personal papers contain several "Declarations of Interdependence" – small pamphlets designed by one Otto Tod Mallory containing tenets of international, collectivist humanitarianism. On their covers appears a spoked wheel and the motto, "Each spoke depends on every other."

Morris also went on record against the McCarran Act (also known as the Internal Security Act) of 1950, which legislated anticommunist measures against Moscow and its "world communist movement." This movement's purpose, the act specified,

is, by treachery, deceit, infiltration into other groups (governmental and otherwise), espionage, sabotage, terrorism, and any other means deemed necessary, to establish a Communist totalitarian dictatorship in the countries throughout the world. (quoted in Klingaman 1996, 195)

In May 1952, Morris and Carnap were among the seventy-five prominent signatories of an open letter to congressional candidates. Sponsored by the New York–based National Committee to Repeal the McCarran Act, the letter turned the charge of "thought control" back against anticommunists as it complained that the act

has already led to serious infringements of the Bill of Rights [and] is responsible for the intolerable situation in which Government agencies, in a manner all too reminiscent of Nazi Germany, are already preparing concentration camps, are holding thought-control hearings, are denying passports to citizens, and are deporting and refusing admission to aliens.[8]

In the first half of the 1950s, it appears, Morris signed at least two other open letters to national legislators encouraging the act's repeal and condemning the "climate of fear," the "atmosphere of repression," and the

---

[8] Press Release, National Committee to Repeal the McCarran Act, May 5, 1952, CMP.

attack on "the civil liberties guaranteed to us by the Bill of Rights" that it legitimated.[9]

The Open Self

Morris's work and political stances did not earn him a leftist reputation like those of Harvard's Harlow Shapley or MIT's Dirk Struik. In one sense, Morris was not intellectually disposed to radicalism because he regularly approached arguments with the goal of synthesizing or reconciling traditionally opposing points of view. This is why Morris's activities and professional memberships reflect elements of both intellectual leftism as well as 1950s-style celebrations of individualism and anticollectivism. He participated, for example, in Paul Mandeville's "Human Destiny conferences," which promoted "democratic processes at grass roots levels" and defended against

habits [that] have been set up during about four years of national dangers that required obedience to top planning and sometimes excessive regimentation in order to execute emerging national policies. . . . The free world is now looking to the United States for long term leadership. In this situation an urgent need is a healthy body politic.

Morris was by 1951 a board member for Mandeville's ongoing "Library Education Conference," whose goals included promoting libraries as tools for "individual and community self help."[10]

Morris's most visible gesture toward anticommunism and anticollectivism was his book *The Open Self* (1948) the title of which drew on the "open society" revered by critics of planning and totalitarian, "closed" societies. (Karl Popper added to his antitotalitarian credentials with his book *The Open Society and Its Enemies.*[11]) Still, not much more than the title of Morris's book would have appealed strongly to Cold War libertarians. Morris praised openness and criticized totalitarianism but remained fully

---

[9] Open Letter to the Chairman and Members of the Senate Judiciary Committee (dated May 1955 by Morris), CMP; Open Letter to Senator Thomas C. Hennings, Chairman, Senate Judiciary Subcommittee on Constitutional Rights, undated, CMP.

[10] Documentation of Morris's participation in the Human Destiny Conferences is contained in the Charles Morris Papers.

[11] Morris appeared with Popper in a colloquium at Stanford University's center for Advance Study in the Behavioral Sciences in 1954 when he was at the center for the academic year. The colloquium series was titled "Competition for Men's Minds," and Popper and Morris both spoke about "The Appeals of Liberalism" (Center for Advanced Study in the Behavioral Sciences, circular, March 27, 1954, CMP).

committed to planning and scientism. Drawing on Neurath's late ideas about "planning for freedom" (Neurath 1942) as well as Dewey's image of a society either drifting or proceeding deliberately and intelligently, Morris argued that "drift is no alternative. To drift is to renounce the human task of man-making." Therefore, planning is inevitable and "*the only relevant issue is whether to plan for the closed or the open society*" (Morris 1948b, 151). Those who fear "serfdom" and reject planning altogether, Morris wrote as he nodded to Hayek, "fail to distinguish between the kind of planning which leads to the closed society, and the kind . . . which issues in the open society" (ibid., p. 152). *The Open Self* called for comprehensive social planning that would yet "have no over-all blueprint for the future":

Planning for freedom is diametrically opposed to planning for the closed society. Where the latter will plan to control the agencies of communication, the supply of food, and the sources of anxiety so as to produce fearsome, conforming, and constricted selves, the open society will plan to extend knowledge and skill and means of subsistence so that each member of society becomes to the limit of his or her ability a mobile, autonomous, spontaneous initiator. For the closed society persons are instruments to be used in order to gain power; for the open society persons are ends and social agencies are means to personal growth. . . . Its planning is the preparation of the social conditions which nourish and sustain the creative individual. (pp. 153–54)

Science and a well-developed "science of man" were important parts of Morris's vision. Not only would the community of scientists be an exemplar for the open society. "The methods of science will themselves be methods by which the open society improves and corrects its existing institutions" (p. 154).

Because Morris's program drew themes from the Unity of Science movement – specifically, the use of science for human enrichment, the role of scientists as partial exemplars in the larger, international, and more unified world, and the kind of deliberate yet open-ended management of future possibilities that Neurath opposed to Kallen's totalitarian vision of unified science – the liabilities of his presentation help to indicate the difficulties that the movement faced in these years. In the eyes of the many postwar intellectuals who had rejected scientism and decided that science was unable to illuminate, much less help to control, the path of history and the often destructive forces driving human actions, Morris's *Open Self* could easily have seemed incoherent. It called for a society that was *both* planned and free, *both* scientific and attentive to human values. If, as Morris wrote, "science provides a major method for controlling the agencies of the open society" (ibid.), many would ask of his book the question Kallen put to Neurath's conception of unified science: How is that different from top-down, totalitarian control?

Preventive Measures and a Sense of Caution

Besides Morris's public stands against the anticommunist "climate of fear" in the 1950s, there is some evidence that he believed he needed to defend himself personally. His student and lifelong friend Howard Parsons may have prompted Morris to reflect about his own situation. In 1948, Parsons was teaching at the University of Tennessee, working on behalf of the Progressive Party and the Wallace campaign, and raising eyebrows because of his leftism and his interests in dialectical materialism. He described his problems to Morris:

> I am counting on being rehired here, unless the anti-Wallace feeling gets too hot. I have been called down twice now – once by the Humanities Dean, once by the head of Counseling – for letting certain facts and convictions be known – though I thought I was cautious to a fault in class! Now I hear the American Legion calling me a Communist and the State legislator planning an investigation.[12]

A few years later, another student's problems surely prompted Morris to think about his own situation. He wrote an affidavit on behalf of Harold Josif, who had been a student and assistant of Morris's who later worked in government foreign service. Apparently, Josif encountered some political trouble and asked Morris for help. Morris wrote "to whom it may concern" that Josif was an "able student and assistant, intelligent, cooperative and responsible." As for the question of communism,

> there was never in his thoughts or in his attitudes the slightest trace of anything that could be called pro-Communist in any degree. Indeed his personality as I know it seems diametrically opposed to the Communist totalitarian mentality.

That was why, Morris implied, he was "very glad when [Josif] entered the Foreign Service, for he seemed to be exactly the kind of person America needed in the field." He concluded the affidavit hoping that Josif "will be enabled to continue his career without interruption."

Then Morris's thoughts turned to himself. In the margin of this one-page letter, he wrote,

> Perhaps it may be relevant to state that I have opposed the totalitarian attitude ... in my book *Paths of Life....* My last work, *The Open Self,* is a defense

---

[12] Parsons to Morris, April 26, 1948, CMP.

of American democracy against the forces that threaten it from within and without.[13]

Though neither *Paths of Life* (1942) nor *The Open Self* could have been recognized as an antitotalitarian manifesto, Morris's clearly wished to appear antitotalitarian and believed it prudent to go on record this manner.

The reputation of the Unity of Science movement as pink or "communistic" was probably one reason Morris felt vulnerable. In 1957, Morris received a letter from Horace Kallen. Surprising as it may have seemed, since Kallen had so vociferously attacked Neurath and the Unity of Science movement years before (as discussed in chapter 9), Kallen now asked Morris to help him to arrange translation and publication of Neurath's writings in English. Otto's widow, Marie, supported the plan but worried that it might be necessary in America's anticommunist climate to selectively edit the translations. "She is afraid," Kallen told Morris, "of having people charge Otto with being a Communist."[14] As usual, Morris agreed with both sides. He agreed with Kallen that there would be no good reason to censor Otto's writings but he also upheld Marie's worries: "I share Marie's sense of caution," he replied to Kallen. "I have met people who think Otto, and indeed the whole unified science program, is communistic."[15] Morris's comment is as much of a smoking gun as one needs to establish some causal influence between Cold War anticommunism and the professional decline of the Unity of Science movement. If even some persons in the field sincerely believed that "Otto, and indeed the whole unified science program" were "communistic," many who had once participated in the movement in the 1930s would have tempered their enthusiasm, if not moved in other professional directions altogether. It was well known, after all, that links to suspiciously pink organizations or individuals that might pique the interest of anticommunist investigators did not need to be substantial. As an editorial in *The Nation* explained when defending some of McCarthy's suspects, "[I]t is enough that at one time or another they addressed groups that were subsequently labeled subversive or that their names appeared on letterheads along with others thought to be Communists."[16] Almost all the names of leading postwar

---

[13] Morris to "Whom It May Concern," May 6, 1953, CMP. Morris pointed out that he was "personally attacked in a book by the English Communist, Maurice Cornforth." Cornforth's book and its probable role in helping establish the political reputation of the Unity of Science movement are examined in chapter 15.

[14] Kallen to Morris, September 24, 1957, JRMC.

[15] Morris to Kallen, October 8, 1957, JRMC.

[16] *The Nation*, March 25, 1950, 261.

philosophers of science once appeared on letterheads of the Unity of Science movement along with "Otto Neurath."

Morris himself neither abandoned nor aggressively pursued the movement during the Cold War. With Carnap, he shepherded the last of the monographs Neurath had commissioned through publication in the 1950s and '60s. But he did not rally his colleagues to revive the *Encyclopedia* and he left to Philipp Frank the job of running the postwar Institute for the Unity of Science. In part, this was surely due to the terrible reception of Morris's opus *Signs, Language, and Behavior* (1946) and his own professional movement away from philosophy of science and toward social science (as discussed in chapter 16). It was probably also due to the facts behind another rumor that Morris surely heard in the early 1950s: Carnap and Frank were being investigated as potential communist operatives by FBI agents.

FBI Investigations of Frank and Carnap

Though Frank was neither a communist nor an admirer of Moscow or dialectical materialism, he took them seriously as political and intellectual realities and, like Morris, urged collaboration and cooperation between the movement and Marxist intellectuals. As shown in chapter 11, Frank continued to address matters ideological in his writings as the Cold War began. Whether or not it was due to Frank's nonchalance about "controversial" topics, a rumor crossed J. Edgar Hoover's desk in the summer of 1952. Hoover found it sufficiently worrisome and wrote to the Pentagon's chief of national security that

an allegation has been made that captioned individual [i.e., Frank] came to the United States for the purpose of organizing high level Communist Party activities. It was stated that the source could not be evaluated as to reliability and no indication of the date of subject's visit to the United States was mentioned.[17]

Following usual practices, Hoover then instructed his agents to search indices of the bureau's files for mention of Philipp Frank in other investigations. They soon learned that Frank had already been noted as author of a biography of Einstein (1947) that was favorably reviewed in

---

[17] Hoover to Pentagon, August 13, 1952. This and the following quotations from FBI correspondence and reports are taken from FBI files on Frank and Carnap, obtained by the author under Freedom of Information Act Requests.

the communist newspaper the *Daily Worker*.[18] Hoover then instructed the local Washington Field Office to examine records of the Immigration and Naturalization Service (INS), whereupon they learned that Frank was teaching at Harvard. That office then sent this information to the Boston Field Office, along with the discovery that Frank was mentioned in William Malisoff's file as a participant in intellectual discussions Malisoff hosted at his New York apartment.[19] Since Malisoff was a known confidant of the KGB, it could easily have appeared that this allegation about Frank was true.

At the Boston office, suspicious information about Frank continued to pile up. From their own files, agents learned that he was a colleague and friend of Harvard astronomer Harlow Shapley, whom the Communist-turned-informer Louis Budenz (once manager of the *Daily Worker*) had fingered as a "concealed communist." Examination of Shapley's mail revealed that Frank and Shapley had corresponded and that Frank went to hear Shapley speak at a 1945 meeting of the Massachusetts Council of American Soviet Friendship. Another informant described Frank as a member of the Young Progressives of Massachusetts. The Young Progressives, like the American-Soviet Friendship, was reputed to be a front group secretly funded and controlled by Moscow.[20]

One particular piece of data from the INS addressed the question raised in the original allegation: When did Frank enter the United States? According to the INS data, he entered first as a tourist in 1938 and then again with an immigration visa he obtained in Cuba in August 1939. But a different – and much more alarming – answer to the question came from the Harvard Corporation Office, where agents interviewed secretaries about Frank and his employment history. According to them, Frank was touring the country from 1932 to 1938 while lecturing at "approximately 20 (unnamed) colleges and universities in the United States."[21] Were the allegation about Frank true, this discovery revealed Frank's modus operandi: While ostensibly lecturing about physics and philosophy, he was actually making contacts and organizing "high level Communist Party activities" around the country.

---

[18] The file cites the *Daily Worker* of March 19, 1947, page 11, but does not specify to which of the several local editions it refers.

[19] New York to Hoover, August 31, 1953. Frank probably met with Malisoff's group when he was in New York for meetings of the annual Conference on Science, Philosophy and Religion which he regularly attended. See Frank 1950.

[20] Boston to Hoover, April 14, 1953.

[21] Ibid.

Hoover was nearly convinced. He informed the director of the CIA that Frank had been an associate of a known subversive (Malisoff) and of a suspected subversive (Shapley) and that he participated in known communist front groups. Presented with contradictory information about the dates of Frank's emigration, Hoover apparently believed the allegation more than the evidence from the INS. Instead of having his agents check the information they obtained from Harvard, he requested that they return to the records of the INS and, while they were at it, to "determine, if possible, the length of the subject's residence at Havana, Cuba." Frank was now the subject of an official "security-type" investigation.[22]

In about three weeks, this suspicious picture of Frank began to fall apart. Boston agents learned that the dates reported by Harvard were incorrect and that Frank had been in the country only since 1938.[23] In addition, interviews with Frank's colleagues and with former communists turned up nothing. Those who knew Frank were sure he was not enamored with communism or with Moscow. The Boston office's secret contacts inside the communist cells at Harvard and MIT had never even heard of him. According to the popular logic of anticommunism, however, absence of evidence is itself strongly suggestive. Boston agents figured that Frank might simply be concealing his activities and working in the party underground, unbeknownst to official party members. What they needed, therefore, was information about Frank's life and activities in Prague prior to his emigration and before Frank assembled whatever cover he might be working under in America. This led them to set up interviews in Boston and Bloomington, Indiana, with individuals who had known Frank in Europe.[24]

The Boston interviews again turned up nothing. Though one informant took Frank for a "muddle-headed" intellectual who might well be a communist because "he is Jewish and the Jews have a penchant for Communism," the Boston office was less impressed with this slur than with the fact that Frank was an associate of Malisoff. For that reason, at least, they urged Hoover to instruct the New York Office to once again pursue Czechoslovakian sources available in New York. The Indianapolis interviews did not clarify matters, either. Only one informant knew Frank years before in Prague, and he claimed that Frank was "very anti-Nazi." For that reason, he suggested, Frank "might possibly have joined the Communists

[22] Hoover to CIA, May 8, 1953.
[23] Boston to Hoover, May 28, 1953.
[24] Boston to Hoover, August 18, 1953.

in fighting against Nazism," though he had no knowledge that this was so.[25]

The Logical Empiricist Network

In the first months of 1954, Frank's investigation continued to wind down. Still more communist informants in Boston had no knowledge of Frank, and still more of his colleagues reported that he was "anticommunist." One altogether dismissed the FBI's interest in Frank: At this late point in his life – Frank was then about seventy – he was merely "a talkative old man who could not possibly be considered dangerous to the security of the United States."[26] By April of 1954, almost two years after Hoover received the original allegation, the investigation of Frank was closed.

Though the investigation failed to connect Frank with active communist operatives, it did reveal his connection to another philosopher whose views and writings piqued the FBI's interest. The informant in Indianapolis who had known Frank in Prague mentioned that Frank was then in contact with Rudolf Carnap. The agents took down this name, learned of Carnap's whereabouts, and then informed Hoover that Carnap was a University of Chicago professor now at the Institute for Advanced Study in Princeton, New Jersey.[27] Hoover instructed the nearby Newark field office to prepare to interview Carnap about Frank. They could not simply approach Carnap directly, however, for were he a part of Frank's alleged communist network, the investigation would be compromised. Agents first had to investigate Carnap and learn whether he himself was a communist.

Newark agents confirmed that Carnap was living in Princeton and checked with local police and a credit bureau to see whether he was known as a communist or agitator. They learned nothing incriminating from these sources, but they did find Carnap's name in an issue of a local New Jersey newspaper, the *Evening Press*. Along with many other intellectuals, Carnap declared himself in favor of clemency for Julius and Ethel Rosenberg. Next to statements by others, Carnap published his own: "I feel that the severity of the sentence is out of proportion to the

---

[25] Unknown to Boston, October 15, 1953; Boston to Hoover, October 22, 1953; Indianapolis to Hoover, January 15, 1954.
[26] Boston summary file, February 19, 1954; Unknown to Boston, January 19, 1954; Boston to Hoover, January 1, 1954.
[27] Indianapolis to Hoover, January 15, 1954.

actual damage which could possibly have been done."[28] In the Manichean
metaphysics of Cold War anticommunism, a published remark such as
this, expressing sympathy with the Rosenbergs, counted as evidence that
Carnap might indeed be an active communist. Newark urged Hoover
to have the Washington office search their indexes for "Rudolf Carnap"
and advise whether they still wished to interview him about Frank. They
correctly suspected, it would appear, that Carnap was about to become a
subject in a separate investigation.

The Carnap Investigation

"Authority to interview Rudolf Carnap concerning the Captioned [i.e.,
Frank] is denied," Hoover replied.[29] It was not only Carnap's support of
the Rosenbergs that raised suspicions, however. One of Hoover's high-
level officers at the bureau had also received, apparently coincidently, a
report about Carnap from the American Legion.[30] The report, which had
already passed through both the House Un-American Activities Committee and the Senate's Subcommittee on Internal Security, listed eighteen
instances (including the Rosenberg statement) where Carnap's name appeared publicly in support of or as an official sponsor of leftist causes,
organizations, or individuals. Ten of these public signatures, the report
made plain, appeared in the communist *Daily Worker*.[31]

The FBI determined that these items were already noted in their indices of other investigations. (Agents routinely scanned the *Daily Worker*
for names of those mentioned in connection with communist causes.)
Still, the American Legion report put Carnap in focus and led them to cull
yet more – and more worrisome – information about him from their files.
Interception of Malisoff's mail from years before, for example, revealed
that Carnap was one of Malisoff's correspondents. They also learned that

[28] Quoted in Newark to Hoover, March 9, 1954. The Rosenbergs were electrocuted in
June 1953 at Sing Sing Prison. Carnap's views were not unique. Some 10,000 protesters
gathered in Union Square in New York City on the day of the execution (Klingaman
1996, 325).
[29] Hoover to Newark, April 7, 1954.
[30] The officer was one A. H. Belmont, whose name sits three (of fourteen possible) spaces
from the top of the routing list for memos in Hoover's office. The list appears to measure
proximity to Hoover himself, since the top name is Tolson, presumably Clyde Tolson,
known to have been inseparable from Hoover during their joint tenure at the FBI.
On the American Legion's practice of forwarding names of suspected communists to
investigators and university administrators, see Schrecker 1986, 68.
[31] HUAC to Jenner, October 19, 1953.

a State Department memo from 1939 included Carnap's name in an anonymous informant's list of "some Communist liaison agents in various countries of the world." Carnap in fact lived in Prague from 1931 to 1935, and his name was given by this informant with a Prague address.[32]

As the investigation began, several offices around the nation were involved. These included Hoover's office as well as the field offices in Washington, D.C., Newark, Chicago, and Indianapolis. In addition, since Carnap had just taken a position at UCLA, overall control of the investigation was given to the Los Angeles office. At nearly all of these offices, September 16, 1954, might have been called Rudolf Carnap Day at the FBI. In a flurry of memo writing, agents across the country shared information that continued to build a picture of Carnap as a potential political subversive. The Newark office wrote to Hoover (sending carbon copies to Chicago and Los Angeles) with some additional information about Carnap's ill health during his stay at the institute; the New York office reported that Carnap was seen with Harlow Shapley (at a dinner party); and the Chicago office announced that Carnap had donated five dollars to MIT mathematician Dirk Struik's defense committee after Struik was indicted in 1951 under the anticommunist McCarran Act. Chicago also documented Carnap's support of Wallace and the Progressive Party in the election of 1948.[33]

The FBI's list of activities and causes that Carnap supported grew longer and more detailed as these memos shuttled across the country. Beginning in 1949, Carnap's name appeared on two public statements on behalf of the International Workers Order (as a signatory to a brief amicus curiae and to an open letter to Thomas Dewey, governor of New York State, where the IWO was being prosecuted). It appeared twice in the *Daily Worker* and once in the *New York Herald Tribune* against the anticommunist McCarran Act. Carnap also supported the American Committee for Protection of the Foreign Born in the *Daily Worker* and signed an open telegram to the Attorney General in support of deportees. Carnap also called for official recognition of the People's Republic of China and gave his name five times in support of the American Peace Crusade. Referred to by anticommunists as Moscow's "peace offensive" (allegedly designed

---

[32] Hoover to Newark, April 7, 1954. This State Department item is mentioned elsewhere in Carnap's file with a date of 1949. It is not clear which is in error. This item did not elicit much concern in Carnap's investigation, most likely because the informant's reliability was suspect.

[33] Newark to Hoover, November 24 and September 16, 1954; Chicago to Hoover, September 16, 1954.

to ease a Soviet takeover by lulling Americans into pacificism), the crusade consisted of various conferences and letter-writing campaigns in support of international peace, disarmament, nonmilitary treatment of tensions in Korea, as well as free-speech rights for citizens wishing to promote these causes (see Fig. 6).

To obtain testimony they might need should they choose to prosecute Carnap, Agents interviewed his friends, colleagues, and neighbors in Chicago, Princeton, and Los Angeles. During the summer and fall of 1954, agents made roughly ten visits to Princeton.[34] Asking whether anything about "the subject" suggested "communist activity," they learned that Carnap was a famous, dedicated, and stereotypical intellectual ("highly impractical, eccentric, and very engrossed in his subject") who often worked in bed due to his chronic back pain. None reported "anything indicating that subject was a Communist sympathizer of any kind." One who claimed to have known Carnap first in Prague in the early 1930s said, "the subject is interested '99% in scholastic matters and has little or no interest in politics of any kind.'" Given the many causes that Carnap supported in the pages of the *Daily Worker*, however, that statement may have seemed plainly false and perhaps deliberately misleading.

In September, an agent from the Chicago bureau interviewed two informants, most likely faculty at the University of Chicago. Both supported the emerging picture of Carnap. One said he was "an introvert who was not well known by any persons in the university circles," and another said that he "appeared to be completely wrapped up in his capacity as Professor . . . and very rarely deviated from his study habits."[35] Several months later, agents in Chicago interviewed the Chicago police department, two landlords from whom Carnap and his wife had rented apartments, and the Chicago Credit Bureau, where Carnap maintained an account. In Los Angeles, agents confirmed Carnap's arrival at the university by interviewing payroll and personnel workers.[36]

Case Closing

By the spring of 1955, the FBI had no proof that Carnap was engaged in subversive political activity. Still, his file reads in places like a study in

---

[34] Interviews in Carnap's file are dated May 17, June 10 and 25, August 3 and 28, September 8, October 27, November 8, 9, and 30, 1954.

[35] Newark Summary Report, June 3, 1955.

[36] Chicago report, June 29, 1955; Los Angeles to Hoover, October 14, 1954.

2,000 In Sciences and Arts Ask for a Big 5 'No War' Pact

Two thousand persons in the cultural and professional fields have already signed "An Appeal for Peace," addressed to the President and the Congress, Professor Henry Pratt Fairchild, secretary of the National Council of the Arts, Sciences and Professions, announced in a preliminary report. The signers come from many sections of the United States and represent a wide variety of pursuits in American culture and science.

The appeal urges "our government to initiate negotiations among the United States, the United Kingdom, France, the Soviet Union and China, in order to arrive at agreements which will end the threat of war and lay the plans for an enduring peace."

Among the eminent signers of the appeal are the following scholars and educators: Prof. Derk Bodde, Prof. Rudolf Carnap, Dr. W. E. B. DuBois, Prof. Robert Morss Lovett and Prof. Pitirim Sorokin. Among the religious leaders are Rabbi Abraham Cronbach, Dr. Fleming James, Sr., and Rev. John Paul Jones. In the world of the arts are composers Ernest Bloch, Charles Ives and Wallingford Riegger; painters, Peter Blume and Raphael Soyer; film and theater people, Peter Lawrence, Howard Koch, Victor Samrock and Martin Wolfson. The scientists and physicians who have joined in the appeal include Prof. Anton Carlson, Prof. I. M. Kolthoff, Dr. Leo Mayer, Dr. John P. Peters, Prof. Theodore Rosebury and Dr. Edward L. Young.

The Appeal for Peace states that "the record of history has shown that an armaments race has never led to peace . . . and that there is a way for all nations to live side by side in peace, notwithstanding the difference in their social systems. That way lies in negotiations among the major powers."

A partial list of additional signers of the appeal follows:

Scholars and educators: Dr. Charles C. Davis, Prof. Harl R. Douglas, Dr. Arnold Dresden, Prof. Royal Wilbur France, Prof. Albert Guerard, Sr., Dr. W. A. Hunton, Prof. Oliver S. Loud and Dr. F. L. Marcuse

Also Dr. Scott Nearing, Prof. Erwin Panofsky, Dr. Vida S.

Scudder, Randolph Smith, Dr. Paul Sweezy, W. Lou Tandy, Prof. Leland H. Taylor, and Prof. Eda Lou Walton.

Scientists and physicians: Dr. Carlton Goodlett, Dr. George Hindes, Dr. Sol Lone, Dr. Clifford Sager and Prof. Frank Weymouth.

Religious leaders: Rev. Charles B. Ackley, Rabbi Robert Goldburg, Rev. Spencer Kennard and Dr. Harry F. Ward.

Lawyers: Harold Cammer, Stanley Faulkner, Hon. Clemens J. France, Samuel Menin and Judge Edward P. Totten.

Writers and journalists: James Aronson, Millen Brand, Angus Cameron, Lester Cole, Shirley Graham, Henry Kraus.

Also Albert Maltz, Sam Moore, Shaemas O'Sheal, William Reuben and Ira Wallach.

Artists, musicians, Film and theater people: Herbert Biberman, Serge Chemayeff, Harry Gottlieb.

Chicago Quakers Asks Fight To Defeat the McCarran Act

CHICAGO, March 24.—Public support for a campaign to repeal the McCarran Act is urged in a letter to the Chicago Daily News by Charles Hait, chairman of the Social Order Committee, Meeting of Friends (Quakers). Hait's letter is topped (March 19) by a three-column headline: "Alarmed by 'Concentration Camps' Set Up in U. S. Under McCarran Act."

Hait writes in part:

"Instead of requiring that the prosecution prove a criminal act beyond a reasonable doubt, it will only be necessary (under the McCarran Act) to assert that a reasonable doubt exists about the individual's intent.

"The imprisoned need not be confronted by his accusers; the evidence need not be revealed to him. Such practices are contrary to time-honored democratic traditions and directly disregard the constitutional safeguards of the civil liberties of those holding unpopular opinions. The camps are concentration camps.

"An act which provides for the imprisonment of thousands in the absence of a criminal act is opposed as immoral and unjust. Of no less consequence is the fact that the climate of conciliation for which we strive is further repressed.

"The appearance in fact of concentration camps in the United States and the emergence of extralegal procedures which undermine basic rights can only be cause for grave concern.

"You are urged to share this concern. The moral issues are quite clear. Lest we be charged with 'guilt by acquiescence' appropriate actions are required. First among these is an effort to repeal the McCarran Act."

NIEMOLLER SAYS SOVIET PEOPLE STRIVE FOR

By DON WHEELDIN

LOS ANGELES, March 24.—In the Soviet Union people are striving for peace, but in Germany there is a revival of Nazism, says Pastor Martin Niemoller, German Lutheran minister jailed by Hitler during World War II.

Dr. Niemoller has been here four days addressing various church congregations before returning to Germany with his wife. Else. A world at peace under "the Fatherhood of God and the Brotherhood of Man" is his simple yet forceful theme.

A recent visitor to the Soviet Union, Dr. Niemoller told one of his audiences, 2,000 persons at Immanuel Presbyterian church:

"While I was in Moscow, I didn't find a single person who did not shrink from the idea of war. And we have to take the slightest possibility as Christians to help build peace."

At a press conference, Dr. Niemoller spoke of a revival of Nazism.

"In the last few years there has been a revival of a militaristic, Nazistic type in Germany, but I should not say a revival in the sense that they have come back to life.

"They are coming up from the past. They have come back to life not as somebody who is raised from the dead, but as somebody who comes back from the dead as a ghost. It is something ghastly."

Pastor Niemoller made his own anti-Communism amply clear. But he stressed often the sincerity of the peace desires of the Russian people and people elsewhere in the world—and the fears they have of American aggression.

"I was invited to come (to Russia) for eight weeks by the Patriarch of Russia, but could only accept for six days," he said. "In that short time I had to rely upon my intuition and what my eyes could observe.

"I took my 23-year-old daughter on the trip to act as an interpreter. I wanted to be sure of what was being said.

"One of the main reasons for my going was in the interest of peace. The last World Ecumenical Conference held in Holland said World War III can be avoided. It was in furtherance of that aim that I made the trip.

"I cannot speak for Americans or the American church but for my nation, Germany, half of which is behind the iron curtain and who live in hope of liberation, fear that if war comes none will survive. That is why I pray every night that this cold war doesn't turn into a hot war."

Dr. Niemoller refuted the anti-Soviet canard about religious persecution. He told of more than 60 churches operating in Moscow alone.

"There are more than 20,000 churches operating through out the Soviet Union today and those

FIGURE 6. One of several articles from the early 1950s in the Communist newspaper the *Daily Worker* listing Rudolf Carnap as a sponsor of events or open letters to government officials calling for peace. In this case, his name is the second listed (in the first column, third paragraph) in a description of 2,000 supporting an "An Appeal for Peace" sponsored by the National Council of the Arts, Sciences and Professions and calling for international negotiations to reduce military tensions.

incriminating circumstantial evidence. Like the stereotype of the secret communist agent popularized in literature and film, Carnap was mysteriously private and often unknown to those around him. At the same time, he was demonstrably connected to Malisoff and suspicious groups. In his INS file from 1941, Carnap declared that he belonged to eleven organizations, three of which came to be widely regarded as communist front groups. These were the Consumers Union, the American Association of Scientific Workers, and the Chicago Civil Liberties Committee. As if wondering whether that the last might be the main venue by which this shadowy scholar contributed his skills to the coming revolution, the Newark office pointed out that, according to a reliable informant, this last group published "a large number of pamphlets and books" in support of the Communist Party and in "defense of all radical groups."[37]

Fortunately for Carnap, the Los Angeles office was not impressed by circumstantial evidence. Specifically, they told Hoover, there was "no activity on the part of CARNAP that would merit his inclusion in the Security Index."[38] The "Security Index" was the successor to Hoover's "Custodial Detention List" of suspicious individuals about whom, in Hoover's words,

> there is information available to indicate that their presence at liberty in this country in time of war or national emergency would be dangerous to the public peace and the safety of the United States Government. (Hoover, quoted in Klingaman 1996, 182)

Were World War III or a communist revolution to commence, that is, the Los Angeles office did not feel that Carnap should be among those targeted for immediate arrest and detainment. Almost three years after Frank's investigation began, the Carnap case had also led to a dead end. There was no solid evidence that Carnap was an active subversive, and there was no good reason to interview him. His file would remain on FBI shelves, however, and his name would remain in FBI indexes where agents might be led to it in the course of future investigations. In effect, Carnap and Frank were well known to the FBI, and they were being watched.

Loyalty Oaths and the Wrath of Hook

Carnap was the leader of the Unity of Science movement who most directly experienced all three kinds of anticommunist scrutiny mentioned

---

[37] Newark Summary Report, June 3, 1955.
[38] LA to Hoover, June 20, 1955.

in the beginning of chapter 12. Shortly before the FBI began to take an interest in him, he became indirectly involved in the loyalty oath controversies at the University of California. Much as Morris protested the University of Illinois's oath requirement, Carnap protested California's by turning down UCLA's Flint Visiting Professorship as well as two speaking invitations at Berkeley. He was especially angered because his friend psychologist Richard Tolman was among those dismissed from their jobs (though most were later rehired) on the grounds that they refused to sign the oaths. Carnap wrote to Berkeley's President Charles Sproul that his decisions in each case were

expressions of solidarity with the dismissed colleagues, and of protest against the violation of the principle that scholarship, teaching ability, and integrity of character should be the only criteria for judging a man's fitness for an academic position.[39]

Carnap's uncharacteristic use of this vague phrase "integrity of character" – one that nods to the common stereotype of communists as psychologically unstable and susceptible to manipulation and control – suggests a strategic choice of words as he spoke about these matters to a powerful university president (in whose system he wished eventually to teach). For when Carnap wrote about this matter to Berkeley's chair in philosophy, William Dennes, the phrase did not and logically could not appear: "I am opposed in principle to the idea that any but academic considerations should qualify a man as fit or unfit for teaching." Carnap admitted to Dennes, however, that were he subject to the pressures of the situation, he too would have sacrificed his principles and signed the oath "under protest in order to protect my livelihood."[40] Thus Carnap joined Morris's colleague Charles Osgood to make the point that faculty usually did not have the luxury to uphold their conscience or principles in these volatile matters of loyalty oaths.

Four years later, Carnap finally arranged to teach at UCLA, where he would succeed Hans Reichenbach, who had died a year before. Reichenbach, also a native German, did not have an easy time in California during the war. He and his family were classified as enemy aliens and subject to strict curfews and travel restrictions.[41] Carnap surely considered the controversy over the state's loyalty oaths as he made his decision to accept

[39] Carnap to Robert Gordon Sproul, October 22, 1950, ASP RC 085-29-11.
[40] Carnap to William Dennes, October 12, 1950, ASP RC 085-29-02.
[41] Reichenbach to Morris, May 24, 1942, CMP.

the offer. After 1952, the oath requirement had been discontinued. Still, Carnap wrote to his friends and colleagues, "the political situation there does not look too good and inspires little confidence. On the other hand, my appointment presumably has not met any opposition on that score."[42]

Carnap's Attack from Hook

It is hard to believe that Carnap was not aware that he was being investigated by the FBI as he wrote this letter to his friends. Almost surely, one or more of his neighbors or colleagues whom the FBI interviewed would have told him. In any case, Carnap was not unknowing about the politics of anticommunism. He knew firsthand that some in the philosophical profession did not care for his politics and his political behavior. However briefly, Kallen attacked Carnap's program in his denouncement of "unity of science" as a totalitarian project. Ten years later, in 1949, shortly after lending his name in support of the upcoming Cultural and Scientific Conference for World Peace, Carnap opened his mail to find himself being scrutinized by Sidney Hook.

Known as the Waldorf conference after the Manhattan hotel where it was held, the conference was organized mainly by Harlow Shapley. Not surprisingly, Shapley and his co-organizers did not invite Hook to speak. Not only was Hook's bullying and staunch partisanship out of place at a conference dedicated to the cause of peace and disarmament, his support of anticommunist investigations would have won him few friends among those intellectuals gathering to promote world peace.[43] After requesting a chance to speak at the conference and being turned down, Hook decided that the conference was a communist front – merely "another ambitious propaganda event to further the Soviet cause" (Hook 1987, 382) – and then orchestrated an attack. He sent letters to the *New York Times* and magazines denouncing the conference as a front and argued for weeks with Shapley and his assistant organizers in an effort to get on the program.[44] At one point during the conference, Hook, with a newspaper reporter in tow, surprised Shapley in his hotel room to deliver documents allegedly proving that Shapley had lied to him and unfairly prevented him from participating as a speaker.

[42] Carnap to "Friends," March 6, 1954, CMP.
[43] Relevant examples of Hook's writings include Hook 1953, 1954.
[44] See, e.g., the exchange between Hook and editor Freda Kirchwey in *The Nation*, "Waldorf Aftermath," April 30, 1949, pp. 511–12.

This battle between Hook and Shapley was just beginning when Hook learned that Carnap supported the conference. Hook then sat down at his typewriter and wrote a letter to Carnap. As Carnap read it, he underlined certain parts:

Dear Carnap,

I have just learned from Dr. Jacobs of the Voice of America that your name has been cited (presumably abroad) as a sponsor of the forthcoming Cultural and Scientific Conference for World Peace. I hope this is not true. If you actually have enrolled yourself as a sponsor, I am confident that you are unaware of the real auspices of the Conference. It is being run by people whose first act, if they came to power, would be to liquidate you and people like you.... I am enclosing some materials which I earnestly beg you to read very carefully. My experience with this Conference is decisive.

Besides underlining much of the letter, Carnap wrote "?!" in the margin next to Hook's phrase "to liquidate you and people like you." If he was alarmed and puzzled about why the conference's organizers would want to "liquidate" its sponsors, as Hook claimed, Carnap at least had no doubt that Hook was both agitated and serious. He enclosed a reprint of a recent anticommunist lecture he had delivered at Dartmouth in which Hook argued that "the communist ideology, like the Nazi ideology, but for different reasons, impels those who hold this ideology to embark upon a program of world conflict and conquest" (Hook 1949). Though he had argued in the 1930s that the Soviets under Stalin had just recently abdicated their Marxist obligations to nourish and sustain socialism, Hook now charged that "even before the October Revolution of 1917, the Soviet regime has been in a state of undeclared war against the West" (ibid.). Judging from his underlining of the text, Carnap read the reprint thoroughly and learned (or was reminded) that Hook regarded all intellectuals as foot soldiers in this ongoing, epic war between the freedom-hating Soviets and freedom-loving West.

Carnap was not the only signatory of the congress that Hook urged to see the anticommunist light and withdraw his name. Hook also wrote to Einstein, Thomas Mann, and Rex Tugwell. In Carnap's case, at least, Hook felt that the "political education" he was giving to Carnap was not only politically correct but in Carnap's best, personal interest:

I can very well understand how honest people can disagree with American foreign policy now as in 1940. But what would you have thought of anyone who in 1940 lent his name to the German-American Bund who were then agitating for peace? People will draw the same inference about you if you are a sponsor of the Waldorf

Conference and remain a sponsor. If you want to agitate for peace there is no need to associate yourself with <u>agents of a foreign power</u>: you can do it independently.

This business is no ordinary thing, as you will learn by developments in the next few days. <u>Anybody who is still a sponsor by the time the Party-line begins to sound off at the conference, will be marked for life as a captive or fellow-traveler of the Communist Party</u>. You are not a Communist and you are not a political innocent.

Forgive me for writing so strongly. It is a measure of the respect I have for you and your life-work. There are mistakes and <u>mistakes</u>. This one <u>can be positively fatal</u> if not rectified.

Sincerely Yours,
Sidney Hook[45]

The references to liquidation and fatality in this angry letter were not merely for effect. Hook and other anticommunists routinely warned fellow-traveling intellectuals of Soviet purges and the likelihood (as William Henry Chamberlin put it) that they "would be speedily liquidated if they had to live under the Communist regimes for which they like to apologize" (Chamberlin 1955, 20). They were not much more secure in the United States, either. Given the widespread fear of immanent war between America and the Soviets (instanced by Hook's reference to the German-American Bund, circa 1940), the provisions in the McCarran Act for constructing concentration camps, Hoover's "custodial detention list," and the emotional, sometimes violent exchanges over communism taking place at political meetings and rallies, Hook gave Carnap fair warning that his politics could, if it had not done so already, place him at grave risk.

Hook did not keep his fair warning private. A few days later, Carnap received a letter from Carl Hempel, whom Hook had asked to help pressure Carnap to withdraw his name. Hempel replied to Hook that he was skeptical of Carnap's optimism for securing peaceful relations with the Soviets and his "optimistic, long-range view concerning the possibilities of the future." Still, Hempel told Hook, Carnap should be allowed to make his political decisions by himself. Besides, the conference was probably not "a matter of far-reaching political consequence."[46]

Carnap was surely gratified by Hempel's support. But, at least in retrospect, that support was double-edged. Two elements of Hempel's and Hook's letters illustrate how the popular anticommunist hysteria informed their view of Carnap and his politics. First, Hempel continued

[45] Hook to Carnap, March 20, 1949, in Shapiro 1995, 128–29. Carnap's underlinings are preserved in his copy of the letter, ASP RC 088-38-10.
[46] Hempel to Carnap, March 23, 1949, ASP RC 102-46-03.

Hook's use of extremely threatening language. He wrote to Hook, "I certainly hope that a favorable attitude toward [the conference], or even sponsorship of it, would not brand one for life as a political dupe or captive if not as a traitor to the cause of the United States." Hempel sent Carnap a copy of this letter, circled the word "traitor" and then noted that "this term was not actually used by Hook."[47] Hempel, that is, specifically introduced the concept of treason. Nor did he dismiss Hook's claim that sponsorship of the conference could be so dangerous – he only downplayed the possibility. Second, Hook and Hempel both alluded to the popular stereotype of communist operatives as robotic, unthinking automatons who were secretly controlled from without. Hook worried that Carnap could be seen as a "captive" of the Communist Party, while Hempel defended Carnap against that perception: "I can assure you of one thing," he told Hook, "namely that he is not under the influence of any sinister person."[48] Though neither Hook nor Hempel believed that Carnap was a traitor or a robot communist, these stereotypical features of communists, hovering behind and around their correspondence, could only have added to Carnap's worry that, indeed, Hook might be right and that Carnap was possibly sealing his reputation as a radical.

Some twenty years later, Carnap recalled the episode as difficult. "Some of my good friends," he told Cedric Belfrage, then researching his book (Belfrage 1973), "made strong efforts to persuade me to withdraw my name."[49] But Carnap was proud to have resisted Hook, to whom he replied calmly and forcefully. As for Hook's complaint that he was not allowed to speak at the conference, Carnap pointed out that "nobody has a 'right' to be allotted time on a [conference] program." As for Carnap's signature, he explained, "I gave my name because I found myself completely in agreement with Shapley's statement as to the purpose of this Conference," which was to promote dialogue between the superpowers about peace and disarmament.[50] In one respect, Carnap's defense recalled the ethical rationalism he outlined during his radio broadcast with Kirtley Mather some thirteen years before. He upheld clear, scientific thinking

---

[47] Ibid.
[48] Hempel to Hook, March 23, 1949, ASP RC 102-46-06.
[49] Replying to Belfrage, Carnap wrote, "I still feel proud that I gave my name as a sponsor for the peace conference in 1949. All the more so since some of my good friends made strong efforts to persuade me to withdraw my name. Everybody was for peace, of course; but it became suspicious, when an effort was made that seemed to point directly in the line of genuine peaceful coexistence and disarmament. It is sad to observe how the tide is going again today" (quoted from Carnap's dictated correspondence, Carnap to Belfrage, November 23, 1968, ASP RC 027-23-51).
[50] Carnap to Hook, March 24, 1949, ASP RC 088-38-13.

as a tool to help ensure that expectations about other people and nations were coherent and well grounded on available scientific information. That, in turn, would help to avoid "childish reproval of opposing groups in the name of morality." Now, Carnap seemed to observe, such childish reprovals and misleading "exaggerations" about geopolitics had become the norm in Cold War discourse. He continued,

I regard as grossly exaggerated the picture of the "serious threat to democracy" by communism in America as it is drawn by the press (including many articles in the *New Leader*) and by the State Department. Gross exaggerations are repellent to me, and those here are dangerous because they work against peace and better understanding.

By refusing to withdraw his name as sponsor – even though, he admitted, it "might be 'wise' for the moment" to do so – Carnap took his stand not only for peace and dialogue, but also – as Neurath had in his review of Hayek's *Serfdom* – for "good will and cooperation" instead of international mistrust and aggressive posturing:

Since our government is persistently refusing to extend good will and cooperation to the other side, I welcome the getting together of scientists and artists in an attempt to lift the Iron Curtain (which is chiefly caused by our military threats) and to show each other respect and good will.

Finally, Carnap addressed the larger "climate of fear" that had so clouded American life:

Your own prediction as to an indelible smearing of my reputation in consequence of my sponsorship of this conference shows that there is fear and intimidation operating in this country to an extent unprecedented so far. The same is shown by many other facts; e.g. the University of Chicago is just under investigation by an anti-seditious activities committee of the Illinois legislature.[51]

Carnap evidently knew better than to reprimand Hook for himself being one source of this widespread "fear and intimidation." He focused instead on Hook's "prediction" as an empirical claim that, it turns out, was more or less right. Though it was not only because of his support for the Waldorf conference, Carnap did come to be seen by some as a suspiciously radical intellectual. At these hearings, anticommunist witness Howard Rushmore named Rudolf Carnap as a suspicious character. As a result of a similar rumor about Frank, as we saw, Carnap soon came to be investigated by the FBI.

---

[51] Ibid. Hook's reply to Carnap's defense of his sponsorship is discussed in chapter 17.

14

Competing Programs for Postwar
Philosophy of Science

When Morris and Carnap regrouped after Neurath's death and began charting a postwar course for the Unity of Science movement, they were joined by Neurath's old friend Philipp Frank. Morris and Carnap planned to edit the *Encyclopedia* by themselves, and they met with the University of Chicago Press to plan the next section, to be titled *Methods of Science*.[1] Frank, meanwhile, would become leader of the Institute for the Unity of Science that, with Morris's help, he re-established in Boston as a center of the movement's activities. At the end of 1947, however, William Malisoff suddenly and unexpectedly died, adding to this mix of projects a struggle for the control of Malisoff's journal, *Philosophy of Science*, and the young Philosophy of Science Association. This chapter examines three factions that worked variously with and against each other in the wake of Malisoff's death and into the 1950s: Frank and Morris sought to lead the Unity of Science movement in sociological and humanistic directions, while Feigl and Reichenbach pursued more professional and profession-building projects that were independent of both Frank's institute and Malisoff's eventual successor, C. West Churchman. Churchman, finally, positioned himself (and his co-authors) as a critic of logical empiricism who shared some of Dewey's (as well as Neurath's, Frank's, and Morris's) reservations about the formal, "scholastic" future into which logical empiricism seemed to be heading.

As these three groups negotiated for control of the profession's main institutions – the Philosophy of Science Association, the journal *Philosophy of Science*, and the new institute – initial victories went to Frank and

[1] To WTC from MGP, September 9, 1946, UCPP, box 346, folder 4.

Churchman, who seemed happy to work together and who presided over all three. Against the background of the intellectual, institutional, and political pressures of the Cold War examined in previous chapters, however, it is possible to see why and how these victories for the Unity of Science movement were short-lived. As shown in the next chapters, Frank's institute soon became moribund as final victories went to Feigl, Reichenbach, and other philosophers who chose more technical, professional, and apolitical directions for philosophy of science.

Philosophy of Science versus *Erkenntnis*, Redux

As logical empiricism took root in North America in the 1930s and '40s, there remained no journal that functioned as a unifying, organizing forum (as *Erkenntnis* had, for example, in Europe). As we saw in chapter 5, though Malisoff had reached out to logical empiricism and published and translated some logical empiricist writings, there was a professional and intellectual rivalry between Malisoff and the logical empiricist émigrés that prevented *Philosophy of Science* from becoming a unifying institution. By 1939, *Erkenntnis* was on its way to rebirth as the *Journal of Unified Science*, but the war, chronic miscommunication with its original publisher, Felix Meiner, and Neurath's last-minute decision to publish the journal with Basil-Blackwell (and not the University of Chicago Press) kept it unrevived during the war years.[2]

Soon after the war, Reichenbach wished to resuscitate *Erkenntnis* at the University of California. He planned that he and Carnap would resume their prewar duties as chief editors, Reichenbach's student Abraham Kaplan would assist them, and the journal would be supported by clerical staff.[3] Reichenbach's plan needed money. Knowing that Morris was helping Frank to court the Rockefeller Foundation for funds to revive Neurath's institute in Boston, Reichenbach asked Morris whether they might join the two projects in one proposal. Piggybacked on the institute, that is, Reichenbach could perhaps obtain Rockefeller funding for reviving his old journal. Yet Morris discouraged Reichenbach from building an editorial empire on his own by telling him that the foundation was unlikely to be receptive. "They wish to consider their possible interest in the Unity of Science Movement as a whole before considering any special item like the journal," he responded. Morris volunteered Kaplan

---

[2] Reisch 1995; 2003a.
[3] Reichenbach to Morris, October 16, 1946, CMP.

to handle both editorial and clerical duties as he nudged Reichenbach to think in more collective, collaborative terms. Such a role for Kaplan "would fit in with our policy of bringing in young men – we hope to do this for the Institute and the *Encyclopedia*. Frank and Carnap favor the plan."[4]

Morris defended the forthcoming institute as the "coordinating center of all our activities" to prevent the kinds of problems created by Neurath's relatively autocratic style of leadership. Neurath had made contractual and royalty arrangements with *Encyclopedia* authors on his own, for example, without leaving records and forcing the University of Chicago Press to create new contracts for the *Encyclopedia* after his death. The new institute, that is, would help to "decentralize" and, in a way, collectivize the movement's leadership.[5] "We must avoid such complications in the future," Morris insisted, as he told Reichenbach that the institute, and not an individual, would sign a publishing contract for *Erkenntnis* when it was finally revived. "It is important that we unify our projects" and achieve "agreement on plans. I think we will soon be under way on a large scale." After discouraging Reichenbach's plan to control *Erkenntnis* apart from the institute, Morris reassured him that "you will of course be a major force in the Institute."[6]

Churchman versus Logical Empiricism and Scholasticism

In a little over a year, these questions about the future of *Erkenntnis* were complicated by Malisoff's death in November 1947. C. West Churchman had worked closely with Malisoff at *Philosophy of Science* and stood to inherit his leadership. With co-authors (either T. A. Cowan or Russell Ackoff), Churchman wrote multiple editorials in the 1940s whose critical, negative tone stood out from the broadly sympathetic and admiring view of

[4] Morris to Reichenbach, July 9, 1946, CMP.
[5] University of Chicago Press memorandum, September 13, 1946, UCPP, box 346, folder 4. Reporting conversations between Morris and the press, the memo notes that in the wake of Neurath's death, the movement's "plans are to decentralize the organization" by creating the institute, which "will probably be headed by Carnap." Though Philipp Frank led the effort to establish the institute and became its president, this memo suggests that Carnap was planning or willing at the time to help to fill Neurath's shoes in the new "decentralized" leadership. He also approved of the Institute as Morris and Frank were planning it, writing to Morris, for example, "I am entirely in agreement with the point that both research and education should be included among the aims of the Institute, and our activities" (Carnap to Morris, May 18, 1946, CMP).
[6] Morris to Reichenbach, July 9, 1946, CMP.

logical empiricism in *Philosophy of Science.* Although he charged logical empiricists with epistemological and methodological mistakes, Churchman shared a leftist, progressive vision for philosophy of science that variously overlapped with Neurath's and Frank's criticisms of "scholastic" logical empiricism and with complaints from the Marxist philosophical left that logical empiricism fled social engagement by retreating into the study of language.

In July 1945, Churchman and Cowan published an editorial titled, "A Challenge." Their challenge followed criticisms of logical empiricism that Dewey and Arthur Bentley had recently published in the *Journal of Philosophy.* Churchman and Cowan glibly approved "the spankings which Bentley and Dewey are currently administering to Carnap, Cohen, Nagel, Ducasse, Lewis and Morris" on the grounds "of all things, that they display amazing contempt for clear and consistent definitions of the terms they use" (Churchman and Cowan 1945, 219). They excused the transgressions of Cohen and Nagel, for their *Introduction to Logic and Scientific Method* (1934) took justifiable pedagogical liberties that Dewey and Bentley, they explained, probably mistook for imprecision and sloppiness. But "the logical positivists" could not be excused. Their spankings "were long overdue" not only for the reasons Dewey and Bentley cited, but also because of "the logical positivist's predilection for verbal analysis."

One logical positivist upset Churchman and Cowan more than the others:

Carnap, particularly, has long outraged patience as well as common sense with rambling verbal analysis of the meaning of terms used in the exposition of logical positivism.

The problem with this "verbal analysis," Churchman and Cowan explained, stemmed from the logical positivist's "epistemological position":

To the logical positivists investigation into the nature of logic, or as we should prefer to say, formal science, is quite independent of investigation in the nature of experimental science. They regard experimental science as dependent upon logical analysis but *they fail altogether to show in what way logical analysis is dependent upon experiment.* This means that for them logical analysis can proceed regardless of experiment, that its method is wholly non-experimental. The result is that the method of logical analysis becomes discursive, verbal, arbitrary, common-sensical, and scholastic. (Churchman and Cowan 1945, 219)

Like Neurath and Frank, Churchman and Cowan were concerned about relationships (or lack of them) between formal and logical studies of science, on the one hand, and actual scientific practice, on the other.

Unfortunately, as suggested by their view of Carnap's writings as "rambling" and pointless, Churchman and Cowan did not read Carnap closely. His distinctions between syntax, semantics, and pragmatics were well developed by that time and specifically designed to help organize and clarify philosophical (and scientific) problems *prior to* the kind of empirical, psychological claims about logic and experiment – "in what way logical analysis is dependent upon experiment" – that Churchman and Cowan seemed to be after. Even in *Philosophy of Science*'s very first article, Carnap's "On the Character of Philosophic Problems," Carnap acknowledged that formal analysis of science and scientific languages is no substitute for considering the practical strengths of different languages. "The task of the philosophy of science," he concluded, "can be pursued only in a close cooperation between logicians and empirical investigators" (Carnap 1934a, 19). Churchman and Cowan seemed to ignore this collaborative division of labor Carnap proposed and then faulted Carnap (and others) for not producing some undivided, comprehensive theory of science.

One reason they demanded such a theory was political. The effect of this demarcation between formal and empirical science, they argued, "is the same as that which results from the split between *Naturwissenschaft* and *Geisteswissenschaft*." "The social effect in each case is reactionary":

> Those who believe the method of investigation into human behavior is nonscientific end by becoming *mystics*[;] those who split science into two methods end by becoming *scholastic* in the worse sense of that abused term. (Churchman and Cowan 1945, 219–20)

Logical positivists, they charged, were thus assisting the "enemies of science" (such as the neo-Thomists) and allowing philosophy of science to devolve into merely "pre-scientific discussion groups which dissipate scientific energy in fruitless investigations of the meaning of words" (ibid., p. 220). The solution and the "challenge" they posed was for the logical empiricists to produce the powerful, unified theory of science that the day demanded:

> We insist that they develop a *single* methodology which encompasses both of the inseparable aspects of every scientific investigation, that is, the formal framework which the observer takes to his work and the experiential effect which his observation has on that formal framework.

They demanded of logical empiricism what Neurath demanded of Carnap's semantics, that it be comprehensive and tightly related to scientific practice.

Six months later, Churchman and Cowan resumed their attack in a four-page article, "On the Meaningfulness of Questions," that focused on the charge of "meaninglessness" in logical empiricist discourse. Though not mentioned by name, Carnap and his early syntactic approach to the problem (and probably also Neurath's antimetaphysical proclamations) were set up as a "positivist" strawman:

> If the issue cannot be settled by empirical methods, then our positivist feels that it can never be settled at all, i.e., that the question asked is neither true nor false, and hence is meaningless. (Churchman and Cowan 1946, 20)

The verdict of meaninglessness, they charged, too often served only to evade real problems. Not unlike the socially blinkered logical positivist that Marxists criticized for "social evasion" (as Margaret Schlauch put it), Churchman and Cowan parodied the positivist who trudged "along a narrow path whose vistas are almost everywhere obstructed by his carefully constructed signboards." This time they also urged philosophers of science to treat not only logic and experiment, but also values. For the positivist "has even gone to the point of relegating ethical judgments to the realm of the meaningless. . . . To ask whether there are ultimate values is to ask the empirically meaningless, is it not?"[7]

The logical positivist was not the only villain Churchman attacked. Others such as "nonempiricist" introspectionists, who deny that empirical science can illuminate problems of mind, were also aiding and abetting antiscientific reaction:

> Our dispute with the non-empiricists is based on the fact that they represent a reactionary movement designed to split the ranks of scientific workers into two distinct parts, parts that use discrete and non-related methods. As scientific workers, we cannot tolerate this state of affairs. (ibid., p. 22)

With the phrase "scientific workers" calling out to the well-known American and British associations of Scientific Workers, Churchman and Cowan's agenda also called out to Dewey's characterization of "unity of

---

[7] Churchman and Cowan 1946, 20–21. The same concern with the practical aspects of language fueled another critique of Carnap from Churchman, this time specifically about Carnap's inductive logic. Against the formal manner in which Carnap defines his key terms, Churchman wrote, "There is no freedom from the practical import of terms. . . . The terms of any language, symbolic or otherwise, have a social impact which must be taken into account by the definer. . . . These concepts are far too important in their scientific and social implications to be treated lightly as toys of the positivist at play" (Churchman 1946, 341–42).

science" as a "social problem" (Dewey 1938). Unity in science was one thing, but unity among scientists to defend science against its enemies in contemporary society and culture was equally important. In all these tasks, Churchman and Cowan believed, the profession was obligated to pursue a general, unified theory of science (formal and experimental) and accept that "the persistent problems of philosophy and the sciences come out of the needs of society and its individuals" (Churchman and Cowan 1946, 23).

One year later Churchman revisited some of these issues with Russell Ackoff in an editorial titled "Ethics and Science." Here, he called for collaboration among ethicists and philosophers to produce (and not merely talk about) an empirical "science of value":

There is too much talk and too little action. Action will arise from consideration of such problems as the following:

1. What is an operational (i.e., experimental) definition of value?
2. How do we determine experimentally the value of any act or set of actions?

We need more of a science of value and less talk of the value of science. (Churchman and Ackoff 1947, 271)

In light of the ongoing values debate and the popular view that science and scientific philosophy could contribute little to understanding human values and their roles in history and social life, Churchman was clearly urging the discipline to meet this challenge and demonstrate that philosophy of science was more than an intellectual specialty insulated from ethics and society.

A Neurath-Churchman-Bentley Alliance?

Most likely because Churchman worked closely with Malisoff, and because Malisoff and the Unity of Science movement kept each other at a respectful arm's length, Churchman was not directly involved in the Unity of Science movement. Still, his challenges to logical empiricism aligned roughly with the socially engaged agendas of Frank and Neurath. Churchman's writings also tend to agree with some of the more subtle aspects of Neurath's program. Shortly after Neurath died, for instance, Churchman and Ackoff published a paper, "Varieties of Unification," in which they argued for a nonhierarchical, nonidealized, and collectively pursued unity among the sciences (Churchman and Ackoff 1946).

Churchman also offered something like Neurath's critique of pseudo-rationalism, insofar as he recognized that scientific practice was fueled more by free decisions involving plans and values than by deductions from epistemological theories *about* science. Churchman also acknowledged a kind of reflexivity through which science could gain justification, when possible, only from itself: "Science obviously makes decisions of its own in both theoretical and applied science," Churchman wrote several years after these editorials. "Science must decide to take certain steps in its procedures, and these steps must presumably be evaluated by science.... How does it evaluate its own policies?" Churchman answered that science can evaluate its values only with a "circularity" ("science must accept values to study values") that recalls Neurath's fallibilism and antifoundationalism. "It is necessary," Churchman wrote, "to develop a theory of science in which no scientific conclusion ever has complete validity and in which the methodology used by science is a self-correcting device" (Churchman 1954, 22).

There is no evidence that Neurath and Churchman recognized these or other affinities in their views. Still, Neurath is conspicuously absent in the list of logical empiricist sinners denounced in Churchman's several editorials. There are also tantalizing hints that just before his death Neurath was poised to recognize Churchman as an ally. In early 1945, Churchman sought to establish an "Institute of Experimental Method" designed mainly to facilitate cooperation between empirical social scientists and statisticians.[8] Apparently seeking to connect his proposed institute with the Unity of Science movement, Churchman sent Neurath a copy of his proposal. Though Neurath took several months to respond, he invited Churchman to correspond further and to elaborate his views about "experimental method." In these "I am very much interested," Neurath replied.[9]

Whether or not Churchman recognized and approved aspects of Neurath's distinctive philosophy of science, Dewey and Bentley did. Their critique of logical empiricism was the springboard for Churchman and Cowan's first editorial. Yet Neurath escaped Dewey's and Bentley's wrath. While Carnap and Nagel, for instance, disliked Neurath's *Foundations of the Social Sciences* (1944), Bentley and Dewey found it intriguing. Bentley

---

[8] Churchman to Neurath, January 25, 1945, ONN. Churchman writes that Richard von Mises encouraged him to contact Neurath and the Unity of Science movement with his proposal.

[9] Neurath to Churchman, October 19, 1945, ONN.

sent Neurath comments on the monograph and confessed in his usual acerbic way that he admired Neurath's other writings, as well:

As a matter of fact, I liked your tone very much in all the early statements of yours I read, and I never quite could see how you and Dewey could be mixed up with people who wanted to "unify science" – horrible for me even to think of – and especially when they had nothing with which to do the unifying except "main force and awkwardness" as the colloquial phrase goes.[10]

Bentley suggested organizing a symposium featuring Dewey and Neurath, a possibility made only more intriguing by evidence that Dewey had begun to appreciate Neurath's philosophy of science. Several years before, when Dewey, Carnap, and Neurath discussed logical empiricism's views of value statements, Neurath and Carnap emphasized to Dewey the movement's general programmatic unity. In Neurath's monograph, however, it appears that Dewey noticed some distinctive themes. "I judge from Neurath's last contribution" to the *Encyclopedia*, he remarked to Morris, "that he is pretty definitely breaking away from the Carnapian standpoint."[11] A few months later, Neurath was dead and these conversations and possibilities could not be explored further.

The Logical Empiricists versus Churchman

In the last years of the 1940s, the absence of Neurath and Malisoff – two powerful figures in philosophy of science – created a vacuum that invited the formation of new agendas and new alliances. After Malisoff's death, Frank, Morris, Carnap, Reichenbach, and Feigl corresponded about the situation and opportunities for better relations between logical empiricism and Malisoff's journal, *Philosophy of Science.* One possibility was to resurrect *Erkenntnis* by effectively combining it with *Philosophy of Science.* Another possibility was to gain some degree of editorial control over *Philosophy of Science* in Malisoff's absence. The logical empiricists knew, in any case, that they had to be careful. Plans for *Erkenntnis* and the new Institute for the Unity of Science could easily intimidate Churchman, who was most likely to succeed Malisoff and who was on record as a critic of logical empiricism.

Frank was optimistic. He suggested that the strained relations between Malisoff and the logical empiricists were caused by Malisoff's

---

[10] Bentley to Neurath, September 29, 1945, ONN.
[11] Dewey to Morris, August 31, 1945, CMP.

"personality" – he "suffered from a great many complexes and inhibitions which made him afraid of any cooperation, by which he always feared to 'lose face.' " Frank pointed out, however, that he had "always been, personally, in good relations with him, although (or perhaps because) I never published a paper in his journal."[12] Churchman, Frank believed, would be a better collaborator. He would "certainly not follow Malisoff's example and will be glad if he gets some cooperation from us."[13]

Reichenbach was pessimistic. After Churchman installed himself as acting editor (pending a formal election by the Philosophy of Science Association (PSA), the legal owner of *Philosophy of Science*), Reichenbach wrote,

I understand that Churchman is now acting editor. I think it is impossible for our group to cooperate with him. He is as confused as he is arrogant. I would not object to all sorts of compromise, but there is a limit.[14]

One hope for keeping Churchman at bay was to appeal to the publisher of the journal, Williams and Wilkins, and try to gain some control over the situation through them. Another was to accept Churchman's terms and hope to win some control at the upcoming PSA meeting in Chicago later that month, only two weeks away. Reichenbach wrote to Williams and Wilkins but was offered only what Churchman had already told him: The successor would have to be picked by the journal's subscribers, that is, the members of the PSA. That meant that the result would be determined by "a chance selection of subscribers present," and Reichenbach did not like those odds, especially because he himself could not attend the meeting. Still hanging on to the first option, he asked Morris whether he could pull some strings behind the scenes: "Dear Morris – do you know of a man who could approach Williams and Wilkins and make it *clear to them that Churchman is impossible?*"[15]

Perhaps because Churchman's interests in values and social aspects of science were close to his own, Frank was more diplomatic than Reichenbach. "In order to smooth the situation" between the two groups, he told his colleagues, he had agreed when Churchman asked him to write an obituary for Malisoff to accompany his own. The two testimonies to Malisoff's ambitions appear side by side in one article (Frank and Churchman 1948), as if symbolizing tentative first steps toward the

---

[12] Frank to Morris, November 28, 1947, CMP.
[13] Frank to Morris, December 6, 1947, CMP.
[14] Reichenbach to Morris, Frank, and Carnap, December 13, 1947, CMP.
[15] Ibid.

alliance that Neurath, Churchman, and Bentley had begun to envision. Frank also urged his colleagues to follow his diplomatic lead: The situation must "by no means" come to "the spectacle of two groups of 'fighting philosophers of science.'"[16] Frank believed that Reichenbach's opinions about Churchman were "exaggerated" and nominated himself to lead negotiations about the form that a post-Malisoff *Philosophy of Science* would take.[17] It should not be too difficult, Frank thought, "to get some influence in this group [because] they have few people of any reputation."[18]

I think the best thing to do is to discuss the matter with C[hurchman] directly and to tell him that we shall found a new journal if we are not satisfied by the compromise. I am prepared to start the discussion with C. and later we can arrange a meeting of some representatives of both groups.

Frank did, however, agree with Reichenbach that luck would not be on the logical empiricists' side. Malisoff's successor would be elected at the upcoming PSA meeting in Chicago's Palmer House hotel, and, Frank implied, machine-style politics would be at work: "The people who meet in the Palmer House are probably a handpicked group and will vote as Churchman wants them to."[19]

Frank's diplomacy paid off. The meeting was presided over by Churchman, who first "declared that the positions on the Governing Committee of the Association were vacant and that a new committee would be elected." The new committee was then formed consisting of Cowan, Churchman, and Frank, along with Sebastian Littauer, F. S. C. Northrop, and Gustav Bergmann. Then the committee approved an amendment to the association's by-laws introduced at that meeting by Churchman's friend and co-author Cowan – an amendment granting the Governing Committee the "power to fill vacancies by electing members to hold office until the next annual meeting." Finally, the newly elected committee met privately to appoint the new roster of officers. The result was that Churchman became secretary and editor of the journal, while Frank became president of the PSA. That left two offices vacant on the association's Governing Committee. Making use, it would appear,

---

[16] Frank to Morris, November 28, 1947, CMP.
[17] Frank to Morris, December 22, 1947, CMP.
[18] Frank to Morris, December 6, 1947, CMP.
[19] Frank to Morris, December 22, 1947, CMP. In this letter Frank reported that Feigl had proposed delaying the upcoming PSA meeting, but that tactic was not pursued. It would only have opened up a new issue in the battle between the two groups.

of Cowan's new amendment, the committee appointed Harvard's Clyde Kluckhohn to one vacancy and left the other unfilled.[20]

By the end of the meeting, the Neurath-Frank agenda won some important territory. *Philosophy of Science* and the PSA remained in the control of philosophers who upheld a relatively populist, socially engaged and non-"scholastic" vision of the discipline. Morris was probably pleased, as well, for his hope to make the scientific study of values a core component of the discipline was shared by Churchman and further supported by the appointment of Kluckhohn, whose anthropological studies of culture and human personality connected to Morris's own work in personality and value theory.[21] This agenda was also victorious because Churchman remained committed to Malisoff's original vision for the journal and the association. Though the official report of the meeting ended with a democratic request for members to send Churchman their comments about the "future of the Association and the journal policy," Churchman also chose to reprint an editorial that Malisoff had first published in January of 1944. It called for "a universal amity of the sciences" and, it hoped, "the emergence of a supreme science of amity in a warless world" (Malisoff 1944). In the fine print of a footnote, Churchman informed readers that "in this editorial, Malisoff stated a journal policy which its present editors have every intention of continuing" (Churchman 1948, 81).

The New Institute for the Unity of Science

Shortly after Churchman took the helm of *Philosophy of Science*, Frank assumed presidency of the PSA and presidency of the newly established Institute for the Unity of Science in Boston. With assistance from Morris, Frank persuaded the Rockefeller Foundation to support it with a budget of $5,000 per year. After a year or more of delays created by the need to establish the institute as a nonprofit entity – a requirement met by Cornell's professor of Labor Relations Milton Konvitz, who incorporated the institute in New York State – it came to life in 1949 as a department of the American Academy of Arts and Sciences in Boston. Morris and Carnap negotiated a new contract with the University of Chicago Press that gave royalties from the *Encyclopedia* to the new Institute. Marie Neurath also transferred the rights she had inherited from Otto. Judging from memos

---

[20] The report of the meeting appears in *Philosophy of Science* 15, (April 1948): 176.
[21] See, e.g., Morris 1948b, 46. Morris cites Kluckhohn approvingly in Morris 1946b and 1951.

at the University of Chicago Press, these developments spurred hope for reviving and completing the first installment of the *Encyclopedia*: "This series is about to spring to life again, and there is hope that we may complete it in three or four more years."[22] The institute also renewed the movement's affiliation with the journal *Synthese*, whose postwar issues occasionally included "Communications of the Institute for the Unity of Science" written and submitted by Frank.

The institute's officers were Frank, Morris, Nagel, and Konvitz (president, vice president, vice president, and secretary treasurer, respectively), while the board of trustees included Percy Bridgman, Egon Brunswick, Carnap, Feigl, Hempel, Hudson Hoagland, Roman Jakobson, W. V. O. Quine, Reichenbach, Harlow Shapley, and Stanley S. Stevens. With the exception of Nagel and Konvitz, there is a conspicuous – though not surprising – absence of New Yorkers on the institute's letterhead. In the eyes of an anticommunist such as Hook, the institute would have been suspicious at least because of Harlow Shapley's membership on the board. After their confrontations over the Waldorf Conference (see chapter 13), Hook and Shapley were most likely not on friendly terms. Nor would Kallen have been sympathetic with an institute devoted to a movement whose basic picture of science was "totalitarian."

The institute was chartered to support the *Encyclopedia*, to continue efforts to resuscitate *Erkenntnis* (as the *Journal of Unified Science*), to pursue Neurath's plan for a monographic series to be called the *Library of Unified Science*, and to sponsor additional international congresses. Frank added to the program his plan to sponsor research in sociology of science and, in particular, the ways by which social and cultural values affect scientists' choices of theories when evidence logically underdetermines theory choice. Frank also proposed an institute project that combined his interests in operationism with Neurath's calls for collectively managing scientific terminology. His proposal to the Rockefeller Foundation described the creation and publication of an operational dictionary of 300 scientific terms.

Frank's agenda was also public, educational, and international. He organized two essay contests to popularize the Unity of Science program, and early in the planning of the institute, he discussed with Morris and Harlow Shapley the idea of producing a bibliography of writings about unity of science. Frank excitedly seized the plan as a public-relations

---

[22] Internal press memo, 1949, UCPP, box 346, folder 5.

tool.[23] On another occasion Frank announced that he would organize a project in "international cooperation" involving "some exchange of books and journals and to support the organization of international meetings" in philosophy of science.[24]

Contributions to the Analysis and Synthesis of Knowledge

The Institute had two venues for publication. One was the "Unity of Science Forum" appearing sporadically in *Synthese*; the other was the series *Contributions to the Analysis and Synthesis of Knowledge*, published within the *Proceedings of the American Academy of Arts and Sciences*. In 1950, when asking his colleagues and board members to contribute articles suitable for publication, Frank requested articles for *Synthese* about logical empiricism, its history, its relations to other projects, or "discussions about the validity of [its] chief tenets . . . and modifications which are [made] necessary by new ideas emerging in science and philosophy." They might also

contain any contributions to the problems which are the concern of our Institute. This means mainly contributions to the integration of the sciences on an empirical and logical basis as for instance contributions to the sociology of science, to the logic of science, to psychology of knowledge, etc.[25]

The institute, that is, in contrast to *Synthese*, was primarily about the practical task of unifying the sciences in the contemporary scientific landscape. It was not, that is, a "philosophical" institute if that meant considering unified science and other topics abstractly and outside scientific practice. Frank reinforced this point in the title of its publications: the phrase "analysis and synthesis of knowledge" was a bid to eliminate the word "philosophy" altogether and establish that the institute's goals and values were, broadly speaking, those of science itself (Frank 1951a, 6).

The first issue of *Contributions* appeared in 1951. Frank and Morris articulated a broad, humanistic agenda in the first section of essays, titled "Science and Man." In "The Science of Man and Unified Science,"

---

[23] "I think we should publish every quarter or at least every half-year a survey. . . . At the same time we can give our own views to the public and interest wider groups for this kind of research. . . . I discussed the matter with Shapley who thinks we could place it in some of the popular science journals published by AAAS or Sigma Xi" (Frank to Morris, October 8, 1947, CMP).

[24] Frank to Morris, October 5, 1951, CMP.

[25] Frank to Morris (and Board of Trustees), March 9, 1950, CMP.

Morris revisited his essay "Semiotic, the Socio-Humanistic Sciences, and the Unity of Science" (Morris 1946a), in which he urged his colleagues to acknowledge the importance of studying values and social and cultural sciences.[26] In the intervening years, he reported, interest and research in the new "science of man" was continuing to bring together those areas of research – semiotics, cybernetics, communication theory, and personality theory – that substantiated the "science of man."[27] Importantly, Morris explained, these developments

> will not be of scientific interest alone; they are of importance for the integration of man as man. They help destroy at the root the opposition of man's scientific and humanistic concerns. In integrating science we at the same time integrate and liberate ourselves. (Morris 1951, 43)

Morris's cultural and humanistic ambitions, and his hope that the Unity of Science movement (and now Frank's institute) would share them, had not diminished during the 1940s. The fact that Morris's essay was accompanied by one from Sheldon, whose work and beliefs Morris alone championed within the movement,[28] further indicate that the new institute was, in part, a child of his own.

Most of the institute's agenda, however, belonged to Frank, who spelled out his ambitions in his remarkable essay, "The Logical and Sociological Aspects of Science." Here Frank articulated and defended the formal, sociological, and political functions of science and philosophy of science. Given underdetermination of theories by evidence or facts, he explained, science operates freely to choose and pursue an array of theories, some of which may have more or less social utility or effects than others. He reviewed historical instances where "political or religious authorities" took sides in scientific choices (such as those of Galileo, Darwin, Nazi science, and Soviet science) and cases where scientific developments in

---

[26] See also Morris 1938.
[27] For more on the movement to develop a "science of man" in the 1940s and'50s, see volumes and compendia such as Ruth Nanda Anshen's *Science and Man* (1942), Clyde Kluckhohn's *Personality in Nature, Society, and Culture* (1953), and Roy Grinker's *Toward a Unified Theory of Human Behavior* (1956). In an effort to bridge the perceived gap between science and humanities, advocates of a "science of man" (including Morris) sought to find "the way by which the conception of Man as an element in a moral and spiritual order can be harmonized with the conception of Man as he is influenced by the forces of the material world" (Anshen ed. 1942, 12).
[28] For Neurath's reservations about Morris's early uses of Sheldon's personality theories in his book *Paths of Life* (1942), see Reisch 1995. Sheldon attacked "naturalism" broadly in Sheldon 1945 and 1946 and elicited a joint response from Dewey, Hook, and Nagel (Dewey et al. 1945). This debate is treated in chapter 16.

298   *How the Cold War Transformed Philosophy of Science*

relativity theory or quantum theory were seized by theologians or politicians as justifications for policies or policy-friendly metaphysical doctrines (Frank 1951b, 19, 20, 21–22).

This recurrent pattern of engagement between science and politics, Frank explained, invited two kinds of responses or postures from scientists and scientific philosophers: "humble positivism" and "active positivism." The first was quiescent and passive. For instance, during those historical periods when science failed to determine unambiguously whether heliocentrism or geocentrism was true or whether human freedom has anything to do with quantum theoretic uncertainties, humble positivism allowed nonscientists (or anyone at all) to take up the issues and settle them as they saw fit. For humble positivists,

scientists should simply collect material and leave the undecided questions of science to be resolved by the educators and rulers of men in such a way that the result will support an education for a "good life." (ibid., p. 28)

Humble positivism allowed educators, politicians, or religious authorities to believe, and behave as if, they possess superscientific knowledge that reaches further into the fabric of reality than empirical science.

Active positivism, on the other hand, continued the iconoclasm of the Vienna Circle and Neurath's campaign against metaphysics and pseudorationalism. Level-headed scientists, Frank explained,

do not believe that "superscientific" methods of knowledge can answer the questions concerning which science has reached a stalemate. These scientists believe that where science has no answer, no on else has any answer, either. One has to wait until a scientific solution is reached.

This was the posture that Frank wanted the institute and its patrons to adopt: "not 'humble positivism' or 'passive positivism' but we may call it 'proud positivism' or 'active positivism'" (ibid., p. 28).

For Frank, therefore, philosophy of science and its new institute were set to ring in the 1950s by continuing the battles of the 1930s and early '40s on at least three fronts. First, his rejection of "superscientific" theories and methods continued the naturalistic campaign waged by Hook, Dewey and Nagel against neo-Thomism and its epistemic pretensions. In this way Frank also nodded to Kallen's critique of unified science as "totalitarian." He rejected the Roman Church, the Third Reich, and the Communist Party in the Soviet Union for opposing "active positivism" and trying to assume epistemic control over science from without (p. 29). Second, his "active positivism" rejected the popular antiscientism and postwar ridicule of "intellectuals" who so foolishly believed that science could make a better, more peaceful world. These critics would have preferred that intellectuals adopt something like Frank's "humble positivism" and leave international affairs to others. Yet where scientists can neither predict nor understand the behaviors of individuals or nations, Frank replied, statesmen and "rulers of men" cannot, either.

Finally, Frank implicitly responded to criticism from the Marxist philosophical left holding that logical empiricists were wrong to regard science as "value free" and insulated from social and economic forces. An "active" philosophy of science would adopt Frank's sociological goals and examine the ways that biases of different kinds are typically intertwined within our understanding of scientific theories. Einstein, Whitehead, and Dewey, Frank remarked, had all made the point that

every presentation of science on a high level of abstraction is imbued with a certain philosophical interpretation which is not a result of scientific research. It may be a mechanistic or organismic or materialistic or idealistic interpretation. The real meaning of these interpretations in physics or biology cannot be found in the sciences of physics and biology. (p. 30)

The sciences intertwined with these metaphysical ideas are powerless to understand their "real meaning," Frank reasoned, because these meanings result from features of our minds, social practices, or history that fall outside their relatively narrow, specialized views. Only *other* sciences, namely social and historical sciences, may put them in sufficient relief for us to see. "The real meaning of these generalizations can be found within science if we include also the social sciences in the system of the sciences" (ibid.). Much as William Gruen, John Somerville, and V. J. McGill had called for logical empiricism to join with Marxist theory in pursuing a comprehensive theory of society, Frank's "active positivism" would pursue both the unification of science from within and the use of unified science to understand science's relations to the world without.

Disunity about the Institute for the Unity of Science

In their private correspondence about the new institute, Frank's and Morris's optimism was sincere and palpable. Frank commented once that if they failed "to integrate all the 'logical empiricists' into our Institute, confusion and even quarrels could arise which are completely unnecessary." With such integration, "we can easily discuss with people like Churchman who do not belong to us, but whom we would not like to make our enemies."[29] They figured that their colleagues would more or less support the institute – as much for its solidifying, professional effects as its social and cultural engagements with the wider world.

But publications from the institute indicate that their vision was not long shared by their colleagues. Following their programmatic essays, the remainder of the first issue concerned semantics and "abstract objects in science" with articles by Reichenbach, Hempel, Gustav Bergmann, John Lotz, Quine, Max Black, and Alonzo Church. Frank again led the second issue of "Contributions" with an essay on "The Origin of the Separation between Science and Philosophy," in which he pursued his contention that science had the capacity to analyze and interpret its own concepts in all their historical and sociological richness. The issue was then rounded out with a bibliography of writings on the sociology of science by Bernard Barber and Robert Merton, a bibliography on formal logic by Alonzo Church, and a report on "Some Significant Trends toward Integration in the Sciences of Man" by Morris's student Laura Thompson.

The second issue was shorter than the first (roughly 70 pages versus 95) and, as the series continued into its fourth and final issue, the number of authors declined. The third issue (of 1953) featured short articles by Carnap on inductive logic (Carnap 1953) and Quine on "mental entities" (Quine 1953), a "Bibliography of Cybernetics," and a long article about the Dutch "Significs" program by one D. Vuysje. The fourth issue, of 1954, contained only one, eighty-page article – "Psychoanalysis and the Unity of Science" – by Else Frenkel-Brunswik (1954). Because these more technical articles do not engage or refer to Frank's and Morris's programmatic essays or their agendas, and because the number of authors and articles in each issue diminishes over time, it is hard to escape the impression that the "Contributions" series began to die as soon as it had appeared.

Frank's Calls for Papers

One reason the number of articles declined was that Frank struggled to find papers that would satisfy his superiors at the American Academy of Arts and Sciences. The institute's agreement with the AAAS specified that

we are going to publish mostly such papers which are interesting to scientists in different fields. Above all, we would like to publish such papers which are

---

[29] Frank to Morris, November 28, 1947, CMP.

stimulating scientists of different fields toward mutual cooperation. This means in particular the cooperation between the sciences, the social sciences, and the humanities.[30]

But Frank had trouble getting such papers from his colleagues on the institute's board. "In the case of our *Proceedings* everything isn't all right," Frank once wrote to Nagel. The academy's president, Ralph Burhoe, explained the problem to Frank, who now explained it to Nagel: "As you know, there has been from the beginning certain contrast between the intentions of the Academy and of some members of our board":

In the first issue [of "Contributions"] it has appeared that a great many members of the Academy feel that they can not get out of it as much as they would like. Particularly the articles of Mr. Bergmann and Mr. Hempel contain terminology which is not familiar to the average scientist.

Frank tried to persuade Nagel that the publications of the institute needed to be tailored to meet these expectations of the AAAS:

I see a general feeling going up which you have to take into consideration. It is particularly necessary that the first year contains sufficient material which is in the line of what the Academy expects of our Proceedings.... [They] are particularly interested in the integration of the different fields.... I noticed that they wish that the Academy should do something in this line, and that this wish becomes stronger and stronger.[31]

Much to his frustration, however, Frank's colleagues seemed more inclined to offer him only technical philosophical papers that "don't fit in the program of the Academy"[32] and thus belonged more in *Synthese*.

As deadlines for the 1952 issue grew near, Frank was effectively begging Morris and Nagel for papers. He had already asked Morris to "provide us with such articles either written by yourself or by some of your collaborators"[33] when he put the same question to Nagel:

The most urgent thing seems to be to find a number of say, five or six papers which we could print in our Proceedings in the line which is expected by the Academy.... I would like you to think it over and see whether you could make a suggestion on from whom we could get such papers in the near future.[34]

---

[30] Frank to Morris, October 5, 1951, CMP.
[31] Frank to Nagel, February 5, 1952, CMP.
[32] Ibid.
[33] Frank to Morris, October 5, 1951, CMP. This request most likely led to the paper in *Contributions* by Morris's student Laura Thompson.
[34] Frank to Nagel, February 5, 1952, CMP.

The problem was not confined to Frank's board members. The institute's attempt to reach out to the public and students with its essay contests did not go well, either. The first contest drew only twenty-five submissions (which were "not very good," Frank told Warren Weaver) and the second was won by Neurath's friend, biologist Joseph Woodger (who himself wrote a monograph for the *Encyclopedia*).[35]

The institute was officially connected to the AAAS through its committee on the Unity of Science. From this Committee's perspective, things were not going well for the institute as it neared its third year of existence. At one meeting, Frank, Burhoe, B. F. Skinner, and Stanley Stevens could not manage to solve the several problems and complaints that had arisen. In addition to the complaint that the papers Frank was publishing in *Contributions* were not accessible to nonspecialists, the academy was unhappy that the series was selling poorly. It blamed the institute for a "scattered appearance of issues [that] had handicapped the campaign to build up circulation." Stevens additionally regretted the "rather unexciting format of the Proceedings" and recommended that the series have "an editor with a strong and clear policy and a fairly free hand" as well as a budget. Since there was no stream of articles for Frank to work with, he recommended that the institute should pay authors in order "to get good articles."[36]

In general, the committee felt that Frank's institute had an identity crisis. If its publications were to be mainly "popular" (in a sense that included "consumption by scientists in other fields"), many other magazines and journals had already saturated this market and the institute seemed superfluous. On the other hand, the AAAS could better stand behind the institute were it to publish "primary and/or secondary papers stemming from any of the scholarly or scientific disciplines" which are additionally germane to promoting an "over-all philosophy of science of knowledge." In that case, however, the committee returned to the issue of relations between Frank's institute and Churchman's *Philosophy of Science*. Again, Frank's operation seemed dispensable: Perhaps, Stevens suggested, "the unity of science people [could] get together and unify their own

---

35 "WW's diary," October 27, 1951, RAC RF 1.100, box 35, folder 284. The essay topics were "The Divorce between Science and Philosophy: Its Historical Origins, Its Logical Basis, and Proposals for Its Termination" and "Mathematical Logic as a Tool of Analysis: Its Uses and Achievements in the Sciences and Philosophy" (Frank's report to the RF, November 13, 1956, RAC RF 1.100, box 35, folder 285).

36 American Academy of Arts and Sciences, Committee on the Unity of Science, Meeting Minutes, December 16, 1952, CMP.

organizational structure" and merge their operations with *Philosophy of Science*. The committee closed its meeting without having agreed on "a program for the future of *Contributions to the Analysis and Synthesis of Knowledge*." That program remained "unfinished business" until, as we see in chapter 15, the Academy chose a new journal, *Daedalus*, to promote its interests in the unity of science.[37]

Conferences, "Boston-style" and Other

One glimmer of light in this darkening picture of Frank's institute was a conference in Boston in December 1953 from which emerged the volume of essays, *The Validation of Scientific Theories* (Frank 1957). Frank remained at that time a member of the Governing Committee of the Philosophy of Science Association and helped to arrange the conference in conjunction with the American Association for the Advancement of Science and the new National Science Foundation. Frank led the volume with a programmatic essay that introduced the themes of unified science, science's roles in society, and the sociological factors affecting theory acceptance. Essays by Richard Rudner (1957) and H. Barrington Moore (1957) followed up by addressing values and politics, respectively, in scientific reasoning. Essays dedicated to operationism then posed questions about which theories and concepts legitimately belong to a unified science – questions that became more specific in the third and fourth sections of the book dedicated to psychoanalysis and the concept of "organism" in biology. The last and fifth section, "Science as a Social and Historical Phenomenon," presented articles by two historians of science (Koyré and Guerlac), a psychologist (Edwin Boring), and the philosopher of physics Robert Cohen.

While most of these topics reflected Frank's interests, the 1953 conference more fully represented Frank's agenda than the published *Validation* volume. "Validation of Scientific Theories" was the topic of the leading symposium on December 27. Subsequent topics included "Art and Science," "Recent Developments in Philosophy of Psychology," "The History and Philosophy of Science," and the closing symposium's "Science and General Education." These topics, and the fact that the articles in *Validation* were also published in the popular magazine *Scientific Monthly*, suggest that the movement under Frank had begun to recover its stride and had found both popular and scholarly audiences for its interdisciplinary, humanistic approach to understanding science.

[37] Ibid.

Despite the success Frank was enjoying with the 1953 conference, however, there is evidence that other leaders in the profession did not consider his project fully representative of their goals and methods. It may be only coincidental that Herbert Feigl and Hans Reichenbach did not contribute papers to the *Validation* volume. But it is clear that they were not fully satisfied with the institute, its mission, or Frank's leadership. They wished to conduct professional meetings somewhat differently, as Feigl summed up in a letter he wrote to Frank in 1951:

As we agreed in our brief conversation, it would be very desirable to arrange for a meeting of our Institute on Issues of the Logic of Science during late August or September, for one or two weeks of intensive but informal discussions; of a well-selected group, preferably not larger than a dozen participants (and *not* open to the public). This sort of a conference has long been an idea cherished by Reichenbach and myself.

Feigl provided a list of six possible topics: alternative logics and their application, modal logic and causality, empiricism and conventionalism, alternative types of analysis and reconstruction, confirmability and meaning, and realistic empiricism versus phenomenalistic positivism. Feigl's personal hope was to dedicate the entire meeting to one topic, "induction and probability." He then provided a list of twenty-nine names from which a dozen speakers could be chosen, most of whom were established philosophers. Feigl clearly did not relish watering down the venue with lesser lights unless there were "a few *highly* promising young scholars (or even graduate students)" who could contribute to achieving "a real meeting of minds or . . . a full threshing out of the issues involved." Though he did not name names, Feigl reminded Frank that some combinations of the personalities involved might be explosive and that his leadership skills would have to be well honed. "Under the leadership of a congenial and psychologically insightful chairman," Feigl wrote, "many of the personality-difficulties may be overcome and the meeting made a genuine success."[38]

Two features of Feigl's letter stand out. The first is that Feigl had already conceived and planned this conference for the institute. Besides perhaps trying "to persuade the Rockefeller Foundation (or some other foundation)" to fund the conference, Frank was left with little to do but rubber-stamp Feigl's plans. Feigl had clearly discussed the conference in some detail with Carnap, Hempel, and Nagel; he provided Frank

[38] Feigl to Frank, June 4, 1951, CMP.

with their summer itineraries. "Well, those are my suggestions," Feigl concluded. But they were more than just suggestions. "Many of us have a very real interest" in "realizing this plan," he wrote as he sent carbon copies to Morris, Nagel, and Quine. The conference was held in June 1952 in New York City under the title "Discussions on the Different Theories of Induction."[39]

If Frank's role in planning the conference was small, so was the overlap between the conference and the mission and goals of the institute. Feigl's proposed topics were technical, and they would most likely not lead to publications that could help Frank satisfy his sponsors at the academy. Nor was the kind of conference that Feigl and Reichenbach "cherished" pedagogical, either within the profession or without. Only advanced graduate students could be invited, and it would be closed to the public.

About two years later, this exchange between Feigl and Frank was repeated. This time, Feigl informed Frank of plans he had made for a "Western" conference to be held in June 1953 in Berkeley. Again, Feigl wrote not to consult with Frank but to inform him that "Brunswik, Reichenbach and I have been making intensive preparations." There would be a "public (Boston-style) conference," but it would be preceded "by a restricted ('workshop'-type) conference of about 10 days duration." The topics were set: induction, foundations of physics, and methodological foundations of psychology. Though Frank had offered some funding from the institute, Feigl and Reichenbach were already in negotiation with the National Science Foundation and the Ford Foundation. All that was left for Frank to do, Feigl suggested, was to inquire about funding possibilities with the Carnegie Foundation in case the other possibilities fell through. Judging from Feigl's tone and choice of words, Frank was not even required for this task – it could just as well be done by "you or anyone else of our group or friends of our group."[40]

Out of Step

The troubles with Frank's institute and the manner in which Feigl and Reichenbach kept Frank professionally at arm's length support a picture of Frank increasingly out of step with his profession. This is not to assume that the kind of private, technical, and discipline-building conferences Feigl and others "cherished" excluded or prevented Frank from running

[39] Ibid.
[40] Feigl to Frank, February 7, 1952, CMP.

his institute with more success. Nor is it to assume that an intellectual discipline must choose to cultivate for itself either a professional, technical core or a program more accessible, public, and socially relevant. The point is simply that the institute and the Unity of Science movement could have succeeded in the 1950s only if leading, influential philosophers chose to support it and participate in it. That, however, was not happening. At least in part, subsequent chapters argue, this was because of the manner in which Frank and his institute defied the dominant moods and trends of Cold War culture in ways that Feigl's and Reichenbach's projects did not. At a time when organizers of international conferences often found it difficult to secure visas or security clearances for foreign and especially leftist intellectuals, for example, Frank wished to sponsor international exchange programs. At a time when the journal *Science & Society* was the epitome of radicalism on campuses, Frank urged his colleagues to take up research about relations between science and society, and, as we saw earlier, he chose to reprint an essay by a Soviet dialectical materialist in his contribution to Paul Schilpp's volume on Carnap. At a time when one's political beliefs and activities could lead to rumors, and rumors could lead to investigation by the FBI, Frank appears not to have adopted the "sense of caution," as Morris put it, mandated by the "climate of fear" in the early 1950s. As his colleagues and friends were interviewed by agents inquiring about Frank's politics, they either learned or were vividly reminded that this sense of caution was necessary. Life for intellectuals so out of step with the Cold War's moods and trends could quickly become difficult and unpleasant.

15

Freedom Celebrated

The Professional Decline of Philipp Frank and the Unity of Science Movement

In 1952, Frank's institute sponsored a colloquium on "Science and Human Behavior." The announcement that Frank posted suggests that the meeting was likely to elicit the heat and bluster of the values debate. It specified that

the group will be composed very largely of those interested in science and the logic of science and who will therefore be inclined to accept the working hypothesis that the methods of science can eventually be adapted to any part of nature, including human nature.

Frank clearly did not want the seminar to become a forum for critics of scientism and took a step to prevent that from happening. "Disagreement with that assumption," the announcement stated, "can more profitably be presented elsewhere."[1]

To neo-Thomists, anticommunists, and critics of scientism who exalted human values above the reach of empirical science, Frank was defiant. The goals he pursued with his institute, his campaign to reform and enlighten the Conference on Science, Philosophy and Religion, his proposals for science education, and his continuing interest in discussing dialectical materialism prevented him from joining the patriotic chorus celebrating the transcendence of American or Western social values over those of the Soviets and over merely empirical, scientific knowledge. Frank was out of step with many of his colleagues in philosophy of science, and, as we saw, many of his friends and colleagues fielded questions about him from FBI agents in the early 1950s. As the decade continued, Frank's

---

[1] Institute for the Unity of Science, General Seminar announcement, 1952, CMP.

fortunes did not improve. He observed others receive funding he and his institute no longer received, he saw the replacement of his *Contributions* series by *Daedalus* (over which he did not have editorial control), and was attacked, if not Red-baited, publicly in the pages of *Philosophy of Science.*

The Decline of Frank's Institute for the Unity of Science

Frank's defiance was one of several issues that brought about the decline of the institute. On the financial side, Frank and the institute fell from favor at the Rockefeller Foundation. Its first three-year grant was renewed in 1952 for a second three-year period, but this second grant was terminal. No more Rockefeller funding would be forthcoming largely because Rockefeller officers believed that Frank and other "older men" were dominating the movement and failing to recruit new blood.[2] The institute held (or co-sponsored) several "national" conferences,[3] but made little or no progress toward its other goals.

There is also evidence that Frank was not skilled at leading the institute and the movement. In February 1952, for instance, shortly after Frank announced the institute's second three-year grant, Herbert Feigl wrote to discuss the matter of the conference he was planning and scold Frank for the manner in which he ran things:

I suggest that you discuss the projects of the Institute by corresponding with the entire Board of Trustees. I remember that some of the members expressed some reservations about the value of the prize-essay contests. In any case it would seem very desirable to make our plans for the coming three years in a thoroughly cooperative manner.[4]

---

[2] RF trustee report, January 18, 1952, RAC RF 1.100, box 35, folder 281. Among the handful of officers overseeing Frank's support, assessments of those involved in the movement were sometimes acerbic: "the principals involved are old-timers in the Unity of Science program.... [Brunswik and Feigl] have passed the phase of original work of distinction.... For some time they have been repeating themselves with lack of open-mindedness and confident they have the answers, at least the best line of solution for problems that interest them.... Reichenbach and Feigl tend to be monologists, and show a little too much condescension towards younger men" ([Unknown] to WW, December 9, 1951, RAC RF 1.100, box 35, folder 284).

[3] The conferences held were April 1950, Boston: "Current Issues in the Philosophy of the Sciences"; May 1951, Boston: "Logic and Sociology of Science"; June 1952, New York City: "Discussions on the Different Theories of Induction"; June 1953, Berkeley: "Conference on the foundations of Physics and Psychology"; December 1953, Boston: "Conference on the Validation of Scientific Theories"; and May 1956, Cambridge, Massachusetts: "Science and the Modern World View."

[4] Feigl to Frank, February 7, 1952, CMP.

Though Feigl's comment perhaps should not be isolated from his own assertion of leadership in planning conferences outside of Frank's institute, others noted as well that Frank was sometimes less than exemplary as a colleague. At various times, Carnap, Horace Kallen, Marie Neurath, and Frank's brother Josef criticized him as unreliable or "dilatory."[5]

Frank's enthusiasms were genuine, but many of his projects advanced little beyond planning and proposal writing. As late as 1952, Frank referred to his sociology of science initiative as an embryonic idea: "[W]e should start concretely the project over the 'Sociology of Science,'" he wrote to Nagel. "Perhaps you could again address Professor Merton in this nature. I should also try to get in touch with some sociologists." Two months later, Frank wrote to Morris reporting that "we shall start soon" with the sociology project. Still, Frank seemed tentative as he asked Morris to "make some suggestions for alternative projects."[6]

Frank's proposed operational dictionary of scientific terms met a similar fate. He discussed the project with his Rockefeller Foundation funding officer, Chadbourne Gilpatric, at the 1952 meeting of the CSPR in New York City. Frank had enlisted the historian Karl Deutsch of MIT to be co-editor, and as the three chatted, Gilpatric gained the impression that the project was moving along: "I gathered from you and Deutsch," he later recalled, "that you were quite close to a satisfactory formulation of at least a few selected terms." But when Frank later discussed the project with Warren Weaver at the foundation, he described things differently. When Gilpatric learned what Frank had told Weaver, he was not happy:

I had understood you were prepared to present compact and exemplary operational definitions, which is not by any means the same as "a small essay" for every key term in which principles of finding operational definitions would be discussed. I must also note our disappointment in your intention to begin with such general, vague and ambiguous terms as are given at the end of your list of 300 basic concepts.[7]

---

[5] Carnap to Morris, January 18, 1956, CMP; Kallen to Morris, May 7 and October 19, 1957, JRMC. Marie Neurath reported to Morris that Josef Frank "also complains about Philipp's silences" (Marie Neurath to Morris, May 15, 1958, CMP). Frank once excused himself for not writing to Otto Neurath for over a year: "It seems almost incredible to me that I did not write to you long ago. I found in my desk three letters to you which I had once started and did not continue for some unknown reason" (Frank to Neurath, February 12, 1942, ONN).

[6] Frank to Nagel, February 5, 1952; Frank to Morris, April 29, 1952, CMP.

[7] Gilpatric to Frank, December 1, 1952, RAC RF 1.100, box 35, folder 279.

In a memo to Warren Weaver, Gilpatric criticized not only Frank's and Deutsch's dictionary project but Frank himself:

The value of their proposal in part depends upon how definitions will be published and circulated. Unfortunately Frank says nothing on this point. As it stands, the proposal is confused and certainly not thought through. I would recommend against RF aid.

Compared with Deutsch, with whom Gilpatric was quite impressed, Frank had clearly lost Gilpatric's professional respect:

I would like to note here that Karl Deutsch has a remarkably wide knowledge of scientific concepts, expresses himself clearly, and might be qualified to plan and execute a lexicon of the sort proposed. Frank's collaboration would be a handicap.[8]

Antiscientism and Anticommunism among Philosophers of Science

While Frank and his institute began to lose favor at the Rockefeller Foundation, increasing numbers of his colleagues in philosophy – besides Kallen and Hook, who had led the way in the 1940s – began to voice public sympathy with some of the dominant anticommunist themes and postures of Cold War intellectual life. With regard to the questions of whether and how science (or scientific philosophy) should treat matters regarding values – political, social, or ethical – two general stances were popular. The first called for disengagement and held that philosophy of science should abstain from such debates. The second held that philosophy of science should at least support the view that absolute moral and social values exist, if not also the partisan view that those of the West are plainly superior to those of the Soviets. In the early 1950s, Ernest Nagel and Martin Gardner touched on these issues and illustrated these stances.

In an obituary for Felix Kaufman, Ernest Nagel referred to "the acute issue whether questions concerning public and individual moral values are capable of being settled scientifically and objectively." Nagel approved of Kaufman's position, one that gingerly sided with Frank's: If our "rules of application" of moral terms are made explicit, Nagel explained, then "moral questions are as much capable of objective decision as are questions in physics and biology" (Nagel 1950, 467). When it came to politics and the problem of international peace and stability, however, Nagel was skeptical that science or philosophy had any relevance. Reviewing a book

[8] Gilpatric to Weaver, November 5, 1952, RAC 1.100, box 35, folder 279.

by Lewis Mumford, in which he insisted that we must "learn to participate in a durable universal communion with the whole of humanity," Nagel wrote in the anti-Stalinist *New Leader*,

It is extremely doubtful whether such a reorientation by every human being constitutes a necessary condition for the disappearance of the cold and hot wars that [Mumford] so fervently desires. There are surely respectable grounds for believing that this objective could be realized within the framework of current political organizations of men and on the basis of current moral ideals.

Nagel did not accept Mumford's utopianism because history did not support it:

As for Mr. Mumford's suggestions concerning the establishment of intimate neighborhood organizations, of public-work groups, or of international travel fellowships, there is little evidence to show that such devices have contributed much in the past to the creation of a world community, or that without fundamental political and economic reorganization on both a national and international level they are likely to do so in the future. (Nagel 1951, 22)

Nagel agreed with Kaufman and others (including Frank, Morris, and Dewey) that science and philosophy of science could illuminate values and related questions about human conduct. But with respect to the Cold War, he was resigned to quietism and disengagement. Only political and economic restructuring, and not resources offered by philosophy, could advance a "world community" or reduce international tensions.

Martin Gardner followed Hook and Kallen more closely than Nagel. When he reviewed books in the *New Leader*, he urged philosophers of science take up the cause of anticommunism and not retreat to any apolitical posture. Gardner liked Reichenbach's *The Rise of Scientific Philosophy* (1951) for its technical content. But, he criticized, "the book takes on a flimsiness" when it treated ethical and political matters. "Reichenbach argues that moral laws cannot be considered true or false since they are expressions of personal desire and directives for the behavior of others. 'Science tells us what is, but not what should be.'" Reichenbach's treatment of ethics "does not seem to me to touch fundamentals," Gardner complained, as he prodded Reichenbach to be more partisan: "When a democratic society is in conflict with a fascist society," he wrote with some suspicion, "one wonders how Reichenbach would arbitrate the dispute" (Gardner 1951, 23).

Gardner expressed similar reservations about Frank's *Relativity: A Richer Truth* (1950). In a review titled "Relativity: Hope Chest or Pandora's Box," he approved Frank's distinctions between relativity theory in physics and vague notions about cultural relativity, and he agreed that the former "does not lead, as critics of science often suggest, to a denial of moral and political values." Still, Gardner devoted half his review to articulating, if not searching for, worries about the political implications of the book. "At the level on which this book is written," he wrote, "it is difficult to find areas of disagreement." But that is not the level Gardner cared about: "there are deeper issues that Frank does not touch."

It is one thing to recognize that science never gives us final truth, and quite another to deny that the term "final truth" has meaning. Our knowledge of the center of the earth [to use one of Frank's examples] may alter as investigation proceeds, but this does not mean that the nature of the earth's core is constantly changing. Members of the Vienna Circle have a distressing habit of talking as though the cosmos had no definite structure, or more precisely, that it is meaningless to say it has. (Gardner 1950b, 26)

Like Reichenbach's, Frank's book was unfortunately "flimsy" or "silent" about "questions concerning metaphysical and ethical relativism which still plague us." Frank would have emphatically disagreed, of course, for his book was precisely a treatment of ethical relativism aiming to show how complicated and in need of contextualization most ethical claims really are. But Gardner disliked such subtleties and hermeneutic complications. There was a "definite structure" to the cosmos. Judging from the question he put to Reichenbach, that structure was not only physical and semantic, but also moral and geopolitical. Gardner found it "distressing" that these philosophers of science did not respect all that structure in their writings.

The Congress for Cultural Freedom

While Gardner's book reviews nudged philosophers of science to respect the popular moral and valuational imperatives of Cold War anticommunism, the Congress for Cultural Freedom shows how these imperatives saturated the broader intellectual culture of the 1950s and '60s. The congress was designed by a cadre of intellectuals and government officials from the United States and Europe who were united by their antipathies to totalitarianism and the Soviet Union. These included Hook, Arthur Koestler, Ignazio Silone, Melvin Lasky, James Burnham, Dwight MacDonald (editor of *Partisan Review*), Arthur Schlesinger, Jr., Nicolas Nabokov (the novelist's brother), and military intelligence experts Michael Josselson and Tom Braden. As Frances Saunders recounts it, the seed of their collaboration was Hook's counter-demonstration at the Waldorf Congress in 1949.[9] For Hook and Michael Josselson, an anti-Soviet exile from Estonia, the Waldorf Conference was part of an aggressive Soviet propaganda campaign that was proving to be alarmingly effective at persuading intellectuals that the Soviets sought peace and not totalitarian world domination. Carnap's sponsorship had helped to convince Hook that even leading philosophers were blind to the fact that tensions between Western democracy and Soviet Communism were matters of black and white, good and evil, and that all Soviet causes – even the cause of "peace" – deserved fierce resistance. If the Soviets were going to manipulate the thoughts and allegiances of intellectuals in the United States, Hook and the other leaders of the congress reasoned, they would establish a new forum to counteract Soviet propaganda and promote intellectual freedom and individualism in Europe – preferably as close to the Soviet Union as possible. Josselson had the necessary government connections to make it happen.

The first official Congress for Cultural Freedom was held in Berlin one year after the Waldorf conference. Then the congress became a yearly affair involving artistic exhibitions and intellectual debates throughout Europe and as well as intellectual magazines, including *Preuves* in France, *Encounter* in England, and *Der Monat* in Germany. By the mid-1950s, Saunders writes,

the prolific activity of the Congress . . . had made it a compelling feature of western cultural life. From the platforms of its conferences and seminars, and across the pages of its learned reviews, intellectuals, artists, writers, poets and historians acquired an audience for their views which no other organization – except for the Cominform – could deliver. (Saunders 1999, 340)

Saunders compares the congress to the Cominform, the Soviet propaganda agency, because American journalists in 1967 began to expose the congress's connections to the CIA and private American foundations (ibid., p. 382). Since the congress principally opposed the control of Soviet intellectual and cultural life by Moscow, these revelations that the congress was symmetrically and secretly funded by the CIA and these foundations greatly reduced its credibility. For the many intellectuals and artists who enjoyed junkets to European conferences and stipends or salaries for producing congress publications, the revelations were either shocking (for those who did not know) or highly embarrassing. The

---

[9] See Saunders 1999, esp. chap. 3. Hook eventually had a falling out with the congress and was an important player only in its first few years.

revelations also added to the growing anger and mistrust of the U.S. government among intellectuals and students over the Vietnam War.

Daedalus and the New Unity of Science

The Congress for Cultural Freedom opposed the Unity of Science movement and Frank's institute both conceptually and institutionally. Conceptually, the congress was sympathetic with Kallen's attack on the movement as it embraced the twinned themes of science and freedom. At the first convocation in Berlin, one of the five debates was titled "Science and Totalitarianism." A conference on "Science and Freedom" was held in Hamburg three years later. Institutionally, the congress was linked to Frank's institute via the Rockefeller Foundation, which funded congress-related projects, such as Michael Polanyi's journal *Science and Freedom.* Inaugurated and named after the Hamburg conference, this journal's start-up funds were provided by the Rockefeller Foundation and the Farfield Foundation – the "Far Fetched Foundation," to knowing insiders – which covertly dispensed CIA funds for the congress.[10] The Rockefeller Foundation also supplied the congress with use of its posh Italian retreat, Villa Serbelloni. With a support staff of fifty-three, the villa was as "an informal retreat for [the Congress's] more eminent members – a kind of officers' mess where frontliners in the *Kulturkampf* could recover their energy."[11]

Some of the Rockefeller Foundation officers who supported the congress were also those who had first supported but then lost interest in Frank's institute. The head of its division of humanities in the 1950s was Charles Fahs, a veteran of the CIA's predecessor, the Office of Strategic Services (OSS). His assistant was Frank's contact, Chadbourne Gilpatric, who was a veteran of both the OSS and the CIA. Fahs and Gilpatric were "the principle liaisons for the Congress for Cultural Freedom, and responsible for dispensing large Rockefeller subsidies" to it (Saunders 1999, 145).

In the context of their Rockefeller funding, the contrast between Frank's demise and the success of the congress and its publications is dramatic. In June 1965, the Villa Serbelloni was the site of a Congress seminar titled "Conditions of World Order" that was co-sponsored by the American Academy of Arts and Sciences and the journal *Daedalus*, one

---

[10] Saunders 1999, 77, 214, 357.
[11] Ibid., 346.

of whose founders was Frank's younger colleague at Harvard and in the institute, Gerald Holton. *Daedalus* was also among the journals that the congress supported in the early 1960s by regularly purchasing copies of each issue (500, in this case) with CIA funds and distributing them around the world.[12]

Daedalus emerged from the very frustrations that Frank struggled with when he edited his *Contributions to the Analysis and Synthesis of Knowledge* for the academy. The academy wished to publish a journal with a less awkward title than the *Proceedings of the Academy of Arts and Sciences* and that would better promote interdisciplinary dialogue among specialists in different fields. The result, which Frank had a hand in shaping, was the debut of *Daedalus* in May 1955.[13] In some respects, the new journal signaled an end to the pursuit of the goals that Frank's institute represented. First, as its editor Walter Muir Whitehill made plain, "*Daedalus* does not aim at popularization" (Whitehill 1955, 5). Holton soon succeeded Whitehill as editor and explained that *Daedalus* existed to help "leading scholars in all fields" communicate and focus "attention again on that which does or should make us members of one community" (Holton 1958, 4). Yet the journal would specifically not attempt to connect that community to the public:

Here the specialist can present an aspect of his field … to specialists in other fields, who will demand of the author only that his account be not obscured by undue allowance for his audience. The influence of *Daedalus* on the greater society, on the public at large, will be determined entirely by its impact on this smaller group. (Holton 1958, 4)

The academy and its new journal still upheld the goals of advancing knowledge and enlightening society, but the latter would be only a by-product of its primary concern with communication among specialists.

Second, *Daedalus* promoted views on the unity of science and relations between science and the humanities more similar to Kallen's libertarian view of science than Frank's (and Neurath's) collectivist, coordinated view. Three years after its debut, the journal published papers from a one-day symposium honoring Frank and Bridgman on the occasion of their retirements from Harvard. Frank read a paper, "Contemporary Science

---

[12] Other magazines supported in the late 1950s and early '60s include *Partisan Review, New Leader, Kenyon Review,* and the *Journal of the History of Ideas* (Saunders 1999, 333, 336, 338).

[13] Frank was member of a "special committee" that assisted the academy's Publications Committee in planning the new journal (Whitehill 1955, 4).

and the Contemporary World View" (1958) and revisited the themes of his book *Relativity: A Richer Truth.* He attacked a popular inductivist view of science "as the product of abstraction from our rich and full experience" (Frank 1958, 59) and emphasized scientific creativity and imagination (such as Einstein's). He also revisited his call for pragmatic, operationalist semantics as a tool for understanding science and exposing simplistic popularizations about the political or cultural meanings that science supposedly reveals (inductively) in nature. In a paper citing writers as diverse as Archibald MacLeish, Wittgenstein, and the Bible, Frank urged that science be recognized as a humanistic, creative endeavor, not entirely isolated from art and poetry. Science therefore did not deserve its current reputation as aloof, highly specialized, and not capable of engaging questions about "our way of life":

> Other authors, viewing science as a dehumanized abstraction, conclude that science cannot have any influence on philosophy as a world view which determines our way of life. Evidently this guidance should be left completely to irrational sources of knowledge, for example religion, metaphysics and, implicitly, political philosophy. (ibid.)

Frank found it inexplicable and culturally dangerous that science and philosophy were so specialized, partitioned, and isolated from matters of society and politics.

Physicist Robert Oppenheimer responded to Frank's paper and did not share his regrets about specialization. Science was all about specialization, Oppenheimer explained. He offered a picture of science similar to Kuhn's, according to which sciences mature as they become sufficiently specialized to experience and interact with nature in distinct, individual ways, ways that are "a world apart" from those of other sciences (Oppenheimer 1958, 69).

Oppenheimer's talk also echoed Kallen's attack on the Unity of Science movement over ten years before. Specialization, after all, is a kind of pluralism. Oppenheimer approved this pluralism but recognized that many were intimated by the rapid growth and proliferation of different areas of scientific knowledge. Yet there may remain, he suggested, a comforting "unity of culture" that held the sciences as well as the humanities together. If so, this

> unity, I think, can only be based on a rather different kind of structure than the one that most of us have in mind when we talk of the unity of culture. I think that it cannot be an architectonic unity, in which there is a central chamber into which all else leads, the central chamber which is the repository of the common

knowledge of the world. I think that it cannot have the architectural coherence of a hierarchy. (ibid., p. 75)

Though generalized to embrace both science and the humanities, Oppenheimer's "central chamber" is the image of totalitarian science in which central planners make blueprints to be imposed on the otherwise free thoughts and activities of those ensnared in the system. As he nodded to Kallen's side in his response to Frank's paper, Oppenheimer used the same metaphor that Neurath and Kallen used to debate the expressive and communicative possibilities of physicalist language: He wrote,

We can have each other to dinner. We ourselves, and with each other by our converse, can create, not an architecture of global scope, but an immense, intricate network of intimacy, illumination and understanding. (p. 76)

Those bewildered by specialization and the growth of knowledge, Oppenheimer suggested, could still take comfort in this *new* kind of unity, nonarchitectonic and not imposed from without.

Not every paper in the conference so echoed Kallen's libertarian and anti-architectonic view of knowledge. Presentations by Bridgman, Morris, and Howard Mumford Jones agreed with the possibility of dialogue, even active collaboration, between the sciences and humanities. Theoretical and conceptual links might even be forged among unlikely areas of knowledge and culture. Bridgman called for brain research, guided by insights about art and artistic experience, while Morris promoted a "new synthesis" of knowledge based on a relation of "complementarity." Holton, however, seemed to prefer Oppenheimer's new image of unity and quoted him when he introduced the essays. "There may be real hope for reaching a common understanding" between the sciences and the arts,

but not by the imposition of a new system from without to replace the unified world systems we have lost. Rather, the way to the goal is precisely through discussions such as these. As Robert Oppenheimer says, "We ourselves, and with each other by our converse. . . . " (Holton 1958, 4)

Even Bridgman's, Morris's and Jones's papers, Holton suggested, ultimately pointed in the same direction toward this basic circumstance of "modern thought":

In the last three articles, P. W. Bridgman, Charles Morris, and Howard Mumford Jones look to the future. They reject the opinion, held by a number of influential spokesmen, that the sciences and humanities are in detrimental conflict. But they also hold out little hope for a discovery of some master-plan for fitting into one pattern our separate preoccupations. And this sums up both the triumph of modern thought and its despair. A synthesis on a large scale is possible neither within the individual person nor among the several fields of learning. But as a consequence the need is greater than ever to recognize how small one's own portion of the world is, to view from one's own narrow platform the search of others with interest and sympathy, and so to re-establish a learned community on the recognition that what binds us together is mainly, and perhaps only, the integrity of our individual concerns. (ibid., p. 6)

Just as Kallen had concluded that the real meaning of unity was "nothing else than the congress of the plurality of the sciences for the unified defense of their singular freedoms against the common totalitarian foes" (Kallen 1940, 83), this issue of *Daedalus* confirmed that contemporary knowledge was particular, pluralistic, and servant to no "master plan" imposed from without.[14] *Daedalus*'s version of the unity of science, Kallen would have happily agreed, was hardly "totalitarian."

Questions about the Unity of Science Movement

There is no direct evidence that the Rockefeller Foundation terminated their funding of Frank's institute only or primarily for political reasons. There were real administrative problems with Frank's leadership, as mentioned above, and there were real concerns that the movement was not making contact with scientists engaged in real, borderland research in areas such as biophysics and biochemistry.[15] Nor is there evidence that Frank's officers knew that Frank was, or would be, under investigation by the FBI. But there is evidence for what should not be surprising given the ordinary exigencies of Cold War intellectual life: Officers at the foundation who evaluated the Unity of Science movement and its project that they funded, or considered funding, took politics into account.

Shortly after the war, when Morris and Frank were planning to approach the foundation for help to rebuild Neurath's institute, Morris sent them a letter describing a proposal he had made to UNESCO, for an "Institute for the Study of Man." Morris proposed that UNESCO "assume a central role in coordinating and advancing our knowledge of man" and provide a forum that "could be the center of integration for the various scientific approaches to man, and a meeting place for scientist and

---

[14] For another early essay in *Daedalus* that celebrates the themes of individualism and "unorthodoxy" in both science and statesmanship, see Haskins 1956.

[15] Report on "Philipp Frank and the Unity of Science Movement" by WFL, May 19, 1950, RAC 1.100, box 35, folder 284.

humanist." Morris passed a copy of his letter along to the Rockefeller Foundation because, as he penciled in the top margin, it was a letter "indicating the possibilities before the Institute for the Unity of Science."

In some quarters, UNESCO would later come to be seen as suspiciously pink. Hutchins singled it out as one of the controversial topics that McCarthyism was making taboo (Hutchins 1954, 27). Whether or not it had such a reputation in 1947, Morris emphasized in his proposal that the unified science of man he promoted was on the right side of the Cold War divide: This new, integrated science "does not restrict but rather enhances man's individual and cultural freedom."[16] At least one officer at the foundation was not so sure, however. As Morris's letter circulated through the foundation accumulating signatures and brief comments, one wrote, "sounds good, only I hope there's no faint tinge of pink in it."

Frank's institute could not have seemed excessively pink to the Rockefeller Foundation, for they supported it for six years. But this comment and others suggests that there were reservations about the politics of Frank's project. One officer circulated a three-page report casually ruminating on questions raised in his mind by the Unity of Science movement and its activities. "The most striking word in the Vienna Circle's vocabulary is 'unity.'" The writer, William F. Loomis, promptly walked directly into the question Kallen posed in 1939 and again in 1945. "Frankly," Loomis wrote, "I ask what does it mean?" His answer was not Kallen's answer. But as he explored in his memorandum a couple of different meanings of "unity" that he could imagine, it led him to pose the question of "whether the Vienna Boys have really isolated any one meaning for their 'unified' attack" on the problem. One of the report's readers agreed with the doubts and lack of confidence expressed in the question. "Sometime I would like to know more about just what it is that the U of S Movement seeks to accomplish," he penciled on the first page.[17]

Frank's unpopularity at the foundation, it would appear, was part of a larger suspicion about what "the Vienna Boys" were up to. On one occasion, Frank spelled out to Weaver, head of the Foundation's Natural Sciences Division, some of logical empiricism's political and cultural significance. The institute's mission included sponsoring international conferences in philosophy of science because (as Weaver recorded it) "it is politically important to support activities which emphasize empirical

---

[16] Morris to Mayoux, November 18, 1947, RAC RF 1.100, box 35, folder 281.
[17] Report on "Philipp Frank and the Unity of Science Movement," from WFL, May 19, 1950, RAC 1.100, box 35, folder 284.

and pragmatic philosophies, since Fascism and Communism depend essentially on metaphysical doctrines." Though he didn't say why, Weaver pointed out to his colleagues that he found the argument "very tenuous and unconvincing."[18] In any case, Weaver's colleagues Fahs and Gilpatric in the Humanities Division were already supporting international conferences of the Congress for Cultural Freedom. These conferences, unlike those Frank proposed, were not designed to fight totalitarianism with empiricism, pragmatism, or any kind of scientific philosophy that might sever its metaphysical roots. Instead, they pitted absolute values of the West against those of the Soviets and sought to influence nonpartisan European intellectuals ("neutralists," the Congress organizers called them) before they fell under the spell of Soviet propaganda. Into an ongoing war of values and propaganda, in other words, Frank proposed introducing the tools of intellect and reflective epistemological analysis. It is perhaps less surprising, therefore, that Frank came to be viewed as an outsider in the foundation's circles.

Unity of Science as a Communist Theme

Compared with the Congress for Cultural Freedom, Frank's institute and its projects had weak antitotalitarian credentials. That was not only because the congress was dedicated to anticommunism, but because the goal of unifying the sciences became increasingly associated with radicalism after the war. Communist and radical philosophers did not embrace the goal with more vigor than they had in the 1930s. But their voices may have seemed louder in the 1950s because so many other intellectuals who once supported the unity of science (and Neurath's Unity of Science movement) moved to the political right, refrained from political engagements altogether, or no longer endorsed unifying the sciences because of the "totalitarian" implications it seemed to possess.

In the 1930s, Maurice Cornforth was the among many leftist philosophers who was welcomed by the Unity of Science movement. Herbert Feigl once reminded Neurath to invite Cornforth to the first International Congress.[19] By 1950, however, when Cornforth published his book *In Defense of Philosophy against Positivism and Pragmatism* (1950), any sense of mission shared by Marxist philosophers and leading philosophers of science was greatly reduced. In his new book, much as in his essay in

---

[18] "WW's Diary, Tuesday, September 27, 1951" RAC 1.100, box 35, folder 285.
[19] Feigl to Neurath, December 11, 1934, ONN.

Philosophy for the Future, Cornforth took logical empiricism to be counterprogressive. As he rehearsed Lenin's complaints about positivism, he accused Morris, Frank, Carnap, Reichenbach, and Schlick of subjective idealism and for promoting an excessively formalistic approach that was "alien to the development of science as real knowledge of the external world."[20]

Yet one logical empiricist, in Cornforth's eyes, received a different and much more friendly treatment. In a section titled "Semantics and Sociology," Cornforth largely applauded Neurath's much-maligned monograph for the *Encyclopedia* (Neurath 1944):

> Neurath inveighs against attempts to describe and explain historical events in terms of high-order abstractions. For instance, he says, cases of wars and conquests are often described by historians ignorant of semantics in such terms as these: "Forced by its historical mission, the nation started to spread its civilisation." Here, he says, are three well-nigh meaningless abstractions. The correct account of such an event, Neurath maintains, would be rather as follows: "One human group killed another and destroyed their buildings and books." (Cornforth 1950, 123)

Neurath's criticism "seems justified," Cornforth wrote, because the abstractions he rejected are "idealist abstractions employed by reactionary historians," and his physicalism wisely sided with materialism in seeking always to avoid abstractions and "replace them by bald statements about the [particular] actions of particular men" (ibid., p. 124). Though Neurath sometimes went too far in his rejection of abstractions, Cornforth complained,[21] he singled him out from other logical empiricists. Neurath, it seemed to Cornforth, was on the right track.

One reason Cornforth admired Neurath was that they shared a commitment to unifying the sciences. Unity would strengthen their epistemic and technological powers as tools to modernize and improve the world. A unified science would be "a weapon of enlightenment and material progress" (p. 155). Other logical empiricists, notably Carnap, had turned against these causes, Cornforth argued, and taken refuge in formal studies of language. With Neurath, on the other hand, Cornforth agreed that Marxist socialism and the unity of science were closely related. Neurath,

---

[20] Cornforth 1950, 18. For examples of Cornforth's criticism's of Morris, see 225, 235; of Frank, 141–43; of Carnap, 17; of Reichenbach, 75; of Schlick, 13–14.

[21] Neurath's sociology, Cornforth stated, was powerless "to say why such wars happen, which wars are just and unjust, how wars are determined by economic factors and class interests, and what part the various human institutions play in them" (Cornforth 1950, 124). Note that there is no obvious reason why Neurath's sociology would be incapable of addressing "economic factors and class interests" if that is where empirical sociological inquiry led.

for example, once described "bourgeois science" (Neurath 1928, 296) as uncoordinated and lacking in overall vision:

> The bourgeois formation of thought is indefinite . . . proceeding here in an anthroposophic way, there mathematicising, here psychologising, there pursuing the idea of fate, here in a technical manner, there in an occultist one. The wealth of scientific detail is no longer held together by a unitary approach, and in a certain sense it is left to chance whether a man thinks about some linguistic formations in Chinese or about a medieval legal text, about African Beetles or about wind conditions at the North Pole. (ibid., pp. 294–95)

Neurath hoped that the Unity of Science movement would remedy this condition by encouraging scientists to collaborate across disciplines and collectively to plan the future of science. In an essay promoting his encyclopedia project, he recommended "serious efforts of organisation." "This new encyclopedia has to induce its collaborators to agree among themselves in order to push the unity of the form of their contributions as much as possible" (1936a, 141). This would, he hoped, create a "framework" in which "scientists, though working in different scientific fields and in different countries, may nevertheless co-operate as successfully within unified science as when scientists co-operate within physics or biology." "The maximum of cooperation," Neurath enthused on another occasion, "That is the program!" (Neurath 1938b, 23, 24).

In his new book, Cornforth also applauded the unity of science as an antidote to overspecialization and "bourgeois science" in which

> some branches of science [are] developing in a one-sided way while others lag behind. Science is called upon to answer just those particular problems in which the capitalist monopolies are interested, which is by no means the same as answering the problems which are bound up with the future development of science and with the interests of the people. (Cornforth 1950, 245)

"The real unity of science," Cornforth explained, could be achieved

> by the organized pressing forward of research in all fields of science in accordance with a single plan – directed towards a single practical goal, the enlargement of knowledge in the service of the people. (ibid., p. 156)

Neurath would have largely agreed with Cornforth this far in his discussion. His Unity of Science movement should never lose sight of practical goals such as enlightenment and ameliorating human life.

There remained, of course, an overwhelming difference between Cornforth's and Neurath's programs. Cornforth was committed to dialectical materialism as a kind of superscience that would guide the development of science's unity. This "single practical goal" of attaining a unified science to serve the people would be "informed by a single scientific method, the method of dialectical materialism" (ibid.). Neither Neurath nor Frank accepted dialectical materialism, much less as a central organon for the Unity of Science movement. But this difference appeared to get lost as the unity of science in popular and intellectual discourse came to seem wedded to totalitarian control *over* science.

Nagel reviewed Cornforth's book and summarily dismissed most of it as misrepresentation and non sequitur in the service of Communist propaganda. Yet Nagel took care to make a particular point:

> But there is one feature of the book which is perhaps more important than its general philosophic incompetence – namely, the explicit assumption that science flourishes best under regimentation. Mr. Cornforth finds that the development of "bourgeois" science, like that of capitalism, is anarchic and uncoordinated, and that its theories "essentially distort and mystify our conceptions of the world and of human relationships and activities." (p. 247) And he believes that just as capitalist production must inevitably be replaced by socialist production, so bourgeois science must be succeeded by a planned, unified science. (Nagel 1952, 650)

Nagel then cited Cornforth's description of "the real unity of science" (given above) and took it as self-evident that "planned, unified science" would be impossible to achieve without also killing science itself. "If Mr. Cornforth has his way," he concluded at the end of his review, "the conditions under which modern science can contribute to human welfare and human dignity may disappear from the face of the earth" (ibid.).

In one sense, it would seem, Kallen had won his debate with Neurath. Kallen's premise that science was essentially libertarian, and his claim that any attempt to coordinate or organize the sciences among each other would harm them, were now shared by Nagel's review, much as they were by Oppenheimer's essay and Holton's editorial in *Daedalus*. Neurath was no longer able to offer his opinion, but he would have been right to say that while he had lost his debate, the profession had lost something as well. It had lost appreciation for the possibility that science (and, in turn, society) might benefit from a kind of collective, democratic planning among and across disciplines that was neither prescribed by nor expected to conform to any prior theory of science, society, or history. Kallen's dichotomy – holding that science was either free and good or controlled and corrupted – left no room for Neurath's middle way in much the same way that popular political discourse allowed no middle ground between democracy and totalitarianism. Neurath's notion of a democratically, freely unified science, like noncommunist socialism, was suspect, if not unrecognizable, in a Cold War structured by rigid, exclusive alternatives.

Kegley's Attack

The 1950s were dismal for Frank. The decade began with high hopes for the new institute, but was quickly dragged down by the institute's problems, his investigation by the FBI, his loss of funding from the Rockefeller Foundation, and the celebration of intellectual freedom that cast a suspiciously pink light on the goals and values of his and Neurath's Unity of Science movement. The way that the movement collided with these exaltations of intellectual freedom and individualism is perhaps the reason that some philosophers decided to talk and "name names" of philosophers whose work and beliefs seemed so out of step.

One informant the FBI spoke to in its investigation of Frank was sufficiently familiar with logical empiricism and its history to discuss the political currents within it. The informant who dismissed Frank as a "talkative old man" described him as "an exponent of logical empiricism" who in his younger days was "under the influence of Marxist ideology" at least partly because he was a Jew who had to flee his homeland. But Frank later moderated his leftism, the informant said, and was never as "leftist or radical" as another logical empiricist who was "a former member of a brief Communist regime in Munich, Germany in 1919." The informant seemed to have forgotten the name, but it could only have been Otto Neurath.

To this informant, philosophy of science of the sort that Frank represented ought to be eliminated for the good of the profession. Consider his concluding remarks as the FBI agents recorded them: "Upon further consideration" of Frank, the report reads, the informant

had come came to the conclusion that while subject's [i.e., Frank's] intellectual views could be harmful in their effect on American scholarship, actually [the informant's] personal views referred to academic philosophy rather than political ideology. (April 29, 1954, Boston summary report)

However cryptic the remark or the transcription, the informant unmistakably distinguished the FBI's interest in Frank as a political subversive with his own worries about Frank's role in the profession. He "could not possibly be considered dangerous to the security of the United States," but he was indeed harmful to "American scholarship" and "academic philosophy."

In 1954, this may have been an isolated opinion about Frank or it may have been part of more widely shared criticism of his projects and goals. In 1959, in any case, a similar criticism became public. That year, Charles W. Kegley attacked Frank in a book review published in *Philosophy of Science*. Kegley's essay, "Reflections on Philipp Frank's Philosophy of Science," is an animated and angry review of Frank's book, *Philosophy of Science: The Link between Science and Philosophy* (1957). Kegley argued that Frank was highly confused about the issue of theory acceptance, and he diagnosed that confusion by claiming that Frank had recently converted to neo-Thomism. Only a few years after he helped to create *Daedalus*, which opposed architectonic systems of thought, Frank was now being fingered in the main journal of his profession as a slave to Thomism.

Frank mentions Aquinas several times in his book and attributes to him basic epistemological insights that could as well be attributed to many others. Far from being "beholden to Aquinas," as Kegley charges (Kegley 1959, 39), Frank almost certainly nodded to Aquinas because he was long engaged, as we saw in chapter 11, in a sustained attempt to persuade neo-Thomists that they had nothing to fear and much to learn from understanding philosophy of science (even certain parts of Aquinas's). Indeed, Frank nodded approvingly and politely in many intellectual directions. Beginning with some epistemological preliminaries, he examined aspects of the history of physics and the topic of theory choice and, at the end, he once again plugged for sociology of science and the ways by which social and cultural factors can influence the reasoning and conduct of scientists. Far from a Thomist treatise, the book, as its subtitle declares, is Frank's attempt – his last, it turns out – to lead his profession into those border areas between science and humanities that he considered most important.

Kegley was aware of Frank's strategy, for he began his review by describing the professional growth of philosophy of science. "Scarcely five years ago there were only two or three books that could have served as textbooks in such a course," Kegley wrote. "Now a teacher has a choice of at least five more. One a year is not a bad rate" (ibid., p. 35). Kegley clearly supported all this growth, but not Frank's views about where it should lead. Specifically, he objected to Frank's view that scientific theories are chosen – in part, and when theory choice is undetermined – for practical and social reasons, such as the "guidance of human conduct" they offer (ibid.). Kegley believed that this view dangerously hand[s] over to the scientific chooser between theories the responsibility of making simultaneously a factual and a value judgment, namely: that a certain theory will, in fact, have precisely such and such results in human conduct, and furthermore that these are the results that scientists *ought* to try to produce.

Here Kegley pinned on Frank two opinions that were unpopular during the Cold War: First, with regard to social and economic planning, he resented Frank's suggestion that responsible planning is even possible, that "it is possible ordinarily to predict the results which a given theory will produce." Second, nodding to the values debate and the popular dichotomy between scientific facts and cultural and moral values, Kegley criticized Frank for assuming "a positive answer to one of our more hotly debated questions today, namely, is the scientist qualified or even expected to determine the guidance of human conduct?" (p. 36).

This confusion of facts and values made for a disastrous philosophy of science, Kegley argued. On the topic of theory choice, he wrote, "one should imagine that the clear-headed quest for explanation of facts ought, always and everywhere, to be the final ground of choice. But with Professor Frank this is by no means clearly the case" (ibid.). Of course, Frank did not deny that "agreement with observations" or "agreement with the facts" (Frank 1957, 353) were good criteria for theory choice. What Kegley could not abide – especially when the education of young, impressionable minds was at stake – was Frank's insistence that relations between theory and evidence were more complicated, often by sociological factors.

I fear that the student who uses this book as a text would be in doubt about so fundamental matter as the direction which scientific thinking takes.... What is to become of the student who, in half a dozen courses, including logic, thought he had it straight that one proceeds from observation to generalization on the basis of observed facts? (Kegley 1959, 39)

Kegley believed scientific facts and observations were simple and brute, uncomplicated by the kind of hermeneutical and sociological factors that interested Frank, and that science was a straightforward, inductivist affair. The scientist's job, as Kegley saw it, was merely to attend to the facts of the world, induce some general theories, and emphatically leave questions about values and the conduct of life to be explored by others. If Kegley's philosophy of science was naive, his essay was even more confused, for if Frank really had gone Thomist (in the style of Adler, at least) his book would have trumpeted precisely such a sharp dichotomy between facts and values.

Did Kegley Red-Bait Frank?

Passages in Kegley's review suggest that he aimed to do more than criticize Frank's philosophy of science. Were his goals only intellectual, it remains difficult to explain Kegley's angry tone and his preposterous charge that Frank had capitulated to neo-Thomism. What would explain Kegley's exaggerations, however, is the possibility that he intended it not only as a review of Frank's book but also as a broader denunciation of Frank himself and his project. Several features of his review suggest this reading.

First, Kegley's essay has a dual focus on Frank's book (and the views contained in it) and on Frank himself. Given Frank's "leading role in the Philosophy of Science Association and the Institute for the Unity of Science," Kegley wrote, there is all the more reason to examine the book. It raises an "important question" not only about "trends in thinking in this field" (Kegley 1959, 35), but also about "Frank's thinking." The two foci are blurred at the outset by Kegley's title, "Reflections on Philipp Frank's Philosophy of Science." It does not distinguish the book under review from Frank's overall conception of science, his "philosophy of science." Second, when Kegley refers directly to Frank, his words sometimes call out to popular stereotypes of communist agents and intellectuals. He gives the impression, for instance, that something has happened to this one-time leader in philosophy of science, that something has gone wrong. Frank's colleagues, when they read this book, "are likely to raise their eyebrows in surprise at a number of points," for Frank has "much to say about science, but says it in a somewhat special way and, as I wish to point out, with rather astonishing assumptions" (ibid., p. 35). This "somewhat special" aspect to Frank's recent thinking is this alleged conversion to aprioristic Thomism:

Is philosophy necessarily concerned with first principles? Something has evidently developed in Professor Frank's thinking in recent years which enamors him with this notion, and for some strange reason its association with Aquinas gives the view added reason for being treated with reverence. (pp. 37–38)

The result of all this is "disturbing to say the least," Kegley gasped (p. 38).

The professional message of Kegley's essay, it would appear, is that Frank had fallen into not the usual grip of communism or dialectical materialist orthodoxy but an authoritarian intellectual program. Frank was now "beholden" to a kind of foreign agent, and, third, he was not sufficiently open and honest with his colleagues about his new allegiance. Kegley's most accusatory paragraph leads inexorably to the popular dichotomy between "censorship and thought-control" and "free inquiry":

> Any author who presents a philosophy of science which is beholden to Aquinas and the Thomistic system ought to scrutinize his task and its implications and to state his position forthrightly. Judging from Professor Frank's many and excellent contributions to scientific thought, it is difficult to believe that he now wishes to place himself within this framework. He does no service by advocating Aquinian positions, for, as Bertrand Russell succinctly puts it, Aquinas was not a philosopher in the first place, if one means by philosophy the free inquiry into truths rather than possession in advance of the essential truth. Furthermore, in an age which suffers from censorship and thought-control by secular and religious authorities, the cause of free inquiry is hardly aided by reverence for an age which, after all, burned Giordano Bruno and forced Galileo to recant. Surely our comprehension of the spirit of philosophy and of science, and the strengthening of that spirit throughout the world today, lies in other directions. (p. 39)

Neither this man nor his views, Kegley declared, deserved a position of leadership in the profession. Philosophy of science had a role to play "throughout the world today," he admitted, but not the role Frank wanted it to play.

One year after Kegley's essay was published, Frank's student F. James Rutherford came to his defense with a short article describing the several logical and methodological errors in Kegley's arguments. Rutherford knew that Kegley's attack was no ordinary (or fair) exercise in philosophical criticism. After showing that Frank was not advocating Thomism in substance or method and that Kegley's rhetoric and indignation amounted to nothing, Rutherford turned to the institutional and professional issues that he recognized behind the attack:

> One last point. The philosopher of science must have something to say to intelligent, curious, questioning human beings who may stand all too confused about the meaning of science for their own lives. In particular he must stand ready to assist the students and teachers and practitioners of science itself in coming to grips with the philosophic foundations of their subject. Part of the problem is that the teacher of physics, say, in searching around for books on the philosophy of science suitable for himself and for his students must usually rely upon the judgement of the professionals as they appear in journals such as this one. It would be unfortunate indeed if Frank's *Philosophy of Science* should receive less attention than it deserves because of Kegley's article. Professor Frank has made a notable attempt to bring some important notions to a wide audience. That Charles Kegley thinks otherwise is not nearly as discouraging as the likelihood that (opening and closing pleasantries notwithstanding) he really is out

of favor with the attempt itself. We need more attempts like Philipp Frank's, not fewer.[22]

Kegley's real aim, Rutherford implied, was to tarnish Frank's book and, in turn, Frank's vision of professional philosophy of science as a bridge between disciplines and a bridge between science and the public. Where Kallen had demanded that Neurath's project engage politics and take on an antitotalitarian agenda, Kegley pulled Frank's in the opposite direction and demanded that philosophy of science avoid contact with sociological and valuational complexities. In terms of Frank's call for "active positivism" in his inaugural essay as president of the Institute for the Unity of Science, Kegley urged his colleagues to reject Frank's ambitions and remain content (and culturally diminished) with something like Frank's "humble positivism."

Frank's Legacy

Kegley's attack raises several questions that can only be raised here. Did he write his essay on his own, or was it commissioned? Which editor in the transitional year of 1959, Churchman or Richard Rudner, finally approved its publication? Was it perhaps the Thomist-friendly CSPR (in which Frank regularly participated) that led Kegley to see Frank as some kind of convert? The episode is also ironic because Kegley later became a leading scholar in religious studies with particular expertise in the theology of Reinhold Niebuhr.[23] If Kegley was once an aspiring philosopher of science, it appears that he, not Frank, experienced a theological conversion that dictated his future intellectual and professional trajectory.

However these questions are best answered, the timing of Kegley's attack is consistent with the effect he wanted it to have. Frank's last year of affiliation with Harvard was 1953, after which time his institutional affiliation remained the Institute for the Unity of Science. If, as Robert Cohen put it years later, the institute was only "half-living," Frank's career was also half-living in the late 1950s, and, it would seem, Kegley's

---

[22] Rutherford 1960, 183, 185–86. Rutherford defended Frank's efforts not only in print, but with his subsequent career in the American Association for the Advancement of Science as an expert in science education.

[23] Kegley wrote and edited several books about Niebuhr and other theologians. I thank David Hollinger for information about Kegley.

attack finally killed it.[24] Assuming that Frank's contribution to the much-delayed Carnap-Schilpp volume (Schilpp 1963) was written in the early to mid-1950s, Frank wrote only two articles after he wrote the book Kegley attacked. After Kegley's review appeared, Frank published nothing for the remaining seven years of his life. Though he was seventy-five years old when Kegley's attack was published, this drop-off in activity seems nonetheless severe. Between 1946 and 1957, besides his work as an editor, Frank wrote five books and roughly twenty articles (some of which appeared in these books). Because his personal papers were not preserved after his death, it may never be known whether Kegley's attack was coincidental to this drop-off in Frank's activity or whether it places Frank and his agenda for "active positivism" among other casualties of Cold War anticommunism.

[24] Quoted in Feigl to Institute Trustees, January 7, 1971, CMP. Robert Cohen recently affirmed that Frank's institute is aptly described as a "a nonstarter" (personal communication).

16

The Marginalization of Charles Morris

If the demise of Frank's institute and influence in the profession helps to explain the eventual dominance of the apolitical, professional agenda in philosophy of science, then some explanatory weight must be placed also on the decline of Charles Morris. As discussed in chapter 2, Morris staked his career on outlining a future science of semiotic that would synthesize the best aspects of logical empiricism and pragmatism and function as an organon for the development of a modern, democratic, and scientifically enlightened world culture. While Frank and the institute were declining, therefore, one might expect that Morris would have come to their aid. But Morris's own star had fallen. Because of his enthusiasm for blending Sheldon's somatotype theory with his philosophical research and because of widely recognized problems with his major book *Signs, Language, and Behavior* (1946), Morris had less prestige in philosophy and admiration from his colleagues than he did in the 1930s.

Signs, Language, and Behavior

Morris's goal in *Signs, Language, and Behavior* was to establish a collection of terms for a future science of semiotic that would analyze sign processes in an objective, behavioristic manner. Examples include, obviously, "sign," "behavior," and "denotatum," as well as other terms that Morris defines on their basis. These include "formator" and "formatum," "pathic sign," "descriptor," "designator," and "determinor." One problem with the book was Morris's wooden, passive, and soporific style of writing. Morris's most aggressive critic, Arthur Bentley, could not help but poke fun in this regard:

Consider the following: "For something to be a sign to an organism . . . does not require that the organism signify that the something in question is a sign, for a sign can exist without there being a sign that it is a sign. There can, of course, be signs that something is a sign, and it is possible to signify by some signs what another sign signifies." (Morris 1946b, 16)

The general purport of this statement is easy enough to gather and Gertrude Stein would certainly feel a home with it, but precision of expression is a different matter. (Bentley 1947, 110)

If Bentley unfairly compared Morris's most technical, most ambitious philosophical contribution to modernist poetry, he was not unfair in his choice of quotation. The book is dull and plodding in a way that invited one of the main criticisms it received, namely, that for all Morris's efforts to lay the terminological foundations of a new science, his new vocabulary simply presents old problems about signs and symbols in new words, without gain. Bentley dryly noted that "the degree of salvation" Morris achieved by latinizing many of his terms (with suffixes "um" and "or") "does not seem adequate" (Bentley 1947, 111). As for Morris's explicit goal of remaining behavioristic and observational, he found the effort pointless and, it would seem, offensive:

Semiotic thus takes goal-seeking psychology at the rat level, sets it up with little change, and then attempts to spread the cobwebs of the older logics and philosophies across it. The failure of Morris's attempt does not mean, of course, that future extensions of positive research may not bring the two points of approach together. Better rats and better men in better days may do better things. (ibid., p. 130)

Other reviewers were more polite. Many objected that Morris's crucial notion of "sign," though he had explicitly had not attempted a definition, was nonetheless unacceptably vague and imprecise (Black 1947; Gentry 1947; Smullyan 1947). Dewey and Bentley angrily argued that Morris's use of Peirce's concept "interpretant" was incorrect (Dewey 1946a; Bentley 1947, 126). Most reviewers found more generally that Morris's behaviorism decisively compromised his philosophical efforts. Black wrote,

My judgment is that Morris builds upon an excessively narrow basis of "behavioral" primitive terms; that the inadequacy of these terms as building blocks leads him to introduce metaphorical usages which are neither "behavioral," well understood, nor epistemologically primitive; and that the resulting vagueness of his terminology allows him to make, with unjustified confidence, a series of critical decisions on debatable questions which are presented with no better foundation than his own pronouncements. (Black 1947, 272)

In his essay "The Philosophical Relevance of a 'Behavioristic' Semiotic," Thomas Storer argued in *Philosophy of Science* that there is no such relevance. Philosophical analysis had no part in Morris's behavioristic project. *Signs, Language, and Behavior* "is only incidentally a contribution to philosophy proper; . . . its worth must be determined by its relation to a rather specialized section of behavior psychology" (Storer 1948, 319).

Morris Responds to His Critics

As Bentley commented to Neurath during their brief correspondence, "various people have been lifting Morris's hide lately."[1] Morris replied to most of them all at once in his paper "Signs about Signs about Signs" (Morris 1948c). Unfortunately, his defensive strategies and arguments could not have impressed his critics or colleagues. If Morris was not reminding his critics that his goals were modest –

The biological orientation of *Signs, Language, and Behavior* is therefore *primarily methodological* and does not involve a defense of "behaviorism" as against "mentalism" considered as psychological doctrines (ibid., p. 117) –

then he was admitting that his efforts in the book were preliminary (if not also defective): "These confusions make my terminology appear to be looser than the facts warrant. There are of course a host of legitimate and difficult problems which further advance in semiotic must meet"; "The term 'sign' clearly needs a better grounding than I have given it, or than I know how to give it"; "The axiomatization of semiotic must be carried out with much greater care than is done in *Signs, Language, and Behavior*" (pp. 117, 119, 131). He also excused himself, not unreasonably, by reminding his readers that his project was "scientific" and that some of its vagueness could be remedied only by future input from empirical science.[2]

---

[1] Bentley to Neurath, December 11, 1945, ONN.
[2] "There is an occasional complaint that I have not told *what* the state of the organism is when it is disposed to respond to something because of a sign, i.e., what the interpretant is as an organic state. The fact that I made no attempt to do so was deliberate. I do not know. The problem is an empirical one and its solution awaits the development of semiotic as an empirical science" (Morris 1948c, 128).

Morris seemed so defensive and bewildered by some of these criticisms that he took refuge in association with other philosophers and scientists. Defending his characterization of "sign," for example, he noted that it was "worked out after months of discussion with Clark Hull, Edward Tolman, and some of their co-workers" (p. 118). To Bentley's image of rats and cobwebs, Morris responded, "it is hard to see how an attempt to use and extend the behavioral analyses of Mead, Tolman, and Hull is to employ a rat-level psychology" (pp. 122–23). For those who questioned Morris's use of "disposition" and disposition predicates, he referred them to Carnap's "Testability and Meaning" as well as "the discussion of the place of intervening variables in psychology given by Tolman and Hull" (p. 128). In his strongest reply (to Storer's criticism), Morris wrote that by identifying philosophy with Wittgensteinian linguistic analysis Storer was himself engaging in judgments that could benefit from semiotic analysis. But instead of attacking Storer's programmatic assumptions about what does and doesn't count as philosophy, Morris replied that he didn't share them merely and "largely on factual grounds. In common usage such thinkers as Plato, Aristotle, Buddaghosa, Cankara, Motze, Loatze, Leibniz, Schopenhauer are accepted denotata of the term 'philosopher'" (Morris 1948a, 331).

Dewey and Bentley Attack Morris

The critic who probably did the most damage, intellectually as well as emotionally, was Dewey.[3] Morris was highly respectful of Dewey and his career, and he saw his own project, including his stewardship of the logical empiricists and the Unity of Science movement, as helping to fulfill the promise of Dewey's scientific pragmatism. Yet Dewey himself starkly rejected Morris's *Encyclopedia* monograph *Foundations of the Theory of Signs* (1938a) and his *Signs, Language, and Behavior* (1946b) as if Morris were a dangerous turncoat to the pragmatist cause.

He attacked in 1946. In an article in the *Journal of Philosophy*, he alleged that Morris misrepresented Peircean semiotics in his encyclopedia monograph (Morris 1938a; Dewey 1946a). Morris wrote to the editor and explained that there had been a simple mistake: He never intended his monograph to be read as a presentation or explication of Peircean semiotics, as Dewey's criticism seemed to presuppose (Morris 1946c). Dewey promptly replied with a letter of his own insisting that there had been

---

[3] Morris's student, psychologist Seth Sharpless, suggested this to me in personal communication.

no mistake on his part. His criticism addressed not Morris's historical relationship to Peirce, but rather Morris's "radical misconception of the scope, intention, and method of *Pragmatism.*" Dewey cared very deeply about how this tradition was represented and seemed to resent Morris's presentation of semiotic as some kind of distillation or culmination of it. Pitting James, Peirce, and himself against Morris, he demanded that Morris specify his disagreements with this great triumvirate and declare "that he is engaged in correcting the serious error of those who have previously written in the name of pragmatism" (Dewey 1946b). Dewey wanted Morris to declare that he was on his own attempting to create a new program for pragmatism. He was not a proper member of the club.

At this late time in his life, Dewey was collaborating with Arthur Bentley on a series of papers in the *Journal of Philosophy* that outlined their "transactional" program in philosophy. As Churchman and Cowan had noted in *Philosophy of Science*, Dewey and Bentley had issued broad charges against many logical empiricists (Churchman and Cowan 1945). Morris's main offense, Bentley agreed, was his treatment of Peirce. To Morris's claim that his semiotic "was an attempt to carry out resolutely the insight of Charles Peirce that a sign gives rise to an interpretant and that an interpretant is in the last analysis 'a modification of a person's tendencies toward action,'"[4] Bentley argued that Morris's conception of "interpretant" smuggled into semiotic notions of "thoughts" and "ideas" understood as "psychic substances or as psychically substantial" – precisely those things that Peirce, James, and Dewey "spent a good part of their lives trying to get rid of" (Bentley 1947, 129).

Morris addressed Dewey's and Bentley's criticism in his paper "Signs about Signs about Signs" and offered a deft analysis of Peirce's usage of "interpretant." He convincingly showed that he had not "foisted an Ersatz version of Peirce on an unsuspecting public" (Morris 1948c, 127) and thus effectively, if only implicitly, questioned why Dewey and Bentley would attack so aggressively about this point. Dewey elaborated his philosophical objections in a personal letter, saying that he agreed with Bentley that Morris had grievously distorted Peircean pragmatism and adding that he could not make sense of Morris's nondefining definition of "behavior." Morris had written,

This term is presupposed by semiotic and not defined within it. Roughly speaking, behavior consists of the sequences of responses (actions of muscles and glands) by which an organism seeks goal-objects that satisfy its needs. Behavior is therefore

---

[4] Morris 1946b, 27–28. The quote is from Peirce's *Collected Papers*, V, section 476.

"purposive" and is to be distinguished from response as such and from the even wider class of reactions. (Morris 1946b, 346)

Dewey found it an "extraordinary statement that behavior is a narrower term than response" and could not understand "its identification with actions of 'muscles and glands.'" Morris's first mistake, Dewey suggested, was an "implicit unstated, perhaps quite unconscious, acceptance of the old subject-object dichotomy, transferred over into different linguistic terms." Morris's (non-)definition of behavior presupposed, that is, a dualism dividing "the activity of the 'organism'" from "that of the 'environment.'" The correct view, Dewey explained, holds that organism and environment are – per Dewey and Bentley's transactionalism – only "abstractions within an inclusive activity which can be said to be the sole existence or event." Dewey laid down the law for Morris: "[A]ny treatment of 'behavior' that does not proceed in accordance with the point of view herewith stated is bound to be radically defective." Dewey closed his case with what seemed to him a compelling *reductio ad absurdum*. If this understanding of "behavior" is not sound, "my work for the last thirty years has all been in the wrong direction."[5]

Morris could not have found these objections believable. In all of his philosophical writings – in his dissertation, his book on theories of mind, his efforts to bridge science and humanism, and in semiotic – he rejected metaphysical and methodological dualisms wherever he found them. The main innovation of semiotic, for example, was its inclusion of syntactic, semantic, and pragmatic theory of language not merely as a concatenation, but in the effort to see and understand – just as Dewey suggested in the case of "organism" and "environment" – that these different kinds of analyses are abstractions or, as Morris usually put it, different "phases" of the entire, undifferentiated empirical reality being studied. Morris could only have been startled to read that Dewey, as he closed his letter, diagnosed Morris's approach as futile, that "in your broad eclecticism you are trying to bring together a variety of views that are mutually inconsistent with one another."[6] Reconciling incompatibilities was Morris's most consistent philosophical aim.

Not surprisingly, Morris believed that Dewey mistook compatibilities and complementarity for genuine conflict. He wrote back with an artful synthesis of humility and defiance:

I am sorry that you found Signs, Language & Behavior to be of so little interest, and I regret that you magnify the differences between our positions. It seems

5 Dewey to Morris, July 30, 1946, CMP.
6 Ibid.

to me that it is very important for pragmatism and modern thought in general to develop a comprehensive theory of meaning, and on as scientific a basis as possible. My book has this aim; it does not claim to be a philosophy. Hence I used the "organism" "object" terminology congenial to scientists. The use does not however repudiate the philosophical position of objective relativism. No claim is made in the book that it a presentation of pragmatism. . . . In the meantime I will consider Bentley's forthcoming paper carefully. . . . As to your view that my position is merely an "selective" juxtaposition of incompatible views I must of course dissent. I take the synthetic task of philosophy very seriously, and I have attempted a synthesis of the various views of meaning on the basis of a single behavioral orientation. Whether I have succeeded is of course an open question. I thank you for the vigor of your reaction.

Again, Morris supported his views by deferring to those of others. "History will not find us so far apart as you suppose,"[7] he wrote as he concluded his letter.

It appears that Dewey's criticisms rested, at least in part, on a personal dislike of Morris. This was perhaps because Morris's semiotic project was not, terminological gymnastics aside, so different from Dewey and Bentley's new "transactional" approach to epistemology and meaning. Whatever his reasons, Dewey did not like sharing pragmatism's stage with Morris: "I confess I do not understand how you can write as you seem to do at times as if there were any similarity between my position and yours."[8] Years earlier, Dewey was not receptive to Morris's inquiries about teaching at Columbia,[9] and in his correspondence about *Signs, Language, and Behavior* Dewey's politeness and formality hardly concealed insult. When Morris confessed that he detected "hard feelings" about the matter and proposed that they meet personally in New York, Dewey responded with a polite and formal scolding:

Dear Dr. Morris, It would be of course a pleasure to see you anytime. But I am not at all sure how likely we should be to get anywhere in an oral discussion – not because of "hard feelings" but because as indicated in my letter it seems to be that your position is badly confused and in need of a clarification that only you can provide. I have not seen your new book but judging from an analysis sent to me by Bentley in an article that will come out in the *Journal of Phil[osophy] and Phen[omenological] Research*, it is even more confused than the article about which I wrote my letter in the *Journal*.[10]

If only because he so emphatically condemned a book he had not even seen, Dewey's criticisms seem connected to "hard feelings" of some kind.

7 Morris to Dewey, September 23, 1946, CMP.
8 Dewey to Morris, July 30, 1946, CMP.
9 Seth Sharpless, personal communication.
10 Dewey to Morris, June 13, 1946, CMP.

Bentley, as Dewey reminded Morris, was no fan of his work, either. Nor was Bentley polite about his dislike of Morris. Even after observing that "Morris's hide" had been lifted by so many, Bentley was moved to attack Morris again in 1949. His essay, "Signs of Error" (Bentley 1949), responded to Morris's "Signs about Signs about Signs" and continued to beat the dying horse of philosophical semiotic. His motivation, in part, was revealed in a comment he made to Neurath: "Morris is a moron and an ass. What I have against him is distortion of Peirce and ignorance of everything he talks about."[11]

By the end of 1948, Morris was professionally and intellectually descending. For philosophers of science, his major work in technical philosophy was carefully taken apart in *Philosophy of Science* by Thomas Storer. The broader philosophical profession observed Dewey dismiss him for misunderstanding pragmatism itself, while a general chorus of reviews criticized his *Signs, Language, and Behavior* for being vague, poorly executed, and of questionable value. By trying to please everyone with his general and synthetic program of semiotic, Morris pleased almost no one.

Morris's Alliance with William Sheldon

As Morris's place in professional philosophy was increasingly questioned, his interests in personality theory led him further toward empirical science. But he did not fare well in his scientific aspirations, either, partly because he eagerly adopted and promoted the theories of Columbia University psychologist William Sheldon. As outlined in his book *The Varieties of Human Physique* (1940), Sheldon, then at Harvard, embarked on a research project to survey and classify human physiques (or "somatotypes"). He and his researchers amassed thousands of photographs (sometimes called "posture photographs") of subjects, often students at colleges and universities. His goal was to find correlations between subjects' physiques and psychological temperaments since, as he believed, "temperament is bound in some fashion to the constitutional pattern" (Sheldon 1940, 234). Sheldon acknowledged that environmental factors, and not just physiology and genetics, were at work in the development of human personalities, but he still believed that physique-temperament correlations were observable and measurable.

---

[11] Bentley to Neurath, December 11, 1945, ONN.

Morris readily took up Sheldon's cause and utilized the idea of somatotype-temperament correlations in his books *Paths of Life: Preface to a World Religion* (1942), *The Open Self* (1948b), *Varieties of Human Value* (1956), and in his everyday correspondence with colleagues. For example, when Neurath strongly criticized Morris's *Paths of Life* (being very suspicious of what he saw as arbitrary, highly subjective classificatory schemes), Morris suggested playfully and seriously that Neurath's reaction could be explained away on the basis of his elephantine body: "So much for you!"[12] There is little evidence that other colleagues took much interest in Morris's and Sheldon's somatotype work.[13] Morris became friends with Sheldon himself and, at one point, tried to arrange for him a position at the University of Chicago.[14] On another occasion, when Morris's interest in world religions led him to visit a Hopi reservation in New Mexico, he tried to arrange for Sheldon to visit the reservation, cameras in tow.[15] Morris was probably intrigued the idea that their distinctive cosmological and ethical beliefs somehow corresponded to the Hopi physique.

One reason why Morris accepted Sheldon's somatotype theory was its promise for integrating and synthesizing psychology and the other sciences with pragmatism. Morris's postwar writings about the new "science of man" (including his piece for Frank's institute (Morris 1951)) described an important place for Sheldonian studies of physique and personality. He drafted a proposal in November 1946 for "A Project in the Science of Man" in which 1,000 college students would be measured for "individual differences at every level of personality." As with his proposal for an "Institute of Philosophy" in 1932, Morris's enthusiasm leapt from the pages he typed. With social scientists of all kinds collaborating, he explained,

We can find out in this way what kind of person [has] what kinds of preferences, needs, and behaviors. We can find out who likes what with respect to social organization, professions, courses, arts, religions, philosophies, friends. We can approach experimentally the problem of value, and the role of signs in the lives of individuals. We can help to bring man back into the humanities. . . . We could attempt . . . similar studies in other cultures, and obtain a new basis for cultural comparisons.[16]

---

[12] Morris to Neurath, June 22, 1943, CMP.
[13] Seth Sharpless, personal communication.
[14] Robert J. Havighurst to Morris, April 12, 1944, CMP.
[15] Morris to Sheldon, March 7, 1948; Sheldon to Morris, April 13, 1948, CMP.
[16] Morris, "A Project in the Science of Man," unpublished ms., November 1946, CMP.

Holding fast to these scientific and cultural ambitions, Morris participated in UNESCO-sponsored research in international relations and pursued his own international, intercultural research. In the United States, Europe, and Asia, Morris and his students conducted questionnaire-based studies of value choices (or "ways of life") and aesthetic preferences that he hoped to connect to Sheldon's research on body types. The title of his book *The Varieties of Human Value* (1956) nods to Sheldon's *The Varieties of Human Physique* (Sheldon 1940), and Morris and Sheldon once agreed that Morris found a statistically significant correlation between body types and preferences in philosophical outlook. Sheldon was impressed. He wrote to Morris that "nobody has ever done that before."[17]

Yet by hitching himself to Sheldon, Morris probably further damaged his career and reputation. In an age when issues surrounding marriage, divorce, and sexuality were volatile (and symbolized for academics by the Russell affair at City College years before), it is hard to imagine that Morris's standing at the University of Chicago, already damaged by his divorce,[18] was improved by his fascination with research based on the minimally censored photographs of nude men and women that Sheldon reproduced in his books.

Sheldon's research in somatotypes must have seemed not only morally but also methodologically suspicious. He believed that any individual's body, such as those photographed in his database-like *Atlas of Men* (Sheldon 1954), is specifiable as a composite of three main types, each of which informs a specific body with a strength ranging from one to seven. With so many theoretically permissible bodies (7^3, or 343), Sheldon's classifications – as Neurath suggested to Morris – seem suspiciously subjective and flexible. During and after the 1940s, moreover, the eugenicist aspects of Sheldon's research (fed especially by his hypotheses that criminality was related to somatotype, as he argued in *Varieties of Delinquent Youth* (Sheldon 1949)) could not have seemed impressive to those of Morris's colleagues who had fled Nazism.

When Sheldon directly engaged Morris's colleagues, he was not well received. In 1945, a year before Morris's opus *Signs, Language, and Behavior* appeared, Sheldon jumped into the debate over naturalism between

[17] Sheldon to Morris, October 14, 1955, CMP.
[18] Carnap to Morris, October 19, 1952, CMP. In a portion of this letter written by Carnap's wife, Ina, she reported that Morris, then at Stanford, was not popular with the department's current chair, Charner Perry, "partly . . . for moral reasons: one does not get divorced!"

Dewey, Hook, and Nagel, in the one corner, and Mortimer Adler, in the other. Sheldon charged that the naturalism defended by self-appointed naturalists was indistinguishable from traditional materialism. Thus it hardly counted as the nonpartisan, metaphysically neutral program for philosophy and knowledge it was promoted to be (Sheldon 1945, 268). In their co-authored response, Dewey, Hook, and Nagel argued that Sheldon was unclear about which of many specifiable versions of "materialism" he intended, thus supporting their general view that Sheldon was not only out of his philosophical depth, but also an antinaturalist (though not, perhaps, a Thomist) running from the iconoclastic realities of modern science. Just like Adler and others, they believed, Sheldon suffered from the "new failure of nerve" and clung to certain "human values which [he] wishes to defend against what he believes is the threat of scientific method" (Dewey et al. 1945, 530).

Nor did Morris's alliance with Sheldon go unnoticed by the broader philosophical community. In his *The Open Self* (1948), Morris called for society to acknowledge that different individuals (with their respective physiques) have different, distinctive emotional and intellectual needs. One reviewer found this appeal to "physique-temperament type" unconvincing (Leys 1949, 285), while another worried that Morris's own qualifications about when physique-temperament relations obtain (only under certain conditions, for example) undermined his claim that they obtained at all:

It would seem clear . . . that the various factors, psychological, ethical, and philosophic, that are associated with one body style or another are so associated only under certain conditions which ought to be stated explicitly. The author does not state the conditions. . . . [Thus] the simple fact of the matter is that the list of correlations, to which the author calls attention, have no special status at all . . . Given other conditions, other correlations would presumably be there to list.(Dommeyer 1950, 453)

Readers of *Philosophy of Science* were also apprised of Morris's curious infatuation with Sheldon's theories and research. A review of Ray Lepley's *The Language of Value*, a compendium of essays discussing linguistic analysis of values, dismissed Morris's contribution altogether: "The recurrence of Sheldon's notion of 'somatotypes' puts Morris's own experiment in aesthetic judgement at the level of pseudo-science" (Kaelin 1958, 307–8). Richard Rudner, who would later succeed C. West Churchman as editor, penned a polite, objective, and therefore all the more convincing rejection of Morris's semiotic approach to aesthetic theory. Rudner wrote that, almost regardless of how it could be further specified and applied,

Morris's program "reduces to absurdity" (Rudner 1951, 71). Though Morris surely did not create the problems and complexities of aesthetic theory, Rudner saw him as a philosopher hopelessly mired in them. "There is some evidence that Morris feels uneasy in his predicament" (ibid., p. 75).

Morris's Predicament

Morris's predicament was actually larger than the one Rudner specified and larger than the personal animosities that led Dewey and Bentley to dislike him. His real problem was that his synthetic sensibilities and talents were no longer central and relevant in a philosophical community that was evolving in increasingly analytic directions. When logical empiricism and the Unity of Science movement were tightly wedded and together embraced a wide range of philosophical, scientific, and cultural concerns, Morris's semiotic project, even with its behaviorist orientation, did not seem suspicious or out of place. As late as 1946, even Hans Reichenbach, who later popularized the image of science and scientific philosophy as intellectually distinct from ethics and moral concerns (see, e.g. Reichenbach 1951, 304), heartily applauded Morris's semiotic project for its "social implications." Reviewing *Signs, Language, and Behavior* for *The Nation*, Reichenbach noted that Morris had only scratched the surface of the semiotic problems he raised. But that did not lead him to attack Morris's book. Instead, he warmly introduced Morris to this popular audience as "one of the leaders in the movement of logical empiricism" who had

clear insight also into the social implications of his work. He knows that language, socially speaking, is not only an instrument of progress but also a source of danger; it supplies means of controlling social groups, and its usefulness depends on the way such control is handled.(Reichenbach 1946)

Reichenbach then quoted Morris at length as he discussed semiotic as a tool for empowering citizens to resist commercial, social, and political manipulation:

Great masses of individuals repeat each week what has been digested for their belief, buy things which they approve because they have been shown a pretty girl or a "scientist" using such articles, mechanically repeat actions which they have been assured ought to be performed. . . . Against this exploitation of individual life semiotic can serve as a counter force. . . . If the individual asks himself the kind of sign he encounters, the purpose for which that sign is used, and the evidence of its truth and adequacy, his behavior shifts from automatic responses to critical and intelligent behavior. . . . He becomes an autonomous human being, neither unduly suspicious nor unduly gullible. (Reichenbach 1946, quoting Morris 1946b, 240–41)

Though Morris's work was recognized – at least by Reichenbach – as proudly carrying the torch of philosophical and civic enlightenment, that aspect of his semiotic project probably did not bolster his reputation in the ensuing Cold War climate. In several years at the height of McCarthyism and anti-Soviet patriotism, even such gentle criticism of American social and economic trends could invite Red-baiting and being described as "soft on communism." As a tool for empowering and creating "autonomous" and free human beings, anticommunists would have found Morris's pitch misdirected. It would be, rather, Soviets and their fellow-travelers who, in order to free themselves from Kremlin propaganda, could make use of Morris's theory of language and its coercive, manipulative capabilities.

There was, therefore, a medley of personal, professional, and political circumstances that rendered Morris an unlikely savior for the declining Unity of Science movement in the 1950s. In addition, Morris was not given by temperament and physique (he and Sheldon would presume) to confrontation and doing battle with his colleagues. In the late 1950s, he accepted a research professorship at the University of Florida at Gainesville and – as Dewey rolled in his grave – wrote his last major work, *The Pragmatic Movement in American Philosophy* (Morris 1970). Free from teaching, Morris devoted time to this and other projects (including his life-long devotion to writing poetry (Morris 1966)), to colleagues and former students pursuing the empirical study of values, and to international humanism through his participation in the American Humanist Association. Though Morris's new situation pleased him with its light workload (especially given worsening problems with his eyesight) and pleasant weather, it could not be considered strategic for his goal of keeping pragmatic methods and the study of values at the core of professional philosophy. His new department did not grant Ph.D.'s in philosophy.[19] Morris died in 1979.

[19] Morris to Feigl, January 19, 1962, HFP 03-109-15.

17

Values, Axioms, and the Icy Slopes of Logic

In their manifesto, *Wissenschaftliche Weltauffassung: Der Wiener Kreis*, Neurath, Carnap, and Hahn acknowledged that their philosophical and scientific crusade was both entrenched in the sociological and cultural battles of the day and subject to the vicissitudes of psychology, temperament, and social pressure:

> Thus, the scientific world-conception is close to the life of the present. Certainly it is threatened with hard struggles and hostility. Nevertheless there are many who do not despair but, in view of the present sociological situation, look forward with hope to the course of events to come. Of course not every single adherent of the scientific world-conception will be a fighter. Some glad of solitude, will lead a withdrawn existence on the icy slopes of logic; some may even disdain mingling with the masses and regret the "trivialized" form that these matters inevitably take on spreading. However, their achievements too will take a place among the historic developments. (Neurath et al. 1929, 317)

Neurath, Carnap, and Hahn probably never imagined that "the historic developments" of the twentieth century would include a new kind of war, a cold war, that would influence intellectual, social, and economic life around the world. It was not just a few individuals who would opt for professionalism and social and cultural disengagement in ivory towers or on "the icy slopes of logic" but entire communities of intellectuals, including philosophers of science, who moved farther away from "the masses" as specialization, professionalization, and nonpartisan analysis became the norms of postwar intellectual life.

This chapter examines some of the developments through which logical empiricism was born again, as it were, unencumbered by any suspicious or "pink" connections to philosophical radicalism and ongoing

344

debates about values, "ways of life," and political and economic concerns. The Institute for the Unity of Science, moribund by the end of the 1950s, was later absorbed by the Philosophy of Science Association (PSA), whose charter reflected this postwar professionalism. That professionalism also supported new relationships among leading logical empiricists and military science sponsored by the RAND Corporation.

Intellectually, the content of the discipline evolved in ways suggestive of the climate on university campuses of the 1950s – a climate that rewarded professional, apolitical behavior. Universities were well funded by the federal government for scientific research and by the Servicemen's Readjustment Act of 1944 (known as the GI bill), which paid for college degrees for returning soldiers. With classrooms filled (in part with former soldiers who had defended democracy against foreign totalitarianism) and calls for the unity of science and social and economic planning issuing exclusively from the far left, it is not difficult to see why philosophers of science generally pursued topics such as those championed by Feigl, Reichenbach, and Carnap, and not those championed by Frank and Neurath.

Within the profession, the transition is signaled by the decline of debate over questions about values and the discipline's responsibilities to these questions. Though arguments for and against the relevance of political and valuational matters are as old as logical empiricism itself, the 1950s ushered in a new consensus as influential leaders firmly and programmatically distinguished philosophy of science proper from the study of ethics and normative statements. While these pronouncements signal one kind of depoliticization of philosophy of science, they also signal a different kind of repoliticization. The manner by which values and value statements were excluded from philosophical analysis by Reichenbach, Feigl and others, this chapter argues, involves a conception of values similar to those driving Sidney Hook's and Horace Kallen's campaigns on behalf of freedom and pluralism. In the Manichean political landscape of the Cold War, the social and moral values dividing the West from the Soviets were thought to be absolute and unchangeable in ways that marked them as out of bounds for treatment by philosophers of science.

The End of the Institute for the Unity of Science

After Frank's institute lost its Rockefeller funding, its activity was diminished and it existed, it would appear, mostly on paper until the early 1970s. After Frank's death in 1966, Herbert Feigl succeeded Frank as president, though most of his energy naturally went to his own Minnesota Center for the Philosophy of Science. When Feigl announced his plans to retire from teaching, he also announced to the institute's current trustees – besides himself, Morris, Nagel, Hempel, Milton Konvitz, and Robert Cohen – that he would also resign from the institute. The question, as Cohen put it to the others, became: What to do with the "half-living" institute?[1]

All were agreed that the institute was to be terminated. Cohen suggested that it should be absorbed "quietly, discreetly and gracefully" into the Philosophy of Science Association, where its "intellectual and social value" would be better served.[2] That proposal introduced the question of whether and how the institute's goals could be preserved. Feigl reported conversations with Adolf Grünbaum, soon to become president of the PSA, who agreed that the institute's "affairs and activities should in some form . . . be preserved as a distinct . . . branch or part of the PSA." Indeed, Feigl even spoke of the institute being "revitalized" under these possible arrangements.[3]

Morris was skeptical. If the "*purposes* of the IUS are still in some form to be consciously pursued,"[4] he recommended, the institute should be folded into the Minnesota Center. Despite Grünbaum's offer that each future PSA meeting would feature a session "on a topic relevant to the Institute's concerns,"[5] Morris seemed to worry that the institute's agenda would be lost were it to be adopted by an organization with frequently changing leadership. In either case, however, Morris was not taking a stand. Whatever the outcome, he wrote to his colleagues, he had "no strong convictions on the matter."[6] Feigl would have been happy for his Minnesota Center to absorb the institute and its bank account. His funding from the Carnegie Corporation of New York was soon to be exhausted, while the institute, thanks mainly to royalties from the *Encyclopedia*, had several thousand dollars in the bank. Feigl assured Morris that were the institute to be merged into the Minnesota Center, he had "every reason to believe that the Center will continue with work directly relevant to the unity of science issues in the foreseeable future."[7]

[1] Feigl to Institute Trustees, January 7, 1971, CMP.
[2] Cohen to Feigl, May 10, 1971, CMP.
[3] Feigl to Institute Trustees, January 7, 1971, CMP.
[4] Morris to Feigl, June 29, 1971, CMP.
[5] Cohen to Feigl, May 10, 1971, CMP.
[6] Morris to Feigl, June 29, 1971, CMP.
[7] Feigl to Morris, July 1, 1971, CMP.

In the end, the trustees followed Cohen's suggestion. With the help once again of Milton Konvitz, who handled the institute's life as a legal entity, the trustees transferred its rights and obligations to the Philosophy of Science Association in 1972.[8] Of its assets, however, $1,500 were set aside and, at Feigl's suggestion, sent to Marie Neurath. After having transferred her inherited rights to the *Encyclopedia* to the institute, she was now in need of the money. She was also integral to Otto's life and work, Feigl explained to his fellow trustees, and Otto, of course, "was the '*spiritus rector*' and initiator of our enterprise."[9]

The Institute and the PSA

Neurath certainly had once been the guiding spirit of the "enterprise" that Feigl and these others now guided. But its goals and methods had changed. The petition to dissolve the institute specified that it had been incorporated for

a) encouraging the integration of knowledge by scientific methods, b) conducting research in the logical, psychological, and sociological backgrounds of science, c) compiling bibliographies and publishing abstracts and other forms of literature with respect to the integration of scientific knowledge, d) supporting the International Movement for the Unity of Science and e) serving as a center for the continuation of the publications of the Unity of Science Movement.

Depicting an absorption of one institution by another without loss (perhaps in order to minimize any legal complications the dissolution might raise) the petition stated that the purposes of the PSA were "similar to those of the Institute for the Unity of Science." Yet there are basic differences between the institute's and the PSA's agendas. Under Frank, the institute aimed to blend the study of the logic, psychology, and sociology of science, on the one hand, with the task of promoting the unity of science and "the integration of knowledge" on an international scale, on the other. Unified science or one of its cognates appears in four of the statements' five points.

According to its constitution, the petition explained, the PSA existed for

a) the furthering of studies and free discussion from diverse standpoints in the field of philosophy of science, b) the publishing of a periodical devoted to such studies in this field, and c) the publishing of essays and monographs in this field

---

[8] Petition to dissolve the Institute for the Unity of Science, CMP.
[9] Feigl to Trustees, October 25, 1971, CMP.

which are too lengthy for publication in a periodical. Said Association holds biennial meetings . . . [and] publishes a scholarly journal, "Philosophy of Science."[10]

In light of the common dichotomy between unified science, on the one hand, and freedom and pluralism, on the other, the allegiances of the PSA were clear. Freedom and pluralism were among its core values, and its statement of purpose made no mention of the unity of science or the integration of knowledge. Philosophy of science was now a professional, bordered discipline whose projects and tasks would be defined and maintained internally. If one did not know what "philosophy of science" was, these points provide no information. The PSA's constitution, of course, did not prohibit its members from pursuing issues and topics related to unified science, but such pursuits were not constitutive of the association's identity and shared mission. The petition also exaggerated the similarities of the institute's and the association's publications. It claimed they were "substantially similar,"[11] yet the monographs of the *Encyclopedia* and the essays in Frank's *Contributions* are similar only to articles appearing in *Philosophy of Science* before the 1960s.

In 1972, when the trustees of the institute circulated their petition among themselves for the necessary signatures, the politicized and leftist orientation that Malisoff (and Churchman) once maintained in the journal and in the association had probably faded from view. According to Malisoff's last mission statement, the association existed for

furthering of the study and discussion of the subject of philosophy of science, broadly interpreted, and the encouragement of practical consequences which may flow therefrom of benefit to scientists and philosophers in particular and to men of good will in general.[12]

The "broadly interpreted" clause effectively invited anyone to join the association, and they did so automatically on subscribing to the journal, a circumstance that ended with the revisions made in 1958.[13] Those revisions accompanied a new mission statement (cited above in the petition to dissolve the institute) in which these humanistic and socialistic overtones – "the benefit" and "practical consequences" of philosophy of science for "men of good will in general" – are absent.

---

[10] Petition to dissolve the Institute for the Unity of Science, CMP.
[11] Ibid.
[12] The constitution appears in *Philosophy of Science* 15: 176–77.
[13] "Revised Constitution of the Philosophy of Science Association – 1958," *Philosophy of Science* 26 (January 1959): 63–66.

These changes coincided with the succession of C. West Churchman by Richard Rudner as editor of the journal. A philosopher of social science at Michigan State University, Rudner was elected to the governing board of the PSA only months before his election as editor on the grounds that the board needed someone whose expertise lay in the social sciences.[14] But Rudner quickly gained enormous power because the association's governing board requested that Churchman's successor, whomever that would be, take "initiative in shaping the future of the journal . . . as regards the contents of the journal."[15] Four years before, Rudner had argued (in Frank's *Validation* and at the related conference) that "scientific method *intrinsically* requires the making of value decisions" (Rudner 1957, 28). But he did not influence "the contents of the journal" in this direction. The opposite appears to be the case. As Don Howard's research has shown, the frequency of articles relevant to the values debates of the 1940s or social aspects of science declined rapidly after Rudner's tenure began (Howard 2003, 66–71).

Logical Empiricists at RAND

By some accounts, the 1950s witnessed not "the end of ideology," but rather the homogenization or normalization of ideology. On this view, as the Congress for Cultural Freedom and firings of communist and suspicious professors might suggest, the academy's new professionalism and vocationalism signaled a retreat not from politics, but rather from dissent and, in turn, the cultivation of a new intellectual style that complemented the anticommunist climate of Cold War America.

Most were agreed that intellectual life was changing. H. Stuart Hughes posed the question "Is the Intellectual Obsolete?" in the pages of *Commentary* (Hughes 1956), while Newton Arvin identified a new "American academic type" of intellectual who had abandoned "wide-ranging, curious, adventurous and humane study" in favor of producing professionally valuable "results" in the spirit of James Burnham's "managerial revolution" (cited in Jacoby 1987, 73). Philip Rieff, who helped to found the

---

[14] Ducasse to Board, May 27, 1958, CMP.

[15] Ducasse to Rudner, quoted in Ducasse to Governing Board, January 13, 1959, CMP. The governing board consisted of Gustav Bergmann, Carnap, Churchman, Theodosius Dobzhansky, Philipp Frank, R. B. Lindsay, Henry Margenau, Ernest Nagel, Stanley Stevens, and Louis Zerby (the journal's managing editor). The board granted most votes to Arthur Burks, Rudner, and Wesley Salmon (in that order) and offered the position to Rudner after Burks declined.

journal *Daedalus* and participated in at least one of the colloquia sponsored by Frank's institute, was likewise disapproving of the "new conservatives" who capitulated to national interests and reshaped their careers as policy analysts:

> With this commitment of the new conservatives to policy, loyalty, not truth, provides the social condition by which the new intellectual discovers his new environment and judges it familiar. To move from the New School [for Social Research] to the RAND corporation is an epitomal decision for the new conservatives. (Rieff 1953, 17)

Rieff targeted intellectuals across the board, but it is likely that he had some logical empiricists in mind. He rubbed elbows with them in Frank's institute and probably knew that some had established working relationships with RAND.[16] A sufficient number of logical empiricists and their students had connections to RAND for one historian to claim that "the professionalization of American philosophy of science in the immediate postwar era grew directly out of the soil of Operations Research" cultivated at RAND (Mirowski n.d., 22). Carnap's student Olaf Helmer worked at RAND much of his professional life, while Abraham Kaplan, Carl Hempel, and Morris's students Paul Kecksemeti and Freed Bales at least briefly worked there in the early 1950s. Others working in or near philosophy of science at RAND or in operations research in the 1950s and '60s include Carnap, Quine, Paul Oppenheim, Alfred Tarski, Olaf Helmer, John Kemeny, J. C. C. McKinsey, Patrick Suppes, Donald Davidson, Nicholas Rescher, and Leonard Savage.[17]

The RAND Corporation (originally an acronym for Research and Development) began as a department within the Douglass Aircraft Corporation in Santa Monica, California. It was created with an endowment from the U.S. Air Force as an independent and nonprofit corporation that would conduct and manage intellectual research "related to national security and public interest" (as its letterhead put it). RAND was the first "think tank" inspired partly by the successes and importance of code breaking in the allied victories in World War II. Often with federal financial support, RAND-like research was also conducted at universities,

---

[16] In 1956, Rieff spoke at the institute under the title, "Science and Politics in a Mass Society" (Frank to Morison, November 13, 1956, RAC 1.100, box 35, folder 285).

[17] Bales to Morris, August 20, 1951; Hempel to Morris, April 17, 1951; Kecskemeti to Morris, July 25, 1961, CMP. Kecskemeti is identified in his contribution to the CSPR conference as "Chief Magazine Servicing Unit, U.S. War Department" (Kecskemeti 1948). For others' connections to Rand, see Helmer and Rescher 1960; Quine 1986, 27–28; Mirowski n.d.

often in departments of "communications theory" such as the one at the University of Illinois whose invitation Morris turned down to protest the state's loyalty oath requirement.

The significance of RAND is not simply that it recruited and supported philosophers for engaging in military-related research. Galileo and Descartes show how old and traditional this kind of alliance has been. During the Second World War, Frank himself worked as a report writer at Columbia University for projects sponsored by the military's Office of Scientific Research and Development (OSRD). Philosophers Herbert Marcuse and Norman Brown, to take two additional examples, worked at the Office of Strategic Services, the predecessor to the CIA.[18] Rather, the significance of these personal connections among RAND, operations research, and logical empiricism is that they continued after the war and thus shaped the profession's view of itself and its public profile during and after the 1950s.

In the late 1950s, RAND advertised itself to a broad range of scientific and technically oriented intellectuals reading *Science* magazine. Alongside advertisements for employment at Raytheon, General Electric, Lockheed, and other military contractors, RAND published a campaign featuring full-page portraits of history's luminaries. Stylized, watercolor renderings of Lucretius, Leonardo, and others floated above paragraph-length quotations from their writings. There is no sales pitch in these advertisements. They suggested simply that RAND was an institution friendly to, and respectful of, history's great, independent, and even radical thinkers. One of the ads featured Thorstein Veblen, whose *Theory of the Leisure Class*, originally published in 1899, was and remains an icon of radical intellectual criticism of American economic life. Others featured were Francis Bacon, William Whewell, Galileo, Henri Bergson, Karl Pearson, and Sir James Jeans. With few exceptions, this pantheon was filled with scientifically oriented intellectuals and philosophers. One of the ads featured Hans Reichenbach, whose portrait appears with a quotation from *Atom und Kosmos* describing physics as a living, dynamic science that nourishes "the truth-seeking spirit" (see Fig. 7).

The appearance of Reichenbach as an advertising emblem for RAND is laden with ironies and symbolism. Unlike the others, Reichenbach had been a paid consultant and researcher for RAND since 1948. But

---

[18] Frank's tenure at OSRD is reported in his FBI file. Brown and Marcuse met when they both worked at OSS in the 1940s (Obituary of Norman Brown, *New York Times*, October 4, 2002).

FIGURE 7. Hans Reichenbach depicted in an advertisement for the RAND Corporation in *Science* magazine (1959).

Reichenbach himself almost certainly had nothing to do with his becoming an advertising icon because he died years before this series of advertisements appeared. It was therefore someone else at RAND or perhaps a hired marketing specialist who seized Reichenbach's image and reputation as an alluring emblem for military research. Reichenbach was once sufficiently vocal in his youthful advocacy of socialism that he had great difficulty obtaining a position at the University of Berlin (Traiger 1984). Now, he was upheld as one face of Cold War military research.

Reichenbach's image also floats conspicuously above his words about science's "truth-seeking spirit." While Reichenbach's probabilism is not out of step with the activity of truth seeking, it denies that science (or any other activity) can secure absolute, indubitable truths about nature. While the advertising designer assembling Reichenbach's entry in the series was almost surely not aware of the letter of Reichenbach's work, he or she was certainly aware of the value and currency of "truth" in Cold War culture, both popular and intellectual. Truths about freedom and the evils of communism were widely seen as a second bulwark against communist invasion, right behind the first bulwark of military strength. Readers of the *New Leader*, for example, were asked to contribute "truth dollars" to help Radio Free Europe broadcast truths about the superiority of democracy into Soviet-controlled countries (see Fig. 8).

The Official Apolitical Stance

If Reichenbach's image in the pages of *Science* suggested a patriotic, anticommunist alliance between North American logical empiricism and military research, the consensus within the profession (according, in part, to Reichenbach himself) was that logical empiricism was apolitical and not concerned with problems and questions about values. To be sure, this consensus was neither sudden nor universally agreed to. Some students and supporters of logical empiricism besides Frank and Morris in the 1950s still pushed for engagement with matters social and ideological. Arguments familiar from Dewey's, Frank's, and Neurath's writings, for example, appear in Abraham Kaplan's article "American Ethics and Public Policy" (Kaplan 1958). Kaplan's contribution to the *Library of Living Philosophers* volume about Carnap took stock of emotivism and, following Frank, outlined "the theory of value to which the various currents in contemporary empiricism and pragmatism – emotivist as well as

With "TRUTH DOLLARS"–*that's how!*

Your "Truth Dollars" fight Communism in it's own back yard—*behind* the Iron Curtain. Give "Truth Dollars" and get in the fight!

Truth Dollars" send words of truth and hope to the 70 million freedom loving people behind the Iron Curtain.

These words broadcast over Radio Free Europe's 29 transmitters reach Poles, Czechoslovakians, Hungarians, Romanians and Bulgarians. RFE is supported by the voluntary, cooperative action of millions of Americans engaged in this fight of good against evil.

How do "Truth Dollars" fight Communism? By exposing Red lies ... revealing news suppressed by Moscow and by unmasking Communist collaborators, the broadcasts are by exiles in the native tongues of the people to whom they are beamed.

Radio Free Europe is hurting Communism in its own back yard. We know by Red efforts to "jam" our programs (so far without success). To successfully continue these broadcasts, even more transmitters are needed.

Every dollar buys 100 words of truth. That's how hard "Truth Dollars" work. Your dollars will help 70 million people resist the Kremlin. Keep the truth turned on. Send as many "Truth Dollars" as you can (if possible, a dollar for each member of your family). The need is now.

FIGHT COMMUNISM

with "TRUTH DOLLARS"

Support Radio Free Europe

Send your *"Truth Dollars"* to CRUSADE FOR FREEDOM c/o your Postmaster

Space donated by People's Educational Camp Society, Inc.

FIGURE 8. A call for readers of *The New Leader* to contribute "truth dollars" to help pay for Radio Free Europe's broadcasting of Western programming into Soviet-controlled Eastern Europe. The advertisement appeared in 1954 and 1955.

cognitivist – appear to be converging" (Kaplan 1963, 854). Kaplan enjoyed a long career at UCLA and wrote several books about topics ranging from epistemology to American society and politics. A similar example is Grünbaum, who contested popular critics of scientism, such as Hayek and the popular anticommunist Whittaker Chambers, in the pages of *Scientific Monthly* (where the essays comprising Frank's *Validation* volume first appeared). Grünbaum argued in his essay "Science and Ideology" that there is no compelling reason why science and a scientific outlook, as many supposed, should be excluded from responsible ideological debate (Grünbaum 1954).

Like echoes of an earlier time, these arguments and calls for the profession to engage itself with politics, values, and ideology faded during the 1950s. Though the view that logical empiricism properly had nothing to do with noncognitive, normative claims of ethics or politics had circulated since the 1920s, there are several reasons to identify Reichenbach's popular book *The Rise of Scientific Philosophy* (1951) as the final victor in this debate. It appeared at the beginning of the decade, when logical empiricism, if it were to grow and thrive in the age of the Red menace, would have to appear – as Reichenbach urged it was – free of ideological contamination. The book was also very popular outside the profession and promoted Reichenbach's well-known distinction between so-called contexts of discovery and justification, first introduced in Reichenbach's Experience and Prediction (1938). The epistemic value of scientific theories lay entirely in their logical relations to accepted evidence, the distinction holds, and has nothing to do with contextual facts of the matter regarding the origin of theories, their inspirations, or the personal, much less political, details about the scientists who create them (Giere 1996, 343–46). These origins, if they are ever to be understood, Reichenbach maintained, will be revealed by scientific psychology, not philosophy of science.

The same holds for topics such as valuation, ethics, and politics, Reichenbach explained:

Those who ask the philosopher for guidance in life should be grateful when he sends them to the psychologist, or the social scientist; the knowledge accumulated in these empirical sciences promises much better answers than are collected in the writings of the philosophers. . . .

Traditional philosophers, Reichenbach argued, chronically confuse logical with sociological and psychological factors and circumstances, while "the scientific philosopher keeps clear of such errors in reducing his contribution to ethics to a clarification of its logical structure" (Reichenbach 1951, 316).

The framework of Reichenbach's account was the "received view" – as it was later called – of scientific theories as formal systems in which concepts and hypotheses are introduced or defined on the basis of the system's premises or axioms. Axioms themselves are not without circularity or triviality justifiable within the system they support. Reichenbach invoked formal geometry as a template to view ethical knowledge and firmly distinguish it from science and philosophy of science: "[T]he scientific philosopher distinguishes between ethical axioms, or premises, and ethical implications, and he regards only the implications as capable of logical proof" (ibid., p. 319). There can be no scientific ethics, consequently, because ethical premises are essentially volitional and subjective. Unlike premises of formal geometry, they cannot be further interpreted as empirical claims (about the structure of space-time, for example) and subject to scientific test. "Ethical axioms, or premises," therefore cannot find within science (or scientific philosophy) justification or substantive criticism. Their logical consequences can indeed be explored and analyzed within the system or framework to which they belong, but they themselves cannot be the focus of philosophical inquiry or evaluation. "For the context considered," Reichenbach explained, "they are not questioned" (ibid.).

The Problem with Social Philosophy

While Reichenbach's localization of ethical knowledge to axioms, or premises, allowed him to draw boundaries around the legitimate activity of a philosopher of science, his broader argument that ethics must be unscientific remains, especially from a pragmatic point of view, unconvincing. It may be, as Reichenbach explained, that "axioms of ethics" considered in isolation "cannot be made cognitive statements at all; there is no interpretation in which they can be called true" (Reichenbach 1951, 319). But he admitted that, on his view, they belong to ethical theories possessing a "logical structure" and from which "implications" can be derived. If so, it would appear that "ethical axioms, or premises," may – as parts of broader ethical frameworks – be examined and evaluated on the basis of objective, empirical consequences they may have in society or individual life. On the other hand, Reichenbach was right to emphasize that if an individual's or a society's choices to accept or reject certain ethical axioms are purely volitional and noncognitive, science and philosophy of science provide no tools or expertise to inform or guide those decisions. But, as a variant of Frank's argument for "active positivism" would suggest (which appeared a year after Reichenbach's book), neither can politicians, business leaders, or any other intellectual field supply such tools or expertise. With respect to ethics and social questions, any principled distinction between philosophy of science and these other institutions remains elusive, as does any reason why philosophers of science in particular must opt out of social and ethical debate.

Reichenbach's division of science and scientific philosophy from ethics was nonetheless popular. The division, and sometimes Reichenbach's motivation for it, appeared often in professional literature specifying what philosophy of science was, and was not, all about. In the volume *Readings in the Philosophy of Science* (1953), in which Herbert Feigl and May Brodbeck collected influential essays, Brodbeck introduced the volume by discussing four popular ideas about just what "philosophy of science," means. These were "the socio-psychological study of science," "the moral evaluation of the scientist's role and knowledge," "the philosophy of nature" (involving, mainly, dialectical materialism), and "logical analysis of science" (Brodbeck 1953, 6). Only the last meaning was correct, Brodbeck explained: Philosophy of science consists in, and is exhausted by, "the ethically and philosophically neutral analysis, description, and clarification of the foundations of science."

Brodbeck's treatment of "philosophy of nature" (which, unlike some forms of speculative philosophy, "is no longer harmless, for it provides the rationalization for a particular ethical or political ideology" (Brodbeck 1953, 6)) anticipated Richard Rudner's introduction to his book *Philosophy of Social Science* (1966) over a decade later. Rudner, then editor of *Philosophy of Science*, specified "the character and scope of philosophy of social science" by contrasting them with the character and scope of "social philosophy." Social philosophy is "concerned with the varying views about the nature of desirable social systems or societies, and sometimes it puts forward its own proposals about what constitutes a good or desirable society." The tradition, Rudner explained, reaches from Plato's *Republic* to Hobbes, Locke, and Rousseau to "voluminous writings of Marxist and non-Marxist socialists and nonsocialists, down to our own epoch" (Rudner 1966, 2). All that had nothing to do with the philosophy of social science, however, which confines itself to "methodological" problems encountered by social science.

It was true, Rudner admitted, that very few works in the field "are unrelievedly neutral or methodological in character – i.e. do not have something to say directly or by strong suggestion about the worth of some social arrangements" (ibid.). In principle, however, the philosopher of social science ignored valuations of this sort and attended only to relations between theoretical statements and observational statements, that is, to *testability* conditions and not the *tenability* or defensibility of social theories:

> To go beyond concern with testability in the direction of ascertaining the tenability of a theory of social phenomena would be for the philosopher of social science to enter the domain of the social scientist or the social philosopher. (p. 3)

Were any anticommunist investigators or university administrators perusing Rudner's or Brodbeck and Feigl's volumes, they would have learned in the first few pages that philosophy of science had no place for ideologues who connected their politics to their philosophy. The profession was made up of conceptual technicians and analysts of theoretical structure.

Nor would they have learned much about the views of Morris, Neurath, and Frank, who opposed such disconnection between technical philosophy of science and values. In Rudner's list of suggested readings, the Unity of Science movement was not visible. Most entries belonged to Hempel, Rudner, Nelson Goodman, and Talcott Parsons; one was given to Hayek (pp. 112–13). Brodbeck and Feigl included Frank's "Philosophical Interpretations and Misinterpretations of the Theory of Relativity," which concludes with a paragraph about the need for a progressive "scientific enlightenment."[19] But neither Frank's writings on relativity and values nor writings by Morris or Neurath were included. By contrast, in the book's section on philosophy of social science, Feigl and Brodbeck included an argument against dialectical materialism by Sidney Hook and a short excerpt from J. A. Passmore, titled "Can the Social Sciences Be Value-Free?" Yes, they can, Passmore answered as he urged philosophers of science to build "theories about the structure of morality which are genuinely theories." Much as Brodbeck rejected "philosophy of nature" for being ideological, Passmore urged that scientific studies of ethics refrain from "social advocacy in disguise" (Passmore 1953, 676). Since "it is part of the task of the social sciences to expose subterfuges of this sort," value-free science, with value-free tools from philosophy of science, could better perform one of its civic duties (ibid., p. 674).

[19] This essay was extracted from Frank 1938.

Feigl on Science and Values

The influence of Reichenbach's axiomatic conception of ethical knowledge is perhaps best marked by its function in the programmatic writings of Herbert Feigl. With his Minnesota Center for the Philosophy of Science, Feigl became an important institutional as well as intellectual representative of logical empiricism. Not unlike the case of Reichenbach, Feigl's programmatic writings reflect a profession moving away from political and social engagement. In 1943, responding to the furor over Mortimer Adler's attack on positivism and naturalism, Feigl sharply criticized Adler's conception of values as "absolute" and disconnected from science and the practical, empirical world:

> The acceptance of an absolute authority or extra-mundane sanction for morality, like the belief in an absolute source of factual truth, manifests a not fully liberated, pre-scientific type of mind. A completely grown-up mankind will have to shoulder the responsibility for its outlook and conduct; and in the spirit of an empirical and naturalistic humanism it will acknowledge no other procedure than the experimental. (Feigl 1943, quoted in Lepley 1944, 9)

Much as other logical empiricists and pragmatists had argued, Feigl agreed that cultural and social values were connected to science and were tested in the push and pull of practice. In its rough outlines, Dewey, Hook, and Nagel (who led the counterattack against Adler), Carnap, Morris, Frank, and Feigl shared this view in the 1930s and early '40s.[20]

In *Readings in the Philosophy of Science*, however, Feigl had begun to conceive of values as more disconnected from scientific theory and practice. He agreed with Brodbeck's austere conception of the field and emphasized that "science cannot dictate value standards" (Feigl 1953, 18). Still, he dwelled on the ways that science and a scientific outlook are connected to humanism's social, educational, and moral concerns. In a comment that Dewey especially would have approved, Feigl wrote that "under the impact of modern science" philosophy's function had been narrowed and refined to "the clarification of the foundations of knowledge *and valuation*" (ibid., p. 9, emphasis added).

If Feigl's comments seem out of place for an essay from 1953, consider that he first published them in an early 1949 issue of *American Quarterly*. Feigl therefore wrote them prior to the highly publicized firings at the University of Washington in 1949, before the loyalty oath controversy at

---

[20] For more on Dewey, Hook, and Nagel, see chapter 3. Carnap endorsed Feigl's view quoted here in Lepley 1944, 9. See also Morris 1938a; Frank 1950; Reisch 2001b).

the University of California, and in the same year that Frank's new institute began its effort to reinvigorate the Unity of science movement. Feigl additionally approved of "social philosophy"[21] and named the values that connected scientific humanism with philosophy of science. These included "human values such as freedom and responsibility, rights and obligations, creative and appreciative capacities," as well as "justice, peace and relief from suffering" (pp. 9, 16). Most important, Feigl's essay promoted activism, not unlike Frank's inaugural essay for the publications of the institute (Frank 1951b). Philosophy of science, Feigl explained, can help science to calculate relevant probabilities in cases "when we have to act, as so often in life, on the highest probabilities available even if these probabilities be low in themselves." Joining Malisoff, Neurath, Morris, and others, Feigl even called for social and economic planning: "[C]ooperative planning on the basis of the best and fullest knowledge available is the only path left to an awakened humanity that has embarked on the adventure of science and civilization" (Feigl 1953, 18).

In 1959, ten years after he first published these remarks, Feigl returned to these issues in his volume *Current Issues in Philosophy of Science*, co-edited with Grover Maxwell. The volume's essays address explanation, induction and probability, the structure of theories, and issues in foundations of mathematics, along with more specialized topics in philosophy of physics, psychology, and social sciences. In his introduction, Feigl again described a connection between philosophy of science and humanism's "aspirations . . . toward a better condition of mankind [as a] universal goal" (Feigl 1959, 17). Since human beings tended to be "such egregious blunderers in the art of living," Feigl suggested, science and philosophy of science had an important role to play: "[S]cientifically enlightened education and politics could help enormously" (ibid., p. 17). More than he had in 1949, however, Feigl emphasized that the profession's embrace of these values was indirect because science "can never, by its very nature, provide a reason for our fundamental obligations or for the supreme goals of life" (p. 16). It can possibly account for "the *facts* of human evaluation," and it can of course empirically "record . . . the ideals and moral codes" of different peoples and cultures and possibly explain their historical development. "But it must not be thought that such description and interpretation . . . could possibly establish their moral justification. Scientific truth is ethically neutral" (ibid.).

---

[21] "Clearly nothing is more urgent for education today than a social philosophy that will be appropriate and workable in an age of science" (Feigl 1953, 9).

Feigl mentioned two developments that appear to have widened the gap between philosophy of science and values. One was the professionalism of philosophy of science and, in particular, a "division of labor" among "a large number of specialists in the various fields of philosophy of science" (p. 2). This naturally encouraged research in narrower and more technical questions and problems. Even though aspects of science's history "are of the greatest interest to the historian as well as the philosopher of science" (p. 2), this professionalism also tended to drive a wedge between philosophy and history of science.

The second and main development leading to this divide between science and values also differentiated philosophy of science from empirical science. It lay in the methodological core of the profession:

Studies in the foundations of mathematics during the last eighty years have pointed the way and established for many philosophers of science an idea that can only be approximated, but never be attained in the logic of the empirical sciences. Axiomatization – that is, the construction of a scientific discipline in the form of a deductive system with precisely formulated sets of postulates, definitions and derivations – has become the order of the day. (pp. 3–4)

This order of the day, Feigl explained, and the "generally accepted account of theories as hypothetico-deductive systems" was not without critics. But that qualification suggested one obvious purpose of the formal "received view" of scientific theories: It focused debate and helped to unify a specializing, growing profession.

The axiomatic model allowed Feigl to reinforce Reichenbach's distinction between science (and, hence, philosophy of science proper) and social and political partisanship. For when Feigl noted that science is "ethically neutral" with respect to "our fundamental obligations or . . . the supreme goals of life," he noted that not only science but all intellectual pursuits were fully independent of ethics and moral concerns:

If science cannot validate moral standards and if theology furnishes no more than verbal and emotional stimulates (or sedatives), where else are we to seek the justification for the ethical principles which are the basis for all moral criticism? (p. 16)

"There is," Feigl answered, "but one answer to this question which is compatible with a clarified scientific outlook." As Reichenbach had suggested in his *Rise of Scientific Philosophy*, our ethical beliefs and values have the status of premises in a formal, deductive system. As such, the answer is that they cannot be logically justified:

> The fundamental ethical principles are themselves the basis for the justification for the more special moral precepts, but – without begging the question or starting an endless regress – cannot themselves be logically justified. (p. 16)

By adopting Reichenbach's axiomatic picture of ethical knowledge, Feigl outlined a firmer isolation of philosophy of science from the study of values and ethical premises. It was not that philosophy of science is specially disengaged from ethical debate, but rather that ethical debate is specially disengaged from logical justification. As much as Feigl respected "moral belief" and its importance for societies and individuals, as much as it involved "sincere and wholehearted commitment to a certain attitude toward the values of life," it nonetheless "*simply* consists in a sincere and wholehearted commitment" that no legitimate intellectual project inside or outside science can illuminate or guide (p. 17, emphasis added). In light of these circumstances, Feigl wrote, to criticize science (or scientific philosophy) for being socially and ethically disengaged "would be like reproaching a weaving loom for its incapacity to produce music" (p. 16).

Ethical Absolutism and Cold War Politics

While this view of ethics and values equipped philosophers of science to say that their profession did not address value issues, it nonetheless reflected the prominent "absolutistic" conception of social and political values undergirding North American anticommunism. One indication of this conception of values and its embrace by some philosophers of science was Martin Gardner's criticism of Frank's *Relativity: A Richer Truth* and Reichenbach's *Rise of Scientific Philosophy* for lacking backbone in regard to ethics and politics, as described in chapter 15. Gardner was right to see that Reichenbach's book excused philosophers of science (qua philosophers of science) from having to adjudicate such disputes, but he was wrong to suggest that these two books were equally permissive. Though Reichenbach qualified his point and denied that "the scientific philosopher" in fact holds that "the so-called axioms are invariable premises, holding for all times and all conditions" (Reichenbach 1951, 319), his argument for the separation of scientific philosophy from ethics treated ethical values as axioms that were not subject to argument. Frank's book about ethical values worked in the opposite direction. For Frank, the only good value was a relativized, contextualized, and revisable value.

It is not accidental that the most aggressively anticommunist philosopher of the 1950s was also the most absolutistic about social and political values. When Carnap refused Sidney Hook's request that he dissociate himself from the Waldorf conference, he explained that he supported the conference on the ground that "the maintenance of peace seems to me the main problem and task today, more important even than civil liberties."[22] Maintaining peace seemed more important, Carnap almost surely reasoned, because civil liberties would be of no value at all were contemporary civilization and its infrastructures decimated by nuclear warfare with the Soviets. Like Frank, Carnap knew that values operate in, and are specified and qualified by, practical contexts.

Hook read this and exploded with anger and sarcasm. Earlier, he had firmly but sympathetically warned Carnap against appearing to be a communist fellow-traveler; now he seemed to have concluded that the appearance was accurate:

If I had known that the "maintenance of peace is more important [to you] than civil liberties," I should never have written you in the first place. For whoever believes that has already left the democratic camp. If peace is more important than civil liberties, then you should urge that the United States apply for affiliation to the Union of Soviet Republics at once. Of course, we could have had peace without civil liberties by capitulating to Hitler in 1939 and saved much bloodshed. My own view as a democrat is that peace *without* the specific freedom of the Bill of Rights is *not* worth having. (in Shapiro 1995, 133)

Hook mistakenly suggested that Carnap opted for "peace *without* civil liberties" because Hook construed values as absolute, noncontextual, and, oddly, exclusive: One could have peace without civil liberties, or civil liberties without peace, he seemed to presume. He did not acknowledge as Carnap did the practical task of balancing and managing *multiple* and potentially conflicting values and goals (such as peace and civil liberties) simultaneously embraced by a complex and diverse society. Carnap described this task again years later in his autobiography as he discussed the needs for preserving individual freedom, avoiding nuclear war, and aggressively pursing state-sponsored social and economic planning (Carnap 1963b, 864). At times, circumstances may require that we compromise certain values for the sake of others. But that does not mean that we abandon them completely or permanently, as Hook took Carnap to recommend in the case of civil liberties.

Despite Hook's contemporary reputation as a fighter for pragmatic instrumentalism in politics, his anticommunist polemics of the 1940s and

---

[22] Carnap to Hook, March 24, 1949, ASP RC 088-38-13.

'50s acknowledged none of the complexities that Carnap had in mind.[23] Freedom, in particular, was for Hook a value that was absolute in the sense that it was to be upheld regardless of other beliefs, values, or contexts and how they may change in the course of events. As in Reichenbach's treatment of ethical commitments, it was an axiom to be taken for granted and not questioned or qualified. Hook admitted his absolutism about freedom and explained how it allowed him to tolerate fellow anticommunists who were usually more socially and economically conservative. Some thirty years after his argument with Carnap, in a letter to William F. Buckley, Hook explained that his devotion to freedom "transcends the differences about a market economy, the specific forms of the welfare state, the validity of the arguments for the existence of God, and other political, economic and educational issues" dividing him from Buckley and others. Nor did the importance and meaning of freedom depend on the well-being of the world or other human beings living in it. Recalling his earlier arguments with Bertrand Russell, Robert Maynard Hutchins, and other Cold War advocates of a world government, Hook reminded Buckley, "I argued that it was both wiser and nobler to announce that we were prepared to go down fighting for freedom even at the risk of world destruction, whatever that meant."[24] When the popular American president Ronald Reagan pinned a Presidential Medal of Freedom on Hook's lapel in 1985, the popularity of Hook's absolutistic approach to values in American popular and political culture was especially clear. Reagan's many admirers defend his approach to geopolitics as, similarly, a noble devotion to absolute, anticommunist ideals.

A Hot Potato on the Icy Slopes of Logic

Philosophy of science made its home on the icy slopes of logic when it adopted a conception of values as absolute and isolated from scientifically informed study and criticism. That conception reduced the

[23] One admirer has written, for example, that Hook was an opponent of absolutism, one who bravely "battled those infected by political fanaticism or double standards and those who condoned abhorrent behavior by appealing to some transcendent political or social ideal" (Shapiro 1995, 2).

[24] Hook to Buckley, May 21, 1982, in ibid., pp. 325–26. Hook also commented publicly that certain social ideals (such as freedom) were of greater value than "bare survival": Liberalism, for Hook, necessarily involved understanding that there are "ideas and ideals without which a life worthy of man cannot be attained. Among them are the strategic freedoms of those American traditions which make the continuous use of intelligence possible" (Hook 1953, 36). Carnap mentions his advocacy of world government in Carnap 1963a, 83.

possibility that the profession could be suspected of political mischief and, at the same time, it marginalized the leaders of the Unity of Science movement who championed a different, competing approach to and understanding of values. Not only must values underpinning moral and political questions be understood pragmatically, contextually, and nonabsolutistically, they believed. Values must also be recognized as operative within scientific theory and practice. As is examined more fully in the next chapter, Frank, Morris, and Neurath fought against any absolutistic conception of values that would pretend that science neither had nor needed values or that would deny scientific tools for understanding, specifying, and revising values. In part, as suggested by his defense against Hook's attack, Carnap shared this agenda and this practical, contextual understanding of values. But Carnap also joined Reichenbach, Feigl, Rudner, and others in distinguishing philosophy of science from politics and thus depoliticizing logical empiricism and the profession that practiced it.

To one observer who prided himself on having a keen eye for the political overtones of philosophical projects, this depoliticization was swift and surprising. When Horace Kallen wrote to Morris about collecting and publishing writings by Neurath (as discussed in chapter 13), he did so because he regretted that logical empiricism had become formal and apolitical. A decade and a half after attacking Neurath's project for its "totalitarian" political implications, Kallen was dismayed to see the ironic result that his attack could only have helped to bring about. Formerly, Neurath's project had the *wrong* political edge. Now, Kallen complained, it had *no* political edge.

Above all, Kallen missed Neurath's nonprofessionalism and his lack of detachment. "My feeling," he wrote to Morris, "is that Otto was so much more than a mere logical empiricist": "He had an enormous amount of compassion, a deep feeling for people as people, and an eagerness to serve their liberation and enrichment through the philosophic and sociological arts." By 1957, the intellectual work of a logical empiricist, Kallen implied, was irrelevant to, and uninspired by, social, political, and humanitarian concerns such as those of Neurath. That is why he championed a collected volume of Neurath's papers. Two years before Feigl noted that "axiomatization . . . has become the order of the day," Kallen wrote to Morris that "such a book could save logical empiricism, etc. etc. from the barrenness into which it seems to me to have fallen."[25]

---

[25] Kallen to Morris, May 7, 1957, JRMC.

Frank and Morris both tried to help Kallen and Marie Neurath in their plan. Frank remained in closest contact with Marie and, visiting her in London in the summer of 1957, helped her and Neurath's son, Paul Neurath, choose articles of Neurath's "specially important and still of interest for young students."[26] The work of finding an American publisher, however, fell to Morris. Morris soon learned that there was little institutional support outside or inside philosophy of science for so commemorating Neurath and his project. That summer, he wrote to Feigl and explained the situation: "I wrote to the Rockefeller people about this, but was turned down." Morris was turned away by Chadbourne Gilpatric, the foundation officer who had soured on Frank's dictionary project a few years before and who probably had a hand in the decision to terminate funding for Frank's institute.[27] Frank's institute, Morris knew, without even hearing it from Frank himself, had no money for this, so he made a pitch to Feigl and his Minnesota Center:

Now it occurs to me that your center might be interested and able to help. Could the book be tied into your publication plans? It might be possible to organize a small conference on Neurath's ideas & publish something of this in Marie's book.[28]

Feigl could not help, either. His center, Morris reported back to Kallen, "now operates 'on a sharply reduced budget' and cannot help in any way." Morris had not yet run out of ideas, however. He recommended that Kallen appeal to the Rockefeller Foundation's social sciences division (where Morris had no contacts) and the design expert Gyorgy Kepes in MIT's architecture program.[29]

At this point, Kallen raised Marie's worry that Otto might be seen as a communist. Morris, his memory perhaps sharpened by his recent conversations about Neurath and his legacy with Feigl and the Rockefeller Foundation (then funding the anticommunist Congress for Cultural Freedom), concurred; indeed, he had "met people who think Otto and indeed the whole unified science program is 'Communistic.'"[30] Morris then turned to the University of Chicago Press, who had already dealt with Neurath through the *International Encyclopedia*. Aiming high, Morris decided that he would meet with the press's director, Roger Shugg. But

[26] Marie Neurath to Horace Kallen, July 18, 1957, JRMC.
[27] Morris reported that he spoke with Gilpatric in Morris to Kallen, June 25, 1957, JRMC.
[28] Morris to Feigl, July 10, 1957, HFP 03-109-03.
[29] Morris to Kallen, August 16, 1957, JRMC; Feigl to Morris, July 23, 1957, HFP 03-109-04.
[30] Morris to Kallen, October 8, 1957, JRMC.

he wanted to do so equipped with letters from prominent members of the profession attesting to Neurath's importance and the importance of such a volume. He asked Carnap, Feigl, Hempel, Kallen, and Nagel to write letters in support of the project, while Marie asked the same of Frank. The volume in question would be substantial. It would include a biographical introduction about Neurath; about 400 pages of his writings in three areas ("society and economy, problems of planning," "unified science, terminology, etc.," and "education, picture language, visual education"); and, finally, a section of "about 50 pages of dates, events, friends, an index, [and] bibliography."

Letters in hand, Morris met with Shugg sometime in the summer of 1958. But their conversation came to naught. If Shugg was interested in publishing a volume on Neurath, that interest did not overcome two practical obstacles: Morris's move to the University of Gainesville in August and, three months later, a case of "acute appendicitis" that prevented Shugg from responding to Morris's inquiry "about the fate of the Neurath volume." Shugg's secretary, Morris reported, said that he would be away from his office for several weeks.[31] Kallen relayed this lack of good news to Marie, who seemed to interpret it as bad news. "I have the impression," Kallen told Morris, "she is whistling to keep her spirits up, not only because of the slowness of our common project, but because her own affairs may not be going as well as they should."[32] After pursuing their project for a year and a half without success, Morris and Kallen were whistling, too. "We must find some way to get the book published if the [University of Chicago] Press does not take it. Neurath was an amazing man," Morris affirmed.[33] Kallen agreed by suggesting that Neurath's contribution to the profession sometimes lay as much in his stubbornness as in his populism:

I have always had that strong feeling about Neurath with whom even disagreement was a happy and creative adventure. His work in making modern modes of thinking communicable to the non-specialist has always seemed to me of prime importance. I do hope that we will get the book we planned published.[34]

Fifteen years later, in 1973, owing largely to the efforts of Robert Cohen, the first volume of Neurath's selected writings appeared in English (see Neurath 1973). Only then, after student activism for civil rights and

[31] Morris to Kallen, November 12, 1958, JRMC.
[32] Kallen to Morris, December 4, 1958, JRMC.
[33] Morris to Kallen, November 12, 1958, JRMC.
[34] Kallen to Morris, December 4, 1958, JRMC.

against the Vietnam War had begun to thaw the McCarthyite, Cold War culture on many campuses, were memories of the Unity of Science movement rekindled. Subsequent generations of philosophers then began to see that the movement's breadth and ambition could never have been contained on the icy slopes of logic.

18

Professionalism, Power, and What Might Have Been

Like any story, the one told here of the rise and fall of the Unity of Science movement in North America is incomplete. Besides the anticommunist pressures described here, other forces and circumstances surely helped to determine the evolution of philosophy of science through the Cold War. Two explored briefly in this chapter are the decline in North America of so-called public intellectuals and the growth of research universities as the main institutions of intellectual life in North America. Both are connected to Cold War anticommunism, and this chapter introduces them as a frame to examine some commonplaces about logical empiricism and the unity of science as well as contemporary interest in the disunity of science. It also addresses two specific questions raised by this story. One concerns the unique role given to Rudolf Carnap, who is depicted both as a leftist philosopher of science in the 1930s and as a professional, apolitical philosopher during and after the Cold War. The other question is one that will have occurred to many readers long before reaching this final chapter: Can't the depoliticization of philosophy of science in the 1950s, described here as the result of multifaceted forces arising from anticommunist powers, be interpreted better as a development or maturation through which twentieth-century philosophy (finally) acknowledged a fundamental and proper disconnection between philosophical research and political partisanship? However clear that conceptual difference between philosophy and politics may now seem, it is suggested, that clarity is itself an artefact of certain historical and institutional conditions. On this note, the chapter concludes by suggesting that the history of the Unity of Science movement invites consideration of the traditional divide between analytic and continental philosophy and their respective concerns with logic and reason, on the one hand, and power, on the other.

The Postwar University and the Decline of Public Intellectuals

If the death of the Unity of Science movement is to be dated, one candidate would be 1955, the year that concluded Frank's second three-year grant for the Institute for the Unity of Science. The year 1959 is also an appropriate date. In that year Feigl followed Reichenbach in delineating a firm, logical gulf between philosophy of science and social and ethical discourse; Frank was cryptically attacked in *Philosophy of Science* as a neo-Thomist; and Reichenbach's image appeared in *Science* to promote an image of logical empiricism in the service of the RAND Corporation and military science. Perhaps the most useful date would be 1966, not only because in that year Frank died and Richard Rudner emphatically demarcated philosophy of science from social philosophy, but because during the 1960s the Unity of Science movement must have been remembered as something of a monster: an awkward patchwork of technical philosophy of science, on the one hand, and public writing contests, vague proposals for studying sociology of science, and interest in society and values of all kinds, on the other. This patchwork not only made little sense in light of the popular distinction between philosophy and politics. It also cut across the walls and borders of the postwar university that rapidly became the dominant forum and institution for intellectual life in America.

Ellen Schrecker made a related point in the title of her monograph *No Ivory Tower* (Schrecker 1986). For an audience of intellectuals and academics reading her book when it was published in the mid-1980s, the very idea that Cold War forces had so affected the lives of individuals and disciplines needed to overcome a popular view that intellectual life has always belonged naturally and exclusively to the university. Inside the ivory tower, the presumption goes, intellectual pursuits driven by reason and the search for truth are insulated and protected from the irrational, unpredictable workings of political power and social fashion that dominate popular culture. Schrecker's book has a shocking effect precisely because it forces this presumption to dissolve. There was no ivory tower, for anticommunists outside easily joined hands with administrators, professors, and trustees inside who were eager to convince senators, journalists, civic groups, and others that their campuses and classrooms were not pink.

Understanding how philosophy of science was affected by Cold War politics therefore goes hand in hand with a larger story about the evolution of the North American university. Keeping classrooms free of leftist ideology, after all, helped to secure federal funds that helped universities to grow during the Cold War. The growth of the universities, in turn, invites consideration of debates concerning so-called public intellectuals. Russell Jacoby helped to frame these debates by documenting that, contrary to the trend of baby-boomer leftists who routinely moved from graduate school to academic life, the "radicals of the early part of the century almost never became college teachers" (Jacoby 1987, 124). Most thrived instead in urban enclaves, the epitome of which was New York's Greenwich Village. Bohemianism secured intellectuals' contact with the public and with realities of political, urban, and civic life. They wrote for journals and magazines that they often founded themselves, were not owned by universities, and were read not only by established intellectuals at Columbia, for example, but also by curious teenagers in Brooklyn and elsewhere.

During the Cold War, however, several trends – some of which are familiar from the story told here – worked to reconfigure the place and role of the intellectual in American society. With the postwar growth of the suburbs and the decline of many major cities, for example, these bohemian enclaves lost their appeal or disappeared. Intellectuals gravitated to stable, better-paying jobs in universities, and, as a result, shop-talk moved from public coffee houses to campuses, seminar rooms, and professional society meetings. Unmoored from its proletarian roots, the tradition of liberal "public intellectuals" that thrived before the war began to atrophy. In the humanities, the rise of specialized jargons in the 1970s and '80s accelerated the trend because popular terms, concepts, and logics of French, German, and Italian theorists, for example, tend to be incomprehensible to those outside the intellectual circles that cultivate them. That circumstance, Jacoby suggests, invited conservative and elite Cold War intellectuals, suspicious of the Marxist roots of literary criticism and the political credentials of the faculty who promote it, to exploit anti-intellectual currents in American culture and thus achieve credibility, if not institutional dominance, over their leftist colleagues in recent decades. Allan Bloom, for example, who wrote the popular *Closing of the American Mind* (1987), found common cause with conservative foundations and publics whose suspicions were confirmed by his account, in relatively clear, jargon-free prose, of the alleged eclipse of Western and American ideals by European theory. In these conservative intellectual circles, as the title of another popular book puts it, the humanities were coopted by "tenured radicals" (Kimball 1990).

One result of these trends is a popular image of the contemporary university as a bastion of variously leftist, feminist, radical thinking about moral, social, and political issues. From the point of view of the Cold War and its effects, however, this image is misleading. It ignores the effects of academic anticommunism and accepts, at most, the popular memory of the McCarthyite hysteria as a momentary, anomalous time in U.S. history during which a handful of scientists and Hollywood screenwriters had some unfortunate professional difficulties. The general transformation of academic life wrought by academic anticommunism, however, was broad and deep. In popular culture, at least, the striking contrast between even the most radical contemporary university intellectuals of the 1990s and their counterparts of the 1920s and '30s has been largely forgotten. Depression-era intellectuals made obligatory trips to the Soviet Union, not just for exposure to Russian culture or art, but also to witness an unprecedented experiment in human society that Soviet intellectuals had actively helped to launch. Many in America and Europe aspired to these ranks, and some succeeded. Dewey's student Max Eastman, after teaching philosophy at Columbia, became confidant to several Bolsheviks. Trotsky's bodyguard and secretary Jean van Heijenoort retired to teach mathematical logic at MIT. Neurath, Hook, Blumberg, Malisoff, and others treated here had similar, if less glamourous, careers that demonstrate how politics, intellectual life, and higher education could be joined before World War II in ways that became impossible after.

The popular image of college professors as highly politicized and leftist ignores the extent to which radical intellectual life in the United States was eviscerated by academic anticommunism. To be sure, students today may yet learn socialist theory and the history of its application during their few years in higher education. But the disconnection between intellectual life inside the ivory tower and popular culture outside reinforces the popular image and critique of leftist intellectuals as peculiar and disengaged from the "real world" and its concerns. In aggressively conservative circles, commentators routinely depict academic leftism as dangerously un-American as they prop up radicalism's corpse for target practice and entertainment.

In part, the Unity of Science movement belongs to this larger history of public intellectuals in the United States. In the late 1930s, especially, there were signs that Otto Neurath was poised for such public recognition, appearing as he did in the pages of *Time*, the *New York Times*, and *Survey Graphic*. Charles Morris also had brief tenure as a public intellectual in the late 1940s following the moderate success of his book *The Open Self.*

But the movement soon took a very different trajectory after the war, as these chapters have shown.

Conventional Wisdom about Logical Empiricism and the Unity of Science

The realities of Cold War anticommunism and their effects on professional philosophy suggest that several items of conventional philosophical wisdom are ripe for revision. Perhaps the most basic is the reputation of logical empiricists as logic-chopping intellects who were blind (as their critics on the far left insisted) to politics, ethics, and urgent questions about the place of philosophy in modern culture. Far from being apolitical, as a snapshot of logical empiricism circa 1960 would suggest, logical empiricism was politicized both when it arrived in the United States and, in a negative sense, after the war when the McCarthyite "climate of fear" disabled the Unity of Science movement and helped to cultivate its apolitical posture. Whether inclined toward socialism in the 1930s or defending itself against anticommunism in the 1940s and '50s, logical empiricism was neither apolitical in its values and ambitions nor an unpolitical community of scholars, somehow insulated from Cold War pressures. Still, it is common to read that logical empiricism was simply "unpolitical" (Kucklick 2001, 232), that it left behind its political and cultural commitments in Europe prior to emigrating, and that it was the agent of its own depoliticization.[1]

Another set of considerations is suggested by contemporary interest in the disunity of science. Recent writings have tended to focus on the unity of science as a *thesis* about science, one that most philosophers view as false and better replaced by its antithesis, the "disunity of science."[2] Hardly a claim for the dialectics of philosophical progress, however, the

---

[1] Thus writes Ron Giere: "Logical empiricism in North America was to a considerable extent a new creation . . . styled for a new audience so that what appeared in public view in North America was something noticeably different from what had existed in Europe" (Giere 1996, 338). Peter Galison, while examining the cultural meanings of Carnap's *Der Logische Aufbau der Welt* (1928) in Europe, similarly writes that "the ideal of an *Aufbau*" was "conjointly architectural, political, and philosophical." But "it did not as such move across the Atlantic" (Galison 1996, 40). Martha Nussbaum, in a tribute to Rawls, depicts Rawls as a politically relevant corrective to the legacy of logical positivism that "had convinced people that there were only two things that it made sense to do: empirical research and conceptual analysis" (Nussbaum 2002).

[2] Relevant literature includes Fodor 1974; Kitcher 1989; Cat et al. 1991; Hacking 1992; Dupré 1993; Rosenberg 1994; Galison and Stump 1996.

suggestion made by most defenders of the disunity of science is that logical empiricists made a simple but colossal mistake. Aside from differences
about what they take "unity of science" to mean, they hold that logical
empiricism was simply wrong to view the sciences as unified or unifiable.

One claim for the disunity of science from the 1990s may seem familiar
to readers of this book. Not only scientific but also social progress, John
Dupré writes, requires us to reject the unity of science thesis in favor of
disunity:

> For if science were unified then the legitimate projects of inquiry would be those,
> and only those, that formed part of that unified whole. On the picture that I am
> presenting, only a society with absolutely homogenous, or at least hegemonic,
> political commitments and shared assumptions could expect a unified science.
> Unified science, we might conclude, would require Utopia or totalitarianism.
> (Dupré 1993, 261)

Though he intends to help "clear away some of the remaining debris of a
philosophical movement that, though largely abandoned some decades
ago, continues to exercise a powerful influence over major areas of philosophy" (ibid., p. 10), Dupré shows that a more powerful and enduring
influence is Kallen's opposition to unified science. His argument dupli-
cates Kallen's stark opposition between unified science and radical pluralism (which Dupré formulates in a metaphysical, ontologized version)
and, in this comment, draws the same negative political conclusions that
Kallen drew about the unity of science.

Another argument for the alleged disunity of science is perhaps better
understood as a substantive contribution to the very line of inquiry that
the Unity of Science movement promoted. Alexander Rosenberg argues
for the disunity of science for reasons of (something like) computational
complexity intervening between conventional object domains of biology
and physics. As Rosenberg summarizes his argument, concepts and laws
of biology simply cannot be reduced to those of chemistry or physics:

> Once matter aggregates beyond the level of the organization of the biologically
> active macromolecule, the level of complexity becomes so great that creatures of
> our cognitive and computational abilities cannot move from models and approx-
> imations to the nomological generalizations governing the biological processes.
> (Rosenberg 1994, 6)

Far from a refutation of "unity of science," however, Rosenberg's project
constructively engages the topic as Neurath and Carnap framed it. Since
bridges among the sciences were to be built within science itself, logical empiricists would generally have welcomed Rosenberg's result that

considerations internal to science cast strong doubt on one kind of unity (the one Carnap dubbed "unity of law") between biology and chemistry and physics. Carnap would take Rosenberg's result seriously, but he would not accept that the result refutes the unity of science. There were, after all, other models of unity available, such as Carnap's "unity of concepts," Neurath's view of a unifying physicalist "universal jargon," and perhaps other conceptions that will only be recognized and formulated by philosophers and scientists in the future.

This last point is important. Since the Unity of Science movement was a collective, collaborative project, 1930s-style arguments for the unity of science are not well joined to recent arguments for the disunity of science. They belong to different modes of philosophical life and practice. Politically, unified science had connections and associations with radical and Communist Party philosophy that intellectuals in the 1950s and '60s would have been wise to avoid. Institutionally, unified science embodied a collaborative, collective, and interdisciplinary impulse that began to lose its sense and purpose in the postwar university. The movement hoped to regularize collaboration with scientists and intellectuals in ways that would diminish, or make more flexible, the traditional borders that separate the sciences from each other, the borders between philosophy of science from science itself, and (as Frank and Morris emphasized) those between science and the humanities. After the war, however, the new professionalism prized disciplinary autonomy and, in growing, well-funded universities, established new borders carving out new specialities and disciplines.

Alongside these political and institutional shifts, the unity of science program transformed from a practical, collaborative goal to a more narrow academic thesis in at least two senses. First, it became an empirical hypothesis *about* science, its languages, and their future historical development viewed, so to speak, from across the quadrangle as an independent intellectual project neither requiring nor requesting input from philosophy. Much as Frank observed philosophers finding an intellectual niche "between and above and below the domain of... isolated special sciences" (Frank 1949b, 269), May Brodbeck explained that philosophy of science hovered discretely between analytic philosophy and science. Because philosophy of science "is about science," she wrote, "it is not quite correct to speak of continuity between these two fields [of analytic philosophy and philosophy of science]." "Yet, because it *is* about science, new problems appear for the philosopher of science as the scientists invent new techniques and extend their knowledge" (Brodbeck

1953, 7). In the next ten years, Carnap referred to his then current ideas about the unity of science as "sweeping extrapolating hypotheses" about the future of science (and philosophical reconstructions of it), and Paul Oppenheim and Hilary Putnam offered an influential view of the unity of science as "a working hypothesis" (see Carnap 1963b, 882–86; Oppenheim and Putnam 1958). Notwithstanding Quine's influential view that science and philosophy are continuous, that "philosophy of science is philosophy enough," this institutional divide remains between "science" as understood and practiced by scientists and its philosophical counterpart.[3]

Second, after it was decoupled from the ideal of active collaboration with scientists, the unity of science thesis was adapted by one of the several specialties within philosophy of science as it became more partitioned and specialized after the 1950s. The "physicalism" and "unity of science" that Neurath and others once championed as a handmaiden to Marxist-inspired progressivism became the "physicalism" that, in its different articulations within philosophy of mind, addresses relations between mind and brain (or psychology and physics) and helps to organize debate and research in that specialty.

The unity of science thesis is an abstraction from what was originally a practical, constructive project. Understood as a tool in that project, the thesis that the sciences can be further unified is no more refuted by the failure of specific models of unity of science or by the historical decline of the Unity of Science movement than failures in physics to reconcile quantum mechanics and general relativity refute the goal of a unified theory. As Neurath regularly reminded his colleagues, our present explications of unity of science are both *our* explications and our *present* explications: "Our thinking is a tool, it depends on social and historical conditions" (Neurath 1930, 46) that limit and shape our conceptions in ways that we cannot always see. We cannot, that is, step outside these particular conditions and see them with some ideal, context-less objectivity. Only in the future may they become visible to our descendants who will be subject, again, to their own social and historical conditions. Thus Neurath concluded his main contribution to the *International Encyclopedia* with a version of his famous nautical metaphor that should have given pause

---

[3] One philosopher arguing that professional philosophy has begun to reemerge toward public engagement from its Cold War formalism and professionalism (again, via Rawls's landmark *Theory of Justice*) nonetheless concedes that there is even within departments "a dangerous fragmentation in the field" and that "most scientists are still not interested in what we do" (Nehemas 1997, 239, 238).

to Kallen, Nagel, and others who variously complained that Neurath and his Unity of Science movement aimed to impose some specific, restrictive philosophical (even totalitarian) plan on Science: scientists and philosophers are in the same boat, like sailors "who, far out at sea, transform the shape of their clumsy vessel":

A new ship grows out of the old one, step by step – and while they are still building, the sailors may already be thinking of a new structure, and they will not always agree with one another. The whole business will go on in a way we cannot even anticipate today. That is our fate. (Neurath 1944)

If we wish to consider the movement on these terms, the question becomes not whether the thesis of the unity of science is true or false but whether this collaborative, collective project in fact had a chance to prove itself and, possibly, succeed. It did not have that chance. After a few optimistic years in North America, the war broke out, and shortly after the war it suffered both the loss of Neurath and the rise of Cold War forces and fashions that soon made the movement's progressive, socialist-friendly values and ambitions a thing of the past.

What about Normativity?

Some philosophers may take history's verdict on the Unity of Science movement to be philosophically correct. Its medley of socialist, internationalist, and populist themes, they may say, has no place in philosophical pursuits. They may accept the apolitical, neutral stance articulated variously by Reichenbach, Rudner, and Feigl during and after the 1950s with the result that, for them, the movement's demise and the obscurity of Neurath's, Frank's, and Morris's agendas for the profession require neither apology nor adjustment. If philosophy of science is logically incapable of treating normative matters of life, society, and politics, then it had no business carrying on before the Cold War as if it could.

There are at least two, related ways to defend the more politically engaged vision of the profession shared by Neurath, Frank, and Morris without violating logic or common sense. From the point of view of 1950s neutralism, the Unity of Science movement's political engagements violated Reichenbach's distinction between cognition and volition. If, as Feigl put it, "there is no way of deducing moral imperatives from the truths of science" (Feigl 1959, 16), then what were Neurath, Frank, Carnap, Hahn, and others doing in 1920s Vienna and 1930s North America? The noncognitive view of ethical statements as purely subjective or expressive, after all, was in full view (and variously endorsed) by Neurath, Carnap, and Hahn when they wrote their politics-laden manifesto, *Wissenschaftliche Weltauffassung*. Yet they did not see themselves struggling to overcome the logical hurdle Reichenbach and Feigl posed because the norms and values they admired and promoted were already established in certain areas of science, politics, education, architecture, and design. The scientific world conception did not need logical justification because it was already, as an empirical fact, alive and poised to dominate modernity:

> We witness the spirit of the scientific world-conception penetrating in growing measure the forms of personal and public life in education, upbringing, architecture, and the shaping of economic and social life according to rational principles. *The scientific world-conception serves life, and life receives it.* (Neurath et al. 1929, 301, 317–18)

What this outlook required, instead, was institutional support (from the Ernst Mach Society and the Unity of Science movement) as well as clarity of conception and vision that logical empiricism could help to provide. With the resources of the new logic and the new scientific philosophy, the movement could "give . . . systematic thought" to these developments, help to clarify the epistemic and humanitarian strengths they shared over competing, reactionary, or regressive forces, and, of course, help to lead the development of unified science (ibid., p. 301).

To the determined neutralist, these considerations are question-begging, for they provide no cognitive justification for the Unity of Science movement's many engagements. In its largest aspect, Neurath's, Frank's, and Morris's crusade to revive the Unity of Science movement after the war aimed precisely to steer professional philosophy of science away from such an outlook, one that echoes Neurath's target, pseudorationalism. It supposes that life's engagements must, to have respectability or validity (in both the logical and colloquial senses of that word), have some place in or connection to ideal philosophical theories of knowledge. For the neutralist critic demanding justification for the Unity of Science movement's ambitions, philosophical conceptions of knowledge come first, and, if derivable from cognitively meaningful premises, political engagements come second. For the left wing of the Vienna Circle that supported the Unity of Science movement, the outlook was quite different. The exigencies of life and philosophical practice were in constant, dialectical dialogue. Life and science inform philosophy of science, while philosophy of science educates and informs our understanding of life and science. As the dialectic continues, we or our descendants may find that we have sailed into strange, presently incomprehensible waters.[4]

This difference helps to illuminate both the political and philosophical marginalization driving Morris's, Frank's, and Neurath's respective campaigns against "absolutism." They opposed, and were out of step with, not only the moral and political absolutism of anticommunism but an institutional and disciplinary absolutism that would isolate philosophy of science from interaction with other disciplines and areas of culture. When Morris urged the movement to pay more attention to the social sciences and, in particular, pragmatic approaches to the study of value, art, and ethics, he rejected what would become the postwar anti-intellectualism and antiscientism that exalted values as transcendent, absolute, and beyond the reach of science and scientific philosophy. As he urged in his lecture on the "cultural significance of science," science needed to be understood as one source of our cultural values and not a servant to supposedly super-scientific or eternal, fixed values. This was also Frank's target at the Conferences on Science, Philosophy and Religion. Absolute moral values are self-defeating, Frank urged, because genuinely absolute values and fixed values must be so abstract and general that they are practically meaningless. When it comes to general and "meaningful principles of human conduct," Frank told his audience that

if we attempt to cling firmly to the general principles without qualification, to the "absolute" meaning of those principles, we shall soon notice that no conclusion at all can be drawn that would be pertinent to an actual life situation. For without using the operational meaning we derive from our abstract principles merely abstract principles. We never get in touch with an actual human problem. (Frank 1950, 42, 43)

In 1950 when Frank published this, the human problems of the day included the epic struggles of capitalism versus communism, individualism versus collectivism, and the problem of "communist professors." By asking his fellow intellectuals to adopt a relativized, contextual understanding of human values and to abandon the popular "absolute" conception, Frank was effectively urging them to remain engaged in these debates, much as he urged philosophers of science to practice an "active positivism," as described in chapter 14. For Frank, a socially and politically engaged philosophy of science could not be one that endorsed the absolute conception of values promoted by Reichenbach, Feigl, and others.

---

4 Frank continued to write about values and science in his last years of life. An unpublished book in the Harvard University Archives addresses science, ideology, and values.

In his arguments with Carnap, Neurath connected his complaints about semantics to the kind of "absolutism in metaphysics and faith" that he believed helped to drive the long, violent history of persecution in the world. Carnap probably never came to share Neurath's view that semantical theory was so directly connected to the history of persecution. But he might have recalled Neurath's more general critique of absolutism during his scuffle with Sidney Hook. When Hook charged that Carnap had "already left the democratic camp" because he diluted Hook's exaltation of freedom with practical considerations about preserving peace and avoiding nuclear war, Carnap may have recognized the absolutist logic that Neurath had so abhorred. Absolutists cannot abide pluralism, Neurath explained to Carnap. "People sometimes cannot bear that we start with many divergent statements, and remain with divergent statements FOREVER, as it were," Neurath wrote. For them, "there HAS TO BE SOMETHING ONE."[5] Freedom, for Hook, was indeed "something one," prior and ultimate. Those who would question that, in Hook's eyes, had already opted out of collaborative, democratic discussion.

Had the profession better accommodated Morris's, Neurath's, and Frank's projects, Reichenbach's and Feigl's arguments for the impossibility of philosophy of science to engage with social and political issues would have seemed less compelling, if only because they would not have seemed unopposed and unqualified. If, with Neurath, Reichenbach were to have emphasized that his axiomatic model of ethical knowledge was actively created by abstraction from a historically evolving field of vague terms and concepts – if, that is, Reichenbach acknowledged something like Neurath's pluralism as an important complement to his formalism – then questions about how, exactly, our ethical knowledge can, and cannot, be usefully and accurately modeled as a formal system could be better focused and organized. Once ethical beliefs have been reconstructed and specified as formal axioms, that is, Reichenbach, Feigl, or other logical empiricists might have followed Frank and posed sociological questions about to what extent these axioms may reflect social influence, accidents of history, or, as Frank was fond of pointing out, bits of antique, now-false scientific knowledge mistakenly seized as indubitable metaphysics or self-evident "common sense." However axiom-like, that is, Reichenbach's ethical volitions could nonetheless be understood as contextualized, analyzable, and no longer treated as absolute and fixed.

[5] Neurath to Carnap, September 25, 1943, ASP RC 102-55-03 (the quote appears on p. 15).

One cannot object that this more fluid and contextual picture of our knowledge is essentially foreign to, or inappropriate for, logical empiricism because, as these chapters have documented, this conception belonged to the mainstream of scientific philosophy before the Cold War. Carnap suggested valuable connections among scientific, ethical, and political knowledge in 1936 when he attributed the unfortunate "conduct of different nations, races, and social classes" to the fact that it is

controlled more often by passions than by reflection upon the facts of psychology and the social sciences. Their expectations, inadequately founded, are usually followed by disappointments in the behavior of other parties: but the failures of their hopes, instead of leading to the correction of erroneous assumptions, frequently become the occasions for a childish reproval of opposing groups in the name of morality. (Carnap 1936a)

Making these comments both publicly and as a visiting professor at Harvard, it would appear that Carnap saw these relations as a legitimate, though perhaps not central, province of logical empiricism. Ernest Nagel similarly attributed "needless sufferings and conflicts" among peoples to excessive reliance on "intuitive insights and passional impulses" that are not grounded in "the firm soil of scientific knowledge." "It is not wisdom," Nagel thundered at the neo-Thomists, "but a mark of immaturity to recommend that we simply examine our hearts if we wish to discover the good life" (Nagel 1943, 54). It was only in the 1950s that the profession's recognition of this intimacy among science, philosophy of science, and "the good life" was downplayed. As he praised the values of scientific humanism, Feigl became more emphatic that those values, like others, could never be logically justified. To those concerned with "how to live," Reichenbach explained, "the scientific philosopher tells him pretty frankly that he has nothing to expect from his teachings if he wants to know how to lead a good life" (Reichenbach 1951, 315).

Those who insist that Reichenbach's sharp distinction between politics and philosophy is correct or more compelling than the alternatives must either believe that Reichenbach suddenly, in the 1950s, saw deeper into this question of relations between politics and philosophy than he and others had before or – as seems more likely – admit that the popularity and seeming inevitability of his sharp distinction owes something (as Neurath would suggest) to the "social and historical conditions" surrounding its rise in the 1950s. Those conditions reduced the stature and influence of those who strongly opposed the distinction and, in the eyes of subsequent generations of philosophers, adds an antique, implausible quality to their alternative visions for philosophy of science. This is not to claim, however, that these alternatives, once recovered and dusted off, are simply superior or even practically available, for such assessments cannot be isolated from the history in question. It is possible, after all, that scientific philosophy survived in North America *only* by purging itself of projects perceived to be "pink" or, as Morris reported about Neurath and the Unity of Science movement, "communistic."

The Special Case of Rudolf Carnap

Like Neurath, Frank, and Morris, Carnap was a leader of the "communistic" Unity of Science movement who took logical empiricism in the 1930s to be relevant to and important for social life and politics. Yet, unlike the others, Carnap suffered no subsequent intellectual or professional decline in the decades after the war. Several considerations explain this circumstance. One is Carnap's talent as philosopher who was able both to pioneer new topics and techniques and to communicate their importance to others. Another is his consistently held policy that philosophy is neutral with respect to social and cultural aims, that it provides no foundation or justification for political views and activities. Like Reichenbach, Feigl, and Rudner, Carnap emphasized that philosophy and politics are two very kinds of enterprises. But his view did not sever connection between them; it rather denied that philosophical research involved or required political beliefs or values. As this quotation from his 1936 radio address makes clear, however, he believed that political thinking and activity indeed *should* make use of methods and insights from science and philosophy of science whenever possible.

Since academic anticommunism was largely driven by the fear that intellectual scholarship would be infected or controlled by Communist Party ideology, Carnap's view of philosophy as independent of politics would hardly have raised eyebrows in the 1950s. Even the FBI agents assessing Carnap's political affiliations seemed to imbibe his neutralism from their interviewees, especially the one who reported that Carnap "is interested '99% in scholastic matters and has little or no interest in politics of any kind.' " That claim was wrong, of course. At the same time, without violating his neutralism, Carnap was highly interested in political issues and took political stands by himself (such as those that aroused the interest of J. Edgar Hoover) and eagerly helped to lead the Unity of Science movement as a point of contact between logical empiricism and science and society.

It would be a mistake, therefore, to take Carnap's postwar work in semantics, probability, inductive logic, and confirmation in the last decades of his life as the result of a substantive change in his philosophical work that was driven by political caution. Rather, any history of Carnap's philosophical work must first contextualize that trajectory within his broad, tripartite conception of philosophy, one that makes room for both his preferred technical work in syntax and semantics as well as pragmatics and its domain of human purposes and needs. Consider Carnap's exchange with Frank in the Carnap-Schilpp volume. Perhaps unwisely revisiting the attacks on logical empiricism from Lenin and other Soviet philosophers, Frank reproduced in his critical essay a largely negative review of Carnap's "The Elimination of Metaphysics through Logical Analysis of Language" by Soviet philosopher V. Brushlinsky. The essay focused on the question, familiar from Lenin's critique of empiricism, of whether and how philosophy is connected to socio-economic structures. As Carnap put it, Brushlinsky held that "I, as a logician and mechanist, and the movement of neopositivism in general, are incapable of understanding the social-economic roots of idealism and of metaphysics which we wish to eliminate" (Carnap 1963b, 867–68). Ignoring the spurious allegation that he was committed to a mechanistic metaphysics, Carnap responded by reiterating that "the Vienna Circle, essentially because of Otto Neurath, did recognize the importance of a sociological analysis of the roots of philosophical movements. But unfortunately a division of labor is necessary." The "division of labor" was Carnap's partition of philosophical analysis into syntax, semantics, and pragmatics. His own work and his talents, he acknowledged, belonged to the first two domains, not to pragmatics. "Therefore," he continued,

I am compelled to leave the detailed work in this direction to philosophically interested sociologists and sociologically trained philosophers. . . . I agree with both Frank and Morris that the pragmatic component has so far not been sufficiently investigated by our movement, although its importance has been acknowledged theoretically by me and by empiricists in general. (ibid., p. 868)

During and after the 1940s, when Carnap's writings and topics became less directly related to the central goals and concerns of the movement (such as the elimination of metaphysics and the Unity of Science), what changed was his personal allocation of labor within philosophy's overall project as he conceived it. He did not change his collectivist, collaborative view of the philosophical enterprise and its divisions, his demarcation of philosophy from politics, nor his acknowledgment of philosophically legitimate work (yet to be done) in the study of values and, more broadly, pragmatics.[6] His general movement away from issues related to "empiricism" and toward formal "scholastic" problems, as Neurath and Frank characterized them, should not be interpreted as some attempt to avoid politics or controversial topics. For even well after the high tide of academic McCarthyism, as illustrated by his late efforts on behalf of persecuted Mexican philosophers, Carnap did not make a point to avoid politics.[7]

Analytic and Continental Philosophy

Almost all of these developments in 1950s philosophy of science – the dominance of a formal model of knowledge, the transition from the Unity of Science movement to the more academic Unity of Science thesis, and the movement toward a politically disengaged professionalism – together offer some perspective on the often-lamented divide between analytic and continental philosophy. John McCumber has argued that this divide is a political imprint from the early years of the Cold War.[8] In the hands of Quine, Carnap, and University of Washington president Raymond Allen, McCumber argues, Cold War exigencies effectively marginalized German and French philosophy within a profession that found safe haven from academic anticommunism in what became modern analytic philosophy. A dispassionate, unpolitical search for scientific truth, McCumber argues, came to dominate professional philosophy because philosophers interested power, dialectics, phenomenology, and other icons of continental philosophy were required to keep these murkier, more political concerns and interests – like a wet dog – outside the ivory tower.

This history of the Unity of Science movement sees Carnap and logical empiricism quite differently – not as agents of Cold War change, but

---

[6] Besides Carnap's remark to Frank, quoted here, see his constructive and sympathetic reply to Kaplan in Carnap 1963b, 999–1013. On Carnap's anti-absolutistic conception of cognitively significant values, see Carnap 1944.

[7] Carnap championed the cause of three Mexican philosophers who had been jailed in Mexico City shortly after student demonstrations in 1968. He met two of them in 1963 during an International Congress of Philosophy and later visited them in a Mexican jail. He published an account in the *Journal of Philosophy*, "My Visit with Two Imprisoned Philosophers in Mexico" (1970). For Carnap's views on world government, the necessity of rational social and economic planning, and the need to avoid nuclear war, see Carnap 1963a, 83–84.

[8] McCumber 2001. For an account that identifies "a parting of the ways" between analytic and continental philosophy in the 1920s, see Friedman 2000.

rather as targets of various political, institutional, and intellectual forces. Still, it agrees with McCumber's account that at least elements of continental philosophy are implicated, perhaps ironically, in an understanding of the history of scientific philosophy in North America. McCumber holds that analytic philosophy adopted Raymond Allen's view that all valid intellectual activity involves a "timeless, selfless quest of truth." Notions about absolute truths were indeed icons of Cold War life. But McCumber's focus on Allen suggests that power, more than truth, was the agent of transformation. Those intellectuals who do not accept this quest of truth, Allen once proclaimed, and who thus fail "to deal in a scholarly and scientific way with controversial questions" will "lose their security" while "the institution from which they come will lose its academic standing" (in McCumber 2001, 40). Allen's warning that a threat of scandal and possible loss of one's job hover over those who treat controversial topics was one instance of the many pressures that helped to depoliticize logical empiricism.

While it may be true, as McCumber says, that most analytic philosophers find writings of continental philosophers merely incomprehensible, "subjective prattle" (ibid., p. 83), it may be that a turn to subjectivity is required to achieve a fuller understanding of both how analytic philosophy and philosophy of science navigated Cold War waters. For these Cold War dynamics present not only questions about what problems, questions, or projects philosophers did or did not pursue in the 1950s, but – as some of these chapters have tried to frame – questions about relations of power and subjectivity in academic life during these pivotal years. Following Foucault's *The Order of Things* (1970) or *Discipline and Punish* (1977), the subjectivity in question is not merely a phenomenological subjectivity that encourages us to ask, What was it like to be a philosopher investigated by the FBI or attacked as promoting "totalitarian" projects? It is also a subjectivity about being *subject to* certain powers and techniques of social control. If, as Foucault claims, criminals, delinquents, and other types of subjectivities and identities were created and maintained by hospitals, schools, the military, and other institutions of modern Europe and North America, then we may ask whether and how the norms and parameters of academic philosophy (or any other profession) were similarly created or affected by anticommunism and its "climate of fear" on North American university campuses.

Such a line of inquiry might well subvert some of the conventional wisdom about the postwar university with which this chapter began. For the ivory tower of the university is widely seen as a kind of sanctuary.

Society exalts and nurtures its intellectuals, on this view, by providing refuge from the sometimes obstructive passions and irrationalities (such as the "Red menace" of the 1950s) of public life outside. Yet a Foucauldian view like this, coupled with the postwar decline of public intellectuals, suggests a reversal: Perhaps the postwar ivory tower functioned rather as a kind of concentration camp into which intellectuals were herded by a largely anti-intellectual society and then permitted to indulge in any sorts of inquiry they liked, as long as their scholarship remained visibly disconnected from radical politics. Radicals may indeed become tenured, that is, but only if they remain either invisible within the ivory tower or incomprehensible to all but insiders and self-chosen initiates or students.

This change of perspective helps to connect academic McCarthyism and the rise of the postwar university to the cogent political and economic fears that, in retrospect, are now being suggested as the foundation of popular hysteria about the "Red menace." The genuine subversive threat, some suggest, was not that communism would be imported from Moscow, but rather that socialism of some kind would emerge indigenously after the end of the war, much as its popularity grew in the early 1930s during and after the Depression early in the century. Many feared, as economist Paul Samuelson put it, that the end of World War II would usher in "the greatest period of unemployment and industrial dislocation which any economy has ever faced" (quoted in Lewontin 1997, 3). Any sudden, peacetime collapse of government spending and purchasing, the argument went, could depress the U.S. economy and potentially revive debates about the intrinsic contradictions and instabilities of capitalism. So understood, Cold War fears of Soviet invasions justified a two-pronged response to this threat of domestic communism: It licensed persecution of domestic leftists (including teachers, importantly, who might enlighten students about alternate forms of social and economic organization), and, as Richard Lewontin emphasizes, it justified high levels of peacetime government defense spending (both within and without universities) that would help to prevent economic depression (ibid., pp. 20–23).

Several different kinds of dynamics and power – institutional, economic, social, and intellectual – were together at work in higher education in the 1950s. They arguably helped to create and institute a new kind of intellectual subjectivity, a self-conscious professionalism that internalized codes that contained and marginalized intellectual radicalism. Though this account of the Unity of Science movement hardly touches on these larger economic and sociological dimensions, there can be no doubt that philosophers of science were among the points of application for anticommunist forces. When philosophy of science distanced itself from the Unity of Science movement and adopted a professional posture that consciously avoided engagement with social and political battles, after all, it did precisely what J. Edgar Hoover, Sidney Hook, and other anticommunists wished *all* leftist and radical intellectual programs would do. That North American intellectuals continue to question their small voice in public affairs (especially when compared with European intellectuals, who were never so attacked) suggests again that this depoliticization did not issue from within intellectual culture.[9] Rather, it was created by a now largely unknown, external set of pressures and agents that, as the Foucauldian theory of subjectivization would have it, concealed its operations.

The result of this professionalization in philosophy of science and other fields is a world very different from the one that most intellectuals of the 1930s could have imagined. For philosophy of science, the differences are made plain by imagining a world in which Neurath's and Frank's projects were accepted as legitimate contenders in the profession as it matured after the 1950s and in which Carnap and Frank were neither investigated by the FBI nor attacked by colleagues with political motives. Neurath's students (had there been more) might have enthusiastically carried his antimetaphysical torch into a world and attacked, for example, Ronald Reagan's infamous speech in which he called the Soviet Union an "evil empire." Perhaps in a public forum, and with the encouragement of his profession and administration, and not only in a private, defensive letter to Sidney Hook, would Carnap have assailed the American government's and American press's "gross exaggerations" about Soviet communism and, by extension, the justifications those exaggerations later made possible for Cold War military interventions (or covert operations) in Africa, South America, the Middle East, and Southeast Asia. Had these mechanisms that transformed philosophy of science been somehow

---

9 A *New York Times* headline, for instance, announces "Intellectuals Wonder Why They're Taken Less Seriously in America" (December 6, 2003). Another *New York Times* report (April 19, 2003) of a public conference on "the future of theory" featuring "academic heavyweights" in the humanities found the event self-parodying. Under the title "The Latest Theory Is That Theory Doesn't Matter," the report depicts Stanley Fish explaining, "I wish to deny the effectiveness of intellectual work" for those who wish to be effective beyond the academy. One panelist privately explained to the puzzled reporter that these theorists "spend so much time talking about current events" because "this particular group of intellectuals has a terror of being politically irrelevant."

deflected, exposed, or counteracted, had the profession not only permitted but encouraged its brightest lights to supplement their technical work in philosophy with analyses of public issues and debates, one cannot but wonder whether scientific philosophy's plans to help to realize a more scientifically and epistemologically informed public, and possibly a more peaceful, economically stable and just world, would not seem as naïve and dreamlike as they seem today.

References

Aaron, Daniel. 1961. *Writers on the Left: Episodes in American Literary Communism.* New York: Harcourt, Brace & World.

Adler, Mortimer. 1941. "God and the Professors." In *Science, Philosophy and Religion: A Symposium.* New York: Conference on Science, Philosophy and Religion in Their Relation to the Democratic Way of Life, 120–38.

Allen, Raymond. 1949. "Communists Should Not Teach in American Colleges." *Educational Forum* 14, no. 4.

Almond, Gabriel. 1954. *The Appeals of Communism.* Princeton: Princeton University Press.

Alston, William P., and George Nakhnikian, eds. 1963. *Readings in Twentieth-Century Philosophy.* New York: Free Press of Glencoe.

Anshen, Ruth Nanda, ed. 1942. *Science and Man.* New York: Harcourt, Brace.

Aron, Raymond. 2001. *The Opium of the Intellectuals.* New Brunswick, N.J.: Transaction.

Ashmore, Harry. 1989. *Unseasonable Truths: The Life of Robert Maynard Hutchins.* Boston: Little, Brown.

Awodey, Steve, and Carsten Klein, eds. 2004. *Carnap Brought Home: The View from Jena.* Chicago: Open Court.

Ayer, A. J. 1936. *Language, Truth and Logic.* New York: Dover.

Ayer, A. J., ed. 1959. *Logical Positivism.* New York: Free Press.

Baumgardt, David. 1947. "Poise and Passion in Philosophy." In Bryson et al. 358–71.

Belfrage, Cedric. 1973. *The American Inquisition, 1945–1960.* New York: Bobbs-Merrill.

Bell, Daniel. 1960. *The End of Ideology: On the Exhaustion of Political Ideas in the Fifties.* Glencoe, Ill.: Free Press.

Bender, Thomas, and Carl Shorske, eds. 1997. *American Academic Culture in Transformation: Fifty Years, Four Disciplines.* Princeton: Princeton University Press.

Bentley, Arthur. 1947. "The New 'Semiotic.'" *Philosophy and Phenomenological Research* 8: 107–31.

Bentley, Arthur. 1949. "Signs of Error." *Philosophy and Phenomenological Research* 10: 99–106.

Bernstein, Jeremy. 1967. *A Comprehensible World: On Modern Science and Its Origins.* New York: Random House.

Beuttler, Fred. 1997. "For the World at Large: Intergroup Activities at the Jewish Theological Seminary." In *Tradition Renewed: A History of the Jewish Theological Seminary,* ed. J. Wertheimer. New York: The Seminary.

Black, Max. 1947. "The Limitations of a Behavioristic Semiotic." *Philosophical Review* 56: 258–72.

Bloom, Allan. 1987. *The Closing of the American Mind.* New York: Simon and Schuster.

Blumberg, Albert. 1958. "Science and Dialectics: A Preface to a Re-examination." *Science & Society* 22: 306–29.

———. 1976. *Logic: A First Course.* New York: Knopf.

Blumberg, Albert, and Herbert Feigl. 1931. "Logical Positivism: A New Movement in European Philosophy." *Journal of Philosophy* 28: 281–96.

Brameld, Theodore. 1938. Letter to the Editors. *The Communist* 17: 381–82.

Brodbeck, May. 1953. "The Nature and Function of the Philosophy of Science." In Feigl and Brodbeck 1953, 3–7.

Bryson, Lyman, L. Finkelstein, and R. MacIver, eds. 1947. *Conflicts of Power in Modern Culture.* Seventh Symposium, Conference on Science, Philosophy and Religion in Their Relation to the Democratic Way of Life. New York: Harper.

———, eds. 1948. *Learning and World Peace.* Eighth Symposium, Conference on Science, Philosophy and Religion in Their Relation to the Democratic Way of Life. New York: Harper.

Buckley, William. F. 1951. *God and Man at Yale: The Superstitions of "Academic Freedom."* Chicago: Henry Regnery.

———. 1976. "The Road to Serfdom: The Intellectuals and Socialism." In *Essays on Hayek,* ed. F. Machlup. New York: NYU Press, 95–106.

Bush, Vannevar. 1960. *Science: The Endless Frontier.* (Originally published in 1945.) Washington, D.C.: National Science Foundation.

Carnap, Rudolf. 1932. "Überwindung der Metaphysik durch logische Analyse der Sprache." *Erkenntnis* 2: 219–41.

———. 1934a. "On the Character of Philosophical Problems." Trans. W. Malisoff. *Philosophy of Science* 1: 5–19.

———. 1934b. *The Unity of Science.* (Originally published in 1932 as "Die physikalische Sprache als Universalsprache der Wissenschaft.") Trans. M. Black. London: Kegan Paul, Trench, Trubner.

———. 1935. "Philosophy and Logical Syntax." In Alston and Nakhnikian 1963, 424–60.

———. 1936/37. "Testability and Meaning." *Philosophy of Science.* 3: 419–71; 4: 1–40.

———. 1936a. Radio interview on the occasion of conferences celebrating Harvard University's tercentenary. NBC broadcast, September 5, 1936 (see Carnap 1937b for Carnap's published contribution to this conference).

———. 1936b. "Über die Einheitssprache der Wissenschaft: Logische Bemerkungen zum Project einer Enzyklopädie." Actes du congrès international de philosophie scientifique, Sorbonne, Paris 1935 [fasc.]. 2. L'Unité de la science *Actualités scientifiques et industrielles* no. 388. Paris: Hermann & Cie, 60–70.

———. 1937a. "Einheit der Wissenschaft durch Einheit der Sprache." In Travaux du IXe congrès international de philosophie, Congrès Descartes [fasc.]. 4. L'Unité de la science: La méthod et les méthodes *Actualitiés scientifiques et industrielles* no. 533. Paris: Hermann & Cie, 51–57.

———. 1937b. "Logic." In *Factors Determining Human Behavior.* Cambridge: Harvard University Press, 107–18.

———. 1937c. *The Logical Syntax of Language.* London: Kegan Paul, Trench, Trubner.

———. 1938. "Logical Foundations of the Unity of Science." In Neurath et al. 1938, 42–62.

———. 1939. *Foundations of Logic and Mathematics. International Encyclopedia of Unified Science,* vol. 1, no. 3. Chicago: University of Chicago Press.

———. 1942. *Introduction to Semantics.* Cambridge: Harvard University Press.

———. 1944. Note on values, quoted in Lepley 1944, 137–38.

———. 1950. "Empiricism, Semantics, and Ontology." Reprinted in *Meaning and Necessity,* 2nd ed. Chicago: University of Chicago Press, 1956, 205–21.

———. 1953. "Inductive Logic and Science." *Contributions to the Analysis and Synthesis of Knowledge, Proceedings of the American Academy of Arts and Sciences,* vol. 80, pp. 189–97.

———. 1956. "The Methodological Character of Theoretical Concepts." In *The Foundations of Science and the Concepts of Psychology and Psychoanalysis,* ed. Herbert Feigl and Michael Scriven. Minnesota Studies in the Philosophy of Science, vol. 1. Minneapolis: University of Minnesota Press, 38–76.

———. 1959a. "The Elimination of Metaphysics through the Logical Analysis of Language." (Originally published in 1932 as "Überwindung der Metaphysik durch logische Analyse der Sprache.") Trans. A. Pap. In Ayer 1959, 60–81.

———. 1959b. "Psychology in Physical Language." (Originally published in 1932 as "Psychologie in Physikalischer Sprache.") In Ayer 1959, 165–98.

———. 1963a. "Intellectual Autobiography." In Schilpp, 1963, 3–84.

———. 1963b. "Replies and Systematic Expositions." In Schilpp, 1963, 859–1013.

———. 1966. *Philosophical Foundations of Physics: An Introduction to the Philosophy of Science.* Ed. M. Gardner. New York: Basic Books.

———. 1969. *The Logical Structure of the World.* (Originally published in 1928 as *Der Logische Aufbau der Welt.*) Trans. R. George. Berkeley: University of California Press.

———. 1970. "My Visit with Two Imprisoned Philosophers in Mexico." *Journal of Philosophy* 67: 1026–29.

Carter, Philip. 1938. "Professor Levy's Approach to Marxism." *The Communist* 17: 667–70.

———. 1939a. "Marxist Philosophy and the Sciences." *The Communist* 18: 572–74.

———. 1939b. "Pitfalls of Pragmatic Logic." *The Communist* 18: 163–69.

Cartwright, Nancy, and Thomas Uebel. 1996. "Philosophy in the Earthly Plane." In *Encyclopedia and Utopia: The Life and Work of Otto Neurath (1882–1945),* ed. Friedrich Stadler and Elisabeth Nemeth. Boston: Kluwer, 39–52.

Cartwright, Nancy, Jordi Cat, Lola Fleck, and Thomas. E. Uebel. 1996. *Otto Neurath: Philosophy between Science and Politics.* Cambridge: Cambridge University Press.

Carus, A. W. 2004. "Sellars, Carnap, and the Logical Space of Reasons." In Awodey and Klein 2004, 317–55.

Cat, Jordi, Hasok Chang, and Nancy Cartwright. 1991. "Otto Neurath: Unification as the Way to Socialism." In *Einheit der Wissenschaften*, ed. J. Mittelstrass. New York: Walter de Gruyter.

Cat, Jordi. 1995. "The Popper-Neurath Debate and Neurath's Attack on Scientific Method." *Studies in History and Philosophy of Science* 26: 219–50.

Chamberlin, William Henry. 1950. "Apologia pro Vita Sua: Why I Once Fellow-Traveled." *New Leader,* May 20, p. 20.

———. 1955. "Communism and the Intellectuals." *New Leader*, June 13, p. 20.

Chambers, Whittaker. 1952. *Witness.* Washington, D.C.: Regnery.

Chislett, Clive. 1992. "Damned Lies. And Statistics: Otto Neurath and Soviet Propaganda in the 1930s." *Visible Language* 26: 298–321.

Churchman, C. West. 1946. "Discussion: Carnap's 'On Inductive Logic.'" *Philosophy of Science* 13: 341–42.

———. 1948. "Editorial." *Philosophy of Science* 15: 81–82.

———. 1957. "A Pragmatic Theory of Induction." In Frank, ed. 1957, 18–24.

Churchman, C. West, and R. Ackoff. 1946. "Varieties of Unification." *Philosophy of Science* 13: 287–300.

———. 1947. "Ethics and Science." *Philosophy of Science* 14: 269–71.

Churchman, C. West, and T. A. Cowan. 1945. "A Challenge." *Philosophy of Science* 12: 219–20.

———. 1946. "On the Meaningfulness of Questions." *Philosophy of Science* 13: 20–24.

Cohen, Morris, and Ernest Nagel. 1934. *Introduction to Logic and Scientific Method.* New York: Harcourt Brace.

Conant, James Bryant. 1951. "Science Views the Future." *New Leader*, September 17, pp. 2–4.

Condon, Edward U. 1950. "The Attack on the Intellect." *The Nation*, March 25, pp. 267–68.

Cooney, Terry A. 1986. *The Rise of the New York Intellectuals: Partisan Review and Its Circle.* Madison: University of Wisconsin Press.

Cork, Jim. 1950. "The Politics of Albert Einstein." Review of A. Einstein, *Out of My Later Years. New Leader,* June 24, p. 22.

Cornforth, Maurice. 1949. "Logical Empiricism." In Sellars et al. 1949, 495–521.

———. 1950. *In Defense of Philosophy: Against Positivism and Pragmatism.* New York: International.

Crossman, Richard, ed. 1949. *The God That Failed.* New York: Harper & Brothers.

Dahms, Hans-Joachim. 1994. *Positivismusstreit.* Frankfurt: Suhrkamp.

Dahms, Hans-Joachim. 2004. "*Neue Sachlichkeit* in the Architecture and Philosophy of the 1920s." In Awodey and Klein 2004, 357–75.

Davis, Wallace Martin. 1952. "Who Listens to the Intellectuals?" *New Leader,* June 16, pp. 19–21.

Dewey, John. 1929. *Impressions of the Soviet Russia and the Revolutionary World, Mexico–China–Turkey.* New York: New Republic.

———. 1938. "Unity of Science as a Social Problem." In Neurath et al. 1938, 29–38.

———. 1939. "Theory of Valuation." *International Encyclopedia of Unified Science,* vol. 2, no. 4. Chicago: University of Chicago Press.

———. 1943. "Anti-Naturalism in Extremis." *Partisan Review* 10: 24–39.

———. 1946a. "Peirce's Theory of Linguistic Signs, Thought, and Meaning." *Journal of Philosophy* 43: 85–95.

———. 1946b. "To the Editors of the *Journal of Philosophy.*" *Journal of Philosophy* 43: 280.

Dewey, John, and Horace Kallen, eds. 1941. *The Bertrand Russell Case.* New York: Viking.

Dewey, John, S. Hook, and E. Nagel. 1945. "Are Naturalists Materialists?" *Journal of Philosophy* 42: 515–30.

Diggins, John Patrick. 1992. *The Rise and Fall of the American Left.* New York: Norton.

Dilling, Elizabeth. 1934. *The Red Network: A "Who's Who" and Handbook of Radicalism for Patriots.* Kenilworth, Ill.: the author.

Dommeyer, Frederick. 1950. Review of Morris 1946a. *Philosophy and Phenomenological Research* 10: 451–53.

Dupré, John. 1993. *The Disorder of Things: Metaphysical Foundations of the Disunity of Science.* Cambridge: Harvard University Press.

Duhem, Pierre. 1954. *The Aim and Structure of Physical Theory.* Princeton: Princeton University Press.

Ebenstein, Alan. 2001. *Friedrich Hayek: A Biography.* New York: Palgrave.

Edel, Abraham. 1939. Review of Bridgman, *The Intelligent Individual and Society. Science & Society* 3: 131–34.

———. 1961. "Science and the Structure of Ethics." *International Encyclopedia of Unified Science,* vol. 2, no. 3. Chicago: University of Chicago Press.

Engel, Leonard. 1947. "Warning All Scientists." *The Nation,* August 2, pp. 117–19.

———. 1948. "Fear in Our Laboratories." *The Nation,* January 17, pp. 63–65.

Faludi, A. 1989. "Planning According to the 'Scientific Conception of the World': The Work of Otto Neurath." *Environment and Planning D: Society and Space* 7: 397–418.

Fast, Howard. 1990. *Being Red.* Boston: Houghton Mifflin.

Feferman, Anita Burdman. 1993. *Politics, Logic, and Love: The Life of Jean Van Heijenoort.* Wellesley, Mass.: A. K. Peters.

Feigl, Herbert. 1943. "Logical Empiricism." In *Twentieth Century Philosophy,* ed. D. Runes. New York: Philosophical Library, 373–416.

———. 1953. "The Scientific Outlook: Naturalism and Humanism." In Feigl and Brodbeck 1953, 8–18.

———. 1959. "Philosophical Tangents of Science." In Feigl and Maxwell 1959, 1–17.

Feigl, Herbert, and May Brodbeck. 1953. *Readings in the Philosophy of Science.* New York: Appleton-Century-Crofts.

Feigl, Herbert, and Grover Maxwell, eds. 1959. *Current Issues in the Philosophy of Science.* New York: Holt, Reinhart and Winston.

Feigl, Herbert, and Charles Morris. 1970. "Bibliography and Index." *International Encyclopedia of Unified Science*, vol. 2, no. 20, 947–1023 in two-volume set.

Feuer, Lewis. 1941. "The Development of Logical Empiricism." *Science & Society* 5: 222–33.

———. 1942. "Ethics and Historical Materialism." *Science & Society* 6: 242–72.

———. 1969. *The Conflict of Generations: The Character and Significance of Student Movements.* New York: Basic Books.

Field, Mike. 1996. "Self Taught and Stubbornly Independent." *Johns Hopkins Magazine* online issue, April.

Fisiak, Jacek. 1984. "Old and Middle English Language Studies in Poland." *Mediaeval English Studies Newsletter*, no. 11 (December): 1–7.

Flewelling, Ralph. 1948. "Philosophy as a Medium of World Understanding." In Bryson et al. 372–83.

Flynn, John. T. 1953. "Twenty-Four Steps to Communism." *American Mercury*, December, 3–6.

Fodor, J. A. 1974. "Special Sciences (Or: The Disunity of Science as a Working Hypothesis)." *Synthese* 28: 97–115.

Folsom, Franklin. 1994. *Days of Anger, Days of Hope: A Memoir of the League of American Writers, 1937–1942.* Niwot: University Press of Colorado.

Foucault, Michel. 1970. *The Order of Things.* New York: Pantheon.

———. 1977. *Discipline and Punish.* New York: Pantheon.

Frank, Philipp. 1938. *Interpretations and Misinterpretations of Modern Physics.* Paris: Hermann & Cie.

———. 1941. *Between Physics and Philosophy.* Cambridge: Harvard University Press.

———. 1946. "Foundations of Physics." *International Encyclopedia of Unified Science*, vol. 1, no. 7. Chicago: University of Chicago Press.

———. 1947. *Einstein: His Life and Times.* New York: Knopf.

———. 1949a. "Einstein, Mach, and Logical Positivism." In Schilpp 1949, 269–86.

———. 1949b. *Modern Science and Its Philosophy.* Cambridge: Harvard University Press.

———. 1950. *Relativity: A Richer Truth.* Boston: Beacon.

———. 1951a. "Introductory Remarks." *Contributions to the Analysis and Synthesis of Knowledge, Proceedings of the American Academy of Arts and Sciences*, vol. 80, pp. 5–8.

———. 1951b. "The Logical and Sociological Aspects of Science." *Contributions to the Analysis and Synthesis of Knowledge, Proceedings of the American Academy of Arts and Sciences*, vol. 80, pp. 16–30.

———. 1957. *Philosophy of Science: The Link between Science and Philosophy.* Englewood Cliffs, N.J.: Prentice-Hall.

———. 1958. "Contemporary Science and the Contemporary World View." *Daedalus* 58: 57–66.

———. 1963. "The Pragmatic Components in Carnap's 'Elimination of Metaphysics.'" In Schilpp 1963, 159–64.

———. 1998. *The Law of Causality and Its Limits.* Ed. R. S. Cohen trans. M. Neurath and R. S. Cohen. Originally published as *Kausalgesetz und seine Grenzen.* Dordrecht and Boston: Kluwer Academic.

Frank, Philipp, ed. 1957. *The Validation of Scientific Theories.* Boston: Beacon.

Frank, Philipp, and C. West Churchman. 1948. "In Memoriam: Dr. William M. Malisoff." *Philosophy of Science* 15: 1–3.

Frank, Richard. 1937. "The Schools and the People's Front" *The Communist* v. 16, 432–45.

Frenkel-Brunswik, Else. 1954. "Psychoanalysis and the Unity of Science." *Contributions to the Analysis and Synthesis of Knowledge, Proceedings of the American Academy of Arts and Sciences*, vol. 80, pp. 271–350.

Friedman, Michael. 1991. "The Re-Evaluation of Logical Empiricism." *Journal of Philosophy* 88, 505–19.

———. 2000. *A Parting of the Ways*. Chicago: Open Court.

Frowen, Stephen. 1997. "Introduction." In *Hayek: Economist and Social Philosopher*, ed. Frowen. New York: St. Martin's Press, xxi–xxvi.

Fuller, Steve. 2000. *Thomas Kuhn: A Philosophical History for Our Times*. Chicago: University of Chicago Press.

Gabriel, Gottfried. 2004. "Introduction: Carnap Brought Home." In Awodey and Klein 2004, 3–23.

Galison, Peter. 1990. "Aufbau/Bauhaus: Logical Positivism and Architectural Modernism." *Critical Inquiry* 16: 709–52.

———. 1996. "Constructing Modernism: The Cultural Location of *Aufbau*." In Giere and Richardson 1996, 17–44.

Galison, Peter, and David Stump, eds. 1996. *The Disunity of Science: Boundaries, Contexts and Power*. Stanford: Stanford University Press.

Gardner, Martin. 1950a. "H. G. Wells: Premature Anti-Communist." *New Leader*, October 7, pp. 20–21.

———. 1950b. "Relativity: Hope Chest or Pandora's Box?" *New Leader*, August 12, p. 26.

———. 1951. "Philosophy as Poetic Speculation." *New Leader*, June 11, pp. 23–24.

Gentry, George. 1947. "Signs, Interpretants, and Significata." *Journal of Philosophy* 44: 318–24.

Giere, Ron. 1996. "From *Wissenschaftliche Philosophie* to Philosophy of Science." In Giere and Richardson 1996, 335–54.

Giere, Ron, and Alan Richardson, eds. 1996. *Origins of Logical Empiricism*. Minneapolis: University of Minnesota Press.

Glazer, Nathan. 1969. "Student Politics and the University." *Atlantic Monthly* 224: 43–53.

Gordon, Elizabeth. 1953. "The Threat to the Next America." *House Beautiful*, April.

Grinker, Roy, ed. 1956. *Toward a Unified Theory of Human Behavior*. New York: Basic Books.

Gruen, William. 1939. "What Is Logical Empiricism?" *Partisan Review* 6: 64–77.

Grünbaum, Adolf. 1954. "Science and Ideology." *Scientific Monthly* 79 (July): 13–19.

Hacking, Ian. 1992. "Disunified Sciences." In *The End of Science*, ed. Richard Q. Elvee. Lanham, Md.: University Press of America, 33–52.

Haller, Rudolf. 1991. "The First Vienna Circle." In Uebel 1991, 95–108.

Haskins, Caryl P. 1956. "Science and the Whole Man." *Daedalus* 86, no. 2: 113–21.

Hayek, Friedrich A. 1944. *The Road to Serfdom.* Chicago: University of Chicago Press.

———. 1999. *The Road to Serfdom: Condensed Version.* Institute of Economic Affairs.

Haynes, James Earl, and Harvey Klehr. 1999. *Venona: Decoding Soviet Espionage in America.* New Haven: Yale University Press.

Heidelberger, Michael, and Friedrich Stadler, eds. 2003. *Wissenschaftsphilosophie und Politik. Philosophy of Science and Politics.* Vienna and New York: Springer.

Helmer, Olaf, and Nicholas Rescher. 1960. "On the Epistemology of the Inexact Sciences" *Management Science* 6: 25–52.

Hempel, Carl. 1942. "The Function of General Laws in History." *Journal of Philosophy* 39: 35–48.

Herberg, Will. 1954. "Why They Became True Believers" (review of Gabriel Almond, *The Appeals of Communism*). *New Leader*, December 13, pp. 12–13.

Hicks, Granville. 1950. "McCarthy and the Homosexuals." *New Leader*, June 17, p. 9.

Hoffer, Eric. 1951. *The True Believer: Thoughts on the Nature of Mass Movements.* New York: Harper.

Hollinger, David. 1996. *Science, Jews, and Secular Culture.* Princeton: Princeton University Press.

———. N.D. "The Unity of Knowledge and the Diversity of Knowers: Science as an Agent of Cultural Integration in the Interwar United States." Unpublished ms.

Holton, Gerald. 1958. "Perspectives on the Issue 'Science and the Modern World View.'" *Daedalus* 58: 3–7.

———. 1993. *Science and Anti-Science.* Cambridge: Harvard University Press.

Hook, Sidney. 1940. "The New Medievalism." *New Republic*, October 28.

———. 1941. "The General Pattern." in Dewey and Kallen 1941, 185–210.

———. 1943. "The New Failure of Nerve." *Partisan Review* 10: 2–23.

———. 1947. "Philosophy and the Police" (review of John Somerville, *Soviet Philosophy*). *The Nation*, February 15, pp. 188–89.

———. 1949. "International Communism." *Dartmouth Alumni Magazine*, March.

———. 1953. *Heresy, Yes – Conspiracy, No.* New York: John Day.

———. 1954. "Robert Hutchins Rides Again." *New Leader*, April 19, pp. 16–19.

———. 1987. *Out of Step: An Unquiet Life in the 20th Century.* New York: Harper and Row.

Horkheimer, Max. 1937. "Der neueste Angriff auf die Metaphysik." *Zeitschrift für Sozialforschung*, vol. 6, pp. 4–51 (translated in Horkheimer 1982, 132–87).

———. 1982. *Critical Theory: Selected Essays.* Trans. Matthew O'Connell et al. New York: Seabury.

Howard, Don. 2003. "Two Left Turns Make a Right: On the Curious Political Career of North American Philosophy of Science at Midcentury." In *Logical Empiricism in North America*, ed. G. Hardcastle and A. Richardson. Minneapolis: University of Minnesota Press, 25–93.

Hughes, H. Stuart. 1956. "Is the Intellectual Obsolete?" *Commentary* 22: 313–19.

Hutchins, Robert M. 1936. *The Higher Learning in America.* New Haven: Yale University Press.

———. 1954. "Are Our Teachers Afraid to Teach?" *Look* 18: 27–29.

Irzik, Gürol, and Teo Grünberg. 1995. "Carnap and Kuhn: Arch Enemies or Close Allies?" *British Journal for the Philosophy of Science* 46: 285–307.

Jacoby, Russell. 1987. *The Last Intellectuals.* New York: Basic Books.

James, William. 1981. *Pragmatism.* Indianapolis: Hackett.

Jerome, V. J. 1937. "Marxism-Leninism for Science and Society" (first of two parts). *The Communist* 16: 1146–63.

———. 1938. "Marxism-Leninism for Science and Society" (second of two parts). *The Communist* 17: 75–91.

Joergensen, Joergen. 1951. "The Development of Logical Empiricism." *International Encyclopedia of Unified Science,* vol. 2, no. 9. Chicago: University of Chicago Press.

Johnson, David K. 2004. *The Lavender Scare: The Cold War Persecution of Gays and Lesbians in the Federal Government.* Chicago: University of Chicago Press.

Kaelin, E. F. 1958. Review of R. Lepley, ed., *The Language of Value. Philosophy of Science* 25: 307–8.

Kaempffert, Waldemar. 1937a. "On Increasing the Span of Life" (review of William Malisoff, *The Span of Life*). *New York Times Book Review,* December 19, p. 4.

———. 1937b. "Science Encyclopedia." *New York Times,* February 14.

———. 1938. "Toward Bridging the Gaps between the Sciences." *New York Times Book Review,* April 7, pp. 2, 18.

———. 1939. "Facts March On – with Neurath." *Survey Graphic,* September, pp. 538–40.

Kallen, Horace. 1940. "The Meanings of 'Unity' among the Sciences." *Educational Administration and Supervision* 26, no. 2: 81–97.

———. 1946a. "An Annotation to the Annotation." *Philosophy and Phenomenological Research* 6: 528–29.

———. 1946b. 'The Meanings of 'Unity' among the Sciences, Once More." *Philosophy and Phenomenological Research* 6: 493–96.

———. 1946c. "Postscript: Otto Neurath, 1882–1945." *Philosophy and Phenomenological Research* 6: 529–33.

———. 1946d. "Reply." *Philosophy and Phenomenological Research* 6: 515–26.

———. 1956. *Cultural Pluralism and the American Idea.* Philadelphia: University of Pennsylvania Press.

Kaplan, Abraham. 1958. "American Ethics and Public Policy." *Daedalus* 87: 48–77.

———. 1963. "Logical Empiricism and Value Judgments." In Schilpp 1963, 827–56.

Kecskemeti, Paul. 1948. "Advice from Philosophy." In Bryson et al. 1948, 319–28.

Kegley, Charles. 1959. "Reflections on Philipp Frank's Philosophy of Science." *Philosophy of Science* 26: 35–40.

Kimball, Roger. 1990. *Tenured Radicals.* New York: HarperCollins.

Kitcher, Philip. 1989. "Some Puzzles About Species." In *What the Philosophy of Biology Is,* ed. M. Ruse. Dordrecht: Kluwer, 183–208.

Klingaman, William K., ed. 1996. *Encyclopedia of the McCarthy Era.* New York: Facts on File.

Kluckhohn, Clyde, ed. 1953. *Personality in Nature, Society, and Culture,* 2nd ed. New York: Knopf.

Koestler, Arthur. 1944. "The Intelligentsia." *Partisan Review* 11: 265–77.

———. 1949. Untitled essay. In Crossman 1949, 15–75.

Korzybski, Alfred. 1933. *Science and Sanity.* Lakeville, Conn.: International Non-Aristotelian Library Publishing.

Kucklick, Bruce. 2001. *A History of Philosophy in America.* Oxford: Oxford University Press.

Kuhn, Thomas. 1962. *The Structure of Scientific Revolutions (International Encyclopedia of Unified Science,* vol. 2, no. 2). Rpt., Chicago: University of Chicago Press, 1970.

Landauer, Carl. 1950. "The Communist Oath at California University." *New Leader* October 14, pp. 12–14.

Lenin, Vladimir I., 1908. *Materialism and Empirio-Criticism.* Trans. 1927. New York: International Publishers, 1972 reprint.

Lepley, Ray. 1944. *Verifiability of Value.* New York: Columbia University Press.

Lewontin, R. C. 1997. "The Cold War and the Transformation of the Academy." In Shiffrin 1997, 1–34.

Leys, Wayne. 1949. Review of Charles Morris, *The Open Self. Philosophical Review* 58: 284–86.

Lipset, Seymour M. 1959. "American Intellectuals: Their Politics and Status." *Daedalus* 88: 460–86.

McCarthy, Mary. 1951. "The Groves of Academe." *New Yorker,* February 3, pp. 28–32.

McCumber, John. 1996. "Time in the Ditch: American Philosophy and the McCarthy Era." *Diacritics* (Spring): 33–49.

———. 2001. *Time in the Ditch: American Philosophy and the McCarthy Era.* Evanston, Ill.: Northwestern University Press.

McGill, J. V., 1936. "An Evaluation of Logical Positivism." *Science & Society* 1: 45–80.

———. 1937. "Logical Positivism and the Unity of Science." *Science & Society* 1: 550–61.

McNeill, W. 1991. *Hutchins' University: A Memoir of the University of Chicago, 1929–1950.* Chicago: University of Chicago Press.

Malisoff, William M. 1930. *A Calendar of Doubts and Faiths.* New York: G. Howard Watt.

———. 1932. *Meet the Sciences.* Baltimore: Williams & Wilkins.

———. 1937. *The Span of Life.* Philadelphia: Lippincott.

———. 1944. "Philosophy of Science after Ten Years." *Philosophy of Science* 11: 1–2.

———. 1945. Review of R. Linton, ed., *The Science of Man in the World Crisis. Philosophy of Science* 12: 228.

———. 1946a. "A Science of the People, by the People and for the People." *Philosophy of Science* 13: 166–69.

———. 1946b. "On the Non-Existence of the Atomic Secret." *Philosophy of Science* 13: 1–2.

———. 1947a. Review of *The Authoritarian Attempt to Capture Education. Philosophy of Science* 14: 103–4.

———. 1947b. Review of N. Cousins, *Modern Man Is Obsolete*. *Philosophy of Science* 14: 171.

———. 1947c. Review of L. Bryson, *Science and Freedom*. *Philosophy of Science* 14: 171.

———. 1947d. Review of J. Baker, *Science and the Planned State*. *Philosophy of Science* 14:, 171–72.

———. 1947e. Review of H. Waxman, *The Source of Human Good*. *Philosophy of Science* 14: 173.

———. 1947f. Review of J. Somerville, *Soviet Philosophy*. *Philosophy of Science* 14: 172.

Marcuse, Herbert. 1939. Review of International Encyclopedia of Unified Science. *Studies in Philosophy and Social Science* 8: 228–32.

———. 1964. *One-Dimensional Man*. Boston: Beacon.

Matthews, J. B. 1953. "Communism in the Colleges." *American Mercury*, (May): 111–44.

Menand, Louis. 2001. *The Metaphysical Club: A Story of Ideas in America*. New York: Farrar, Straus and Giroux.

Meyer, Frank S. 1957. Review of Charles Morris, *The Varieties of Human Value*. *National Review* 3: 118.

———. 1958. "The Bigotry of Science." *The Nation* 5: 234.

Mills, C. Wright. 1956. *The Power Elite*. Oxford: Oxford University Press.

Mirowski, Philip. N.d. "The Social Dimensions of Social Knowledge and their Distant Echoes in 20th-Century American Philosophy of Science." Unpublished ms.

Moore, Barrington. 1957. "Influence of Political Creeds on the Acceptance of Theories." In Frank 1957, 29–36.

Morris, Charles. 1932. *Six Theories of Mind*. Chicago: University of Chicago Press.

———. 1934a. "Introduction: George H. Mead as Social Psychologist and Social Philosopher." In *Mind, Self and Society*, ed. Charles W. Morris. Chicago: University of Chicago Press, ix–xxxv.

———. 1934b. "Pragmatism and the Crisis of Democracy." *Public Policy Pamphlet*, no. 12.

———. 1935. "The Relation of the Formal and Empirical Sciences within Scientific Empiricism." In Morris 1937, 46–55.

———. 1936. "Opening Speech (for the American Delegates)." Actes du Congrès international de philosophie scientifique, Sorbonne, Paris 1935 [fasc.]. 2. Unité de la science *Actualités scientifiques et industrielles* 388. Paris: Hermann & Cie, 22.

———. 1937. *Logical Positivism, Pragmatism, and Scientific Empiricism*. Actualités Scientifiques et Industrielles, 449, Paris: Hermann et Cie.

———. 1938a. *Foundations of the Theory of Signs*. International Encyclopedia of Unified Science, vol. 1, no. 2. Chicago: University of Chicago Press.

———. 1938b. "The Unity of Science Movement and the United States." *Synthese*, (November): 25–29.

———. 1939. "General Education and the Unity of Science Movement." In *John Dewey and the Promise of America*, Progressive Education Booklet no. 14. Columbus, Ohio, 26–40.

Morris, Charles. 1942. *Paths of Life: Preface to a World Religion.* Chicago: University of Chicago Press.

———. 1946a. "The Significance of the Unity of Science Movement." *Philosophy and Phenomenological Research* 6: 508–15.

———. 1946b. *Signs, Language, and Behavior.* New York: George Braziller.

———. 1946c. "To the Editors of the *Journal of Philosophy.*" *Journal of Philosophy* 43: 196.

———. 1948a. "Comments on Mr. Storer's Paper." *Philosophy of Science* 15: 330–32.

———. 1948b. *The Open Self.* New York: Prentice-Hall.

———. 1948c. "Signs about Signs about Signs." *Philosophy and Phenomenological Research* 9: 115–33.

———. 1951. "The Science of Man and Unified Science." *Contributions to the Analysis and Synthesis of Knowledge, Proceedings of the American Academy of Arts and Sciences,* vol. 80, pp. 37–44.

———. 1956. *The Varieties of Human Value.* Chicago: University of Chicago Press.

———. 1960. "On the History of the International Encyclopedia of Unified Science." *Synthese* 12: 517–21.

———. 1966. *Festival.* New York: George Braziller.

———. 1970. *The Pragmatic Movement in American Philosophy.* New York: George Braziller.

———. 1973. "Memories of Otto Neurath." In Neurath 1973, 64–68.

Morris, Charles, and Ferruccio Rossi-Landi. 1992. "The Correspondence between Charles Morris and Ferruccio Rossi-Landi." *Semiotica* $88^1/_2$: 37–122.

Morris, Charles, and Frank Sciadini. 1956. "Paintings, Ways to Live, and Values." In *Sign, Image, Symbol,* ed. G. Kepes. New York: George Braziller, 144–49.

Nagel, Ernest. 1936. Review of A. Lovejoy, *The Great Chain of Being. Science & Society* 1: 252–56.

———. 1939. "Principles of the Theory of Probability." *International Encyclopedia of Unified Science,* vol. 1, no. 7. Chicago: University of Chicago Press.

———. 1943. "Malicious Philosophies of Science." *Partisan Review* 10: 40–57.

———. 1950. "Memorial Address for Felix Kaufmann." *Philosophy and Phenomenological Research* 10: 464–68.

———. 1951. "Mumford on Modern Man." *New Leader,* November 12, pp. 21–22.

———. 1952. Review of Cornforth 1950, *Journal of Philosophy,* v. 49, 648–50.

Nagel, Ernest, et al. 1937. "Four Letters on Ernest Nagel's Review of Lovejoy's 'The Great Chain of Being.'" *Science & Society* 1: 412–13.

Nehemas, Alexander. 1997. "Trends in Recent American Philosophy." In Bender and Schorske 1997, 227–41.

Nemeth, Elisabeth. 2003. "Philosophy of Science and Democracy: Some Reflections on Philipp Frank's *Relativity – A Richer Truth.*" In Heidelberger and Stadler 2003, 119–38.

Neurath, Otto. 1913. "The Lost Wanderers of Descartes and the Auxiliary Motive (On the Psychology of Decision)." In Neurath 1983, 1–12.

———. 1919. "Through War Economy to Economy in Kind." In Neurath 1973, 123–57.

———. 1921. "Anti-Spengler." In Neurath 1973, 158–213.

———. 1928. "Personal Life and Class Struggle." In Neurath 1973, 249–98.

———. 1930. "Ways of the Scientific World Conception." In Neurath 1983, 32–47.

———. 1931a. "Empirical Sociology: The Scientific Content of History and Political Economy." In Neurath 1973, 317–421.

———. 1931b. "Physicalism." In Neurath 1983, 52–57.

———. 1932/33. "Protocol Statements." In Neurath 1983, 91–99.

———. 1934. "Radical Physicalism and the 'Real World.'" In Neurath 1973, 100–114.

———. 1935. "Pseudorationalism of Falsification." In Neurath 1983, 121–31.

———. 1936a. "An International Encyclopedia of Unified Science." In Neurath 1983, 139–44.

———. 1936b. "Encyclopedia as 'Model.'" In Neurath 1983, 145–58.

———. 1936c. "Individual Sciences, Unified Science, Pseudorationalism." In Neurath 1983, 132–38.

———. 1936d. "Physicalism and Knowledge." In Neurath 1983, 159–71.

———. 1937. "Unified Science and Its Encyclopedia." In Neurath 1983, 172–82.

———. 1938a. "Encyclopedism as a Pedagogical Aim: A Danish Approach." *Philosophy of Science* 5: 484–92.

———. 1938b. "Unified Science as Encyclopedic Integration." *International Encyclopedia of Unified Science*, vol. 1, no. 1. Chicago: University of Chicago Press.

———. 1939. *Modern Man in the Making.* New York: Alfred A. Knopf.

———. 1941. "Universal Jargon and Terminology." In Neurath 1983, 213–29.

———. 1942. "International Planning for Freedom." In Neurath 1973, 422–40.

———. 1944. *Foundations of the Social Sciences. International Encyclopedia of Unified Science*, vol. 2, no. 1. Chicago: University of Chicago Press.

———. 1945a. "Germany's Education and Democracy." *Journal of Education* 77, no. 912.

———. 1945b. Review of Friedrich Hayek, *The Road to Serfdom. London Quarterly of World Affairs* (January): 121–22.

———. 1946. "After Six Years." *Synthese* 5: 77–82.

———. 1946a. "For the Discussion: Just Annotations, Not a Reply." *Philosophy and Phenomenological Research* 6, no. 4: 526–28.

———. 1946b. "Orchestration of the Sciences by the Encyclopedism of Logical Empiricism." *Philosophy and Phenomenological Research* 6, no. 4: 496–508.

———. 1973. *Empiricism and Sociology.* Ed. M. Neurath and R. S. Cohen, trans. M. Neurath and Paul Foulkes. Boston: Reidel.

———. 1983. *Philosophical Papers: 1913–1946,* Ed. and trans. R. S. Cohen and M. Neurath. Boston: Reidel.

Neurath, Otto, and J. A. Lauwerys. 1944. "Nazi Text-books and the Future." *Journal of Education* 76, nos. 904 and 905.

———. 1945. "Plato's Republic and German Education." *Journal of Education* 77, nos. 907, 910, and 913.

Neurath, Otto, Niels Bohr, John Dewey, Bertrand Russell, Rudolf Carnap, and Charles Morris. 1938. "Encyclopedia and Unified Science." *International Encyclopedia of Unified Science*, vol. 1, no. 1, Chicago: University of Chicago Press.

Neurath, Otto, Rudolf Carnap, and Hans Hahn. 1929. "The Scientific Conception of the World: The Vienna Circle." In Neurath 1973, 299–319.

Nussbaum, Martha. 2002. "Making Philosophy Matter to Politics." *New York Times,* December 2.

Oppenheim, Paul, and Hillary Putnam, 1958. "Unity of Science as a Working Hypothesis." *Minnesota Studies in the Philosophy of Science* 2: 3–36.

Oppenheimer, Robert. 1958. "The Growth of Science and the Structure of Culture." *Daedalus* 87: 67–76.

Packard, Vance. 1957. *Hidden Persuaders.* New York: David McKay.

Passmore, John. 1953. "Can the Social Sciences Be Value-Free?" In H. Feigl and M. Brodbeck, 1953, 674–76.

Petrilli, Susan. 1992. "Introduction." *Semiotica* 88 $^1/_2$: 1–36.

Popper, Karl. 1935. *Die Logik der Forschung. Schriften zur Wissenschaftlichen Weltauffasung,* vol. 9. Vienna: Springer. Translated as *The Logic of Scientific Discovery.* London: Hutchinson, 1958.

———. 1969. "Science: Conjectures and Refutation." In *Conjectures and Refutations: The Growth of Scientific Knowledge,* 3rd ed. London: Routledge & Kegan Paul, 33–59.

Quine, W. V. O. 1951. "Two Dogmas of Empiricism." *Philosophical Review* 60: 20–43.

———. 1953. "On Mental Entities." *Contributions to the Analysis and Synthesis of Knowledge, Proceedings of the American Academy of Arts and Sciences,* vol. 80, pp. 198–203.

———. 1986. "Autobiography of W. V. Quine." In *The Philosophy of W. V. Quine,* ed. P. A. Schilpp. LaSalle, Ill.: Open Court, 3–46.

Radosh, Ronald. 2001. *Commies: A Journey through the Old Left, the New Left and the Leftover Left.* San Francisco: Encounter.

Rahv, Phillip. 1940. "What Is Living and What Is Dead." *Partisan Review* 7: 175–80.

———. 1948. "Comment: Disillusionment and Partial Answers." *Partisan Review* 15: 519–29.

Rautenstrauch, Walter. 1945. "What Is Scientific Planning?" *Philosophy of Science* 12: 8–18.

Reck, Erich. 2004. "From Frege and Russell to Carnap: Logic and Logicism in the 1920s." In Awodey and Klein 2004, 151–80.

Reichenbach, Hans. 1938. *Experience and Prediction.* Chicago: University of Chicago Press.

———. 1946. "Language as Behavior" (review of Charles Morris, *Signs, Language, and Behavior*). *The Nation,* June 22.

———. 1951. *The Rise of Scientific Philosophy.* Berkeley and Los Angeles: University of California Press.

Reisch, George. 1991. "Did Kuhn Kill Logical Empiricism?" *Philosophy of Science* 58: 264–77.

———. 1994. "Planning Science: Otto Neurath and the *International Encyclopedia of Unified Science.*" *British Journal for the History of Science* 27: 153–75.

———. 1995. "A History of the International Encyclopedia of Unified Science." Ph.D. diss., University of Chicago, 1995.

———. 1997a. "Epistemologist, Economist . . . and Censor? On Otto Neurath's Infamous Index Verborum Prohibitorum." *Perspectives on Science* 5: 452–80.

———. 1997b. "How Postmodern Was Neurath's Idea of Unified Science." *Studies in History and Philosophy of Science* 28: 439–51.

———. 1998. "Pluralism, Logical Empiricism, and the Problem of Pseudo-science." *Philosophy of Science* 65: 333–48.

———. 2001a. "Against a Third Dogma of Logical Empiricism: Otto Neurath and 'Unpredictability in Principle.'" *International Studies in the Philosophy of Science* 15: 199–209.

———. 2001b. "'The Cultural Significance of Science': An Unpublished Lecture by Charles Morris." *RS/SI* 21: 73–99.

———. 2003a. "Disunity within the *International Encyclopedia of Unified Science*." In *Logical Empiricism in North America*, ed. G. Hardcastle and A. Richardson. Minnesota: University of Minnesota Press.

———. 2003b. "On the International Encyclopedia, the Neurath-Carnap Disputes, and the Second World War." In *Logical Empiricism: Historical and Contemporary Perspectives*, ed. P. Parrini, W. Salmon, and M. Salmon. Pittsburgh, Pa.: University of Pittsburgh Press.

Rieff, Philip. 1953. "Are Intellectuals Chained to Policy?" *New Leader*, August 24, pp. 16–17.

Romerstein, Herbert, and Eric Breindel. 2000. *The Venona Secrets: Exposing Soviet Espionage and America's Traitors*. Washington, D.C.: Regnery.

Rosenberg, Alexander. 1994. *Instrumental Biology or the Disunity of Science*. Chicago: University of Chicago Press.

"Rosenberg Clemency Backers Comment on Red Anti-Semitism." 1953. *New Leader*, January 26, pp. 4–6.

Rudner, Richard. 1951. "On Semiotic Aesthetics." *Journal of Aesthetics & Art Criticism* 10: 67–77.

———. 1957. "Value Judgements in the Acceptance of Scientific Theories." In Frank 1957, 24–28.

———. 1966. *Philosophy of Social Science*. Englewood Cliffs, N.J.: Prentice Hall.

Runes, Dagobert, ed. 1960. *Dictionary of Philosophy*. Ames, Ia.: Littlefield, Adams.

Russell, Bertrand. 1929. *Marriage and Morals*. New York: H. Liveright.

———. 2002. *Yours Faithfully, Bertrand Russell*, ed. R. Perkins, Jr. Chicago: Open Court.

Russell, Bertrand, John Dewey, Morris Cohen, Sidney Hook, and Sherwood Eddy. 1934. *The Meaning of Marx: A Symposium*. New York: Farrar and Rinehart.

Russell, Bertrand, and Editors of *New Leader*. 1952. "Is America in the Grip of Hysteria?" *New Leader*, March 3, pp. 2–4.

Russell, Jeffrey Burton. 1991. *Inventing the Flat Earth: Columbus and Modern Historians*. New York: Praeger.

Rutherford, James F. 1960. "Discussion: Frank's Philosophy of Science Revisited." *Philosophy of Science* 27: 183–85.

Saunders, Frances S. 1999. *The Cultural Cold War: The CIA and the World of Arts and Letters*. New York: New Press.

Schilpp, Paul A. 1948. "The Task of Philosophy in an Age of Crisis." In Bryson et al. 1948, 300–10.

Schilpp, Paul A., ed. 1949. *Albert Einstein: Philosopher Scientist*. New York: MJF Books.

Schilpp, Paul A. 1963. *The Philosophy of Rudolf Carnap.* La Salle, Ill.: Open Court.

Schlauch, Margaret. 1942. "Semantics as Social Evasion." *Science & Society* 6: 315–30.

———. 1947. "The Cult of the Proper Word." *New Masses,* April 15, pp. 15–18.

Schlesinger, James M., Jr. 1947. "The Perspective Now." *Partisan Review* 14: 229–42.

Schlick, Moritz. 1985. *General Theory of Knowledge.* (Originally published in 1918 as *Allgemeine Erkenntnislehre.*) Trans. A. Blumberg. La Salle, Ill.: Open Court.

Schrecker, Ellen W. 1986. *No Ivory Tower: McCarthyism and the Universities.* New York: Oxford University Press.

———. 2002. *The Age of McCarthyism: A Brief History with Documents.* New York and Boston: Bedford/St. Martin's.

Schrickel, H. G. 1943. "Philosophy of Science and Social Philosophy." *Philosophy of Science* 10: 208–12.

Scott, Stephen. 1987. "Enlightenment and the Spirit of the Vienna Circle." *Canadian Journal of Philosophy* 17: 695–710.

Sellars, Wilfred, V. J. McGill, and Marvin Farber. 1949. "Foreword." In Sellars et al. 1949, v–xii.

Sellars, Wilfred, V. J. McGill, and Marvin Farber, eds. 1949. *Philosophy for the Future: The Quest of Modern Materialism.* New York: Macmillan.

Shapiro, Edward S., ed. 1995. *Letters of Sidney Hook: Democracy, Communism, and the Cold War.* Armonk, N.Y.: M. E. Sharpe.

Sheldon, William. 1940. *The Varieties of Human Physique.* New York: Harper.

———. 1945. "Critique of Naturalism." *Journal of Philosophy* 42: 253–70.

———. 1946. "Are Naturalists Materialists?" *Journal of Philosophy* 43: 197–209.

———. 1949. *Varieties of Delinquent Youth.* New York: Harper.

———. 1954. *Atlas of Men.* New York: Harper.

Shiffrin, André, ed., 1997. *The Cold War and the University.* New York: New Press.

Singer, Milton, and Abraham Kaplan. 1941. "Unifying Science in a Disunified World." *Scientific Monthly* 52: 79–80.

Smullyan, Arthur. 1947. Review of Charles Morris, *Signs, Language, and Behavior. Journal of Symbolic Logic* 12: 49–51.

"Social Showman." 1936. *Survey Graphic* (November): 618–19.

Somerville, John. 1936. "The Social Ideas of the Wiener Kreis's International Congress." *Journal of Philosophy* 33: 295–301.

———. 1945. "Soviet Science and Dialectical Materialism." *Philosophy of Science* 12: 23–29.

———. 1946. *Soviet Philosophy: A Study of Theory and Practice.* New York: Philosophical Library.

———. 1947. "Ethics and Social Science: Case History of a Sharp Practice." *Philosophy of Science* 14: 345–47.

Stadler, Friedrich. 2001. *The Vienna Circle: Studies in the Origins, Development, and Influence of Logical Empiricism.* Vienna: Springer Verlag.

Stein, Howard. 1992. "Was Carnap Entirely Wrong, After All?" *Synthese* 93: 275–95.

Storer, Thomas. 1948. "The Philosophical Relevance of a 'Behavioristic Semiotic.'" *Philosophy of Science* 15: 316–30.

Sullivan, Kelly. 2000. "Putting the 'Liberal' in Liberal Arts." Editorial essay published at www.heritage.org, February 16.

Suppe, Fred, ed. 1977. *The Structure of Scientific Theories*. Chicago: University of Illinois Press.

"Toward Unity." 1938. *Time*, August 1.

Traiger, Saul. 1984. "The Hans Reichenbach Correspondence: An Overview." *Philosophy Research Archives* 10: 501–11.

U.S. Senate. 1956. *The Communist Party of the United States of America. What It Is. How It Works. A Handbook for Americans.* Washington, D.C.: United States Government Printing Office.

Uebel, Thomas E. 1992. *Overcoming Logical Positivism from Within: The Emergence of Neurath's Naturalism in the Vienna Circle's Protocol Sentence Debate.* Atlanta: Rodopi.

———. 1998. "Enlightenment and the Vienna Circle's Scientific World-Conception." In *Philosophers on Education: Historical Perspectives*, ed. A. Rorty. London: Routledge.

———. 2000. "Some Scientism, Some Historicism, Some Critics: Hayek's and Popper's Critiques Revisited." In *The Proper Ambition of Science*, eds. M. W. F. Stone and J. Wolff. London: Routledge.

Uebel, Thomas. E. 2003a. "History and Philosophy of Science and the Politics of Race and Ethnic Exclusion." In Heidelberger and Stadler 2003, 91–117.

———. 2003b. "Philipp Frank's History of the Vienna Circle: A Programmatic Retrospective." In *Logical Empiricism in North America*, eds. G. Hardcastle and A. Richardson. Minnesota: University of Minnesota Press, 149–69.

———. 2004. "Carnap, the Left Vienna Circle, and Neopositivist Antimetaphysics." In Awodey and Klein 2004, 247–77.

Uebel, Thomas E., ed. 1991. *Rediscovering the Forgotten Vienna Circle: Austrian Studies on Otto Neurath and the Vienna Circle.* Boston: Kluwer.

"Unity at Cambridge." 1939. *Time*, September 18, pp. 72–73.

Viereck, Peter. 1953. *The Shame and Glory of the Intellectuals: Babbit Jr. vs. the Rediscovery of Values.* Boston: Beacon.

Whitehill, Walter Muir. 1955. "A Foreword to *Daedalus*." *Daedalus.: Proceedings of the American Academy of Arts and Sciences* 86, no. 1: 3–5.

Wohlstetter, Albert, and Morton White. 1939. "Who Are the Friends of Semantics?" *Partisan Review* 6: 50–57.

Index

Blumberg, Albert, 21–22, 60, 96–99,
110, 118; FBI investigation of, 119;
Logic: A First Course, 99; political
activity of, 61, 97
Blumberg, Dorothy, 99
Bohr, Niels, 13
Bolshevik revolution, 62
Bolshevism, 126, 138
Boltzmann, Ludwig, 53
Borah, William, 260
Boring, Edwin, 303
Brameld, Theodore, 139, 140; *A
Philosophic Approach to Communism*,
140
Brandeis, Louis, 260
Bridgman, Percy, 22, 129, 295, 315,
317
Broad, C. D., 131
Brodbeck, May, 358, 359, 375;
Readings in the Philosophy of Science,
357
Bronstein, Daniel, 66
Browder, Earl, 136–37, 138,
189
Brown, Norman, 351
Broyles Commission, 260
Brunswik, Egon, xiv, 295
Brushlinsky, A. V., 383
Bryson, Lyman, 160; *Science and
Freedom*, 109
Buckley, William F., 163, 234, 240–42,
248, 256; *God and Man at Yale*, 234,
240
Budenz, Louis, 269
Burhoe, Ralph, 301
Burnham, Daniel, 21–22
Burnham, James, 312, 349
Burtt, E. A., 39
Butterworth, Joseph, 249

Canwell, Albert, 248–49
Carnap, Ina, 67
Carnap, Rudolf, xi, 10, 21, 22, 58, 68,
69, 80, 85, 123, 125, 153, 189, 191,
350; cultural/political interests of,
2, 148; on definitions and
reductions, 87–88; and Dewey,

interaction between, 86–88, 91–92,
93; and the *Encyclopedia*, 12, 18, 35,
67, 123; FBI investigation of, 115,
119, 271–76; as humanitarian, 48;
Introduction to Semantics, 17, 192,
197; on language and pragmatics,
49; and logical empiricism, 8; *The
Logical Syntax of Language*, xiv, 48,
141; *Der Logische Aufbau der Welt*, 4,
48, 49–50, 51, 171; and loyalty oath
issues, 277; manifesto, 27, 126, 163;
on metaphysics, 83, 178; and
Morris, 45; on neo-Thomism, 74;
and Neurath, tension between, xv,
6, 17, 28–29, 38, 48, 191, 192, 201–2;
neutralism of, 49, 365, 369; on
philosophy and politics, 47–48, 382,
383–84; on philosophy of science,
50–52; politics of, 52–53, 363, 369,
382; Principle of Tolerance, 171;
reply to Hook, 281–82; on science
and ethics, 381; as socialist, 27; on
unity of science, 87–88, 376; and
Unity of Science movement, 4, xiv,
10, 48; on values/value statements,
91, 365; and Vienna Circle, 1, 27,
126, 176; and Waldorf conference,
281–82, 363
Carter, Philip, 136–37, 139, 144–46
Cartwright, Nancy, 7
Chamberlin, William Henry, 280
Chambers, Whittaker, 153, 241, 244,
355; *Witness*, 153, 241
Chiang Kai-shek, 255
Church, Alonzo, 300
Churchill, Winston, 237–38
Churchman, C. West, 107, 283,
285–89, 329, 335, 341, 349; on
Carnap, 286–87; on logical
empiricism, 285–89; and Neurath,
289–90
Circle of Wittgenstein, 192
Cohen, I. B., 9
Cohen, Morris, 11, 58, 124;
*Introduction to Logic and Scientific
Method*, 286; response to Hook,
151

Duhem, Pierre, 53, 54–55, 56, 176, 214
Dunham, Barrows, 257
Dupré, John, 374

Eastman, Max, 60, 62, 63, 372
Eckhart, Carl, 41
Eddington, Arthur, 133
Edel, Abraham, 66, 67, 69, 85, 153, 189
Eesteren, Cornelius Van, 32
Einstein, Albert, 53, 55, 77, 131, 157, 212, 215, 220, 268, 279
Eisenhower, Dwight D., 249
Eisler, Gerhard, 245
Encyclopedia. See International Encyclopedia of Unified Science
"end of ideology," 165–66
Engel, Leonard, 156
Engels, Friedrich, 57, 61, 126, 136–37, 144
Enriques, Federigo, 9
Erkenntnis, 10, 44–50; and *Philosophy of Science*, rivalry between, 104–5
Ernst Mach Society, 3
Etc.: A Review of General Semantics, 218

Fahs, Charles, 314
Farber, Marvin, 60, 131
Farrell, James T., 63, 67
fascism, 42–43
Feigl, Herbert, xiv, 2, 8, 96, 97, 99, 204–5, 284, 358, 366; conference planning of, 304–5, 308; *Current Issues in Philosophy of Science*, 360; and the Institute for the Unity of Science, 304–5, 345–46, 347; and politics, 35–36; *Readings in the Philosophy of Science*, 357, 359; on science and values, 359–62, 370
Fermi, Enrico, 106
Feuer, Kathryn, 127
Feuer, Lewis: *The Conflict of Generations*, 127; on logical empiricism, 124–26, 128, 146; move to the right, 127; on the student movement, 127

Finkelstein, Rabbi Louis, 55, 160
Flewelling, Ralph, 162
Flynn, John T., 244
Foundations of the Unity of Science, 11
Fourth International (magazine), 65
Franco, Francisco, 150
Frank, Josef, 32, 309
Frank, Philipp, xv, 10, 21, 32, 44, 77, 93, 105, 119–20, 124, 164, 292–93, 351, 358, 365–66, 370; on absolutism, 222–24, 379; on active positivism, 161, 298; on Aquinas, 325; background of, 53; *Between Physics and Philosophy*, 124; on Carnap, 383; on common sense, as petrifaction, 214–15; and CSPR, 219; on dialectical materialism, 119–20, 226–29, 323; educational proposals of, 226; FBI investigation of, 115, 119, 268–71, 324–25; historicism/contextualism of, xv; holism of, 214; on humble positivism, 298; and Institute for the Unity of Science, 18–19, 147, 294–99, 307–10; Kegley's attack on, 325–30; and Kuhn, comparison between, 225–29, 233; *The Law of Causality and Its Limits*, 212; and logical empiricism, xiv, 2, 8, 53, 209; on Mach, 213–14; on Malisoff, 107, 291–92; and neo-Thomism, 3; and Neurath, alliance between, 24; obscurity of, reasons for, 53; and philosophy of science, 54–55, 213; *Philosophy of Science: The Link between Science and Philosophy*, 325; on politics and science, 297–98, 315–16; as pragmatist, 22–23; as president of Institute for Unity of Science, 294–99, 307–10; *Relativity: A Richer Truth*, 220, 311, 316, 362; on relativity theory, 220–21; and Rockefeller Foundation, 319–20; on science education, 217–18; on science and philosophy/humanities, x–xi, 24,

McCarthy, Joseph, 19, 118, 241,
242–43, 244, 248, 256, 257
McCarthy, Mary, 252–53
McCumber, John, 250, 384, 385
McGeehan, John E., 81
McGill, V. J., 60, 61, 66, 67, 131,
141–42, 299
McKinsey, J. C. C., 350
Mead, George Herbert, 39, 41
Meaning of Marx: A Symposium, The,
151
Meiner, Felix, 11, 284
Menger, Karl, 8
Menshevism, 95, 102, 109, 138
Merton, Robert, 300
Methods of Science, 12
Meyer, Eduard, 32
Meyer, Frank, 163–64
Mills, C. Wright, xii
Minnesota Center for the Philosophy
of Science, 346, 359
Mises, Ludwig von, 235
Modern Man Is Obsolete (Norman
Cousins), 109
Moholy-Nagy, Laszlo, 41
Mont Pelerin society, 236
Moore, H. Barrington, 303
Morris, Charles, 6, xiv, xv, 16, 65, 67,
72, 80, 95, 96, 102, 109, 123, 130,
135, 138, 297, 317, 358, 372; against
absolutism, 379; and Buddhism, 38;
caution of, 266–67; in China, 261;
and Dewey, interaction between,
7–8, 84–85, 86–87, 335–37, 338; as
Deweyan, 38, 40, 43; effort to
publish Neurath, 366–67; and the
Encyclopedia, 12, 34–36, 40, 41, 45;
on fascism, 42–43; *Foundations of the
Theory of Signs,* 334; institute, idea
for, 39–40; and the Institute for the
Unity of Science, 18, 284–85; on
Kallen, 172; and loyalty oath issues,
261–62, 351; marginalization of,
331, 342–43; on Marxist-
communism, 42–43; and Neurath,
collaboration between, 41; *The Open
Self,* 264–65, 267, 339, 341, 372; on

orchestration of science, 173; *Paths
of Life: Preface to a World Religion,* 46,
266, 339; and personality theory,
46, 338; and philosophy of religion,
46–47; and "philosophy of the
heart," 43; political activities of,
263–64; *The Pragmatic Movement in
American Philosophy,* 342–43;
*Pragmatism and the Crisis of
Democracy,* 42, 43, 44, 46; as
pragmatist, 22–23; on science,
social significance of, 44–45; *Signs,
Language, and Behavior,* 47, 130, 132,
268, 331–33, 334, 338, 340, 342,
344; in Soviet Union, 42; and
theory of signs/semiotics, 10, 39,
41; on Thomism, 42–43; and the
Unity of Science movement, 10,
13–15, 41, 45–46, 74–77, 168;
Varieties of Human Value, 339, 340
Muller, H. J., 60
Mumford, Lewis, 311
Murphy, Arthur, 39
Murray, Gilbert, 78
Mussolini, Benito, 45, 52

Nabokov, Nicolas, 312
Nagel, Ernest, 82, 90, 96, 124, 206,
301; on Cornforth, 323; defense of
science, 79–80; and the
Encyclopedia, 67, 69; *Introduction to
Logic and Scientific Method,* 286; and
logical empiricism, 11, 82; on
neo-Thomism, 11; and Neurath,
correspondence between, 65–66;
and New York Intellectuals, 11, 58,
62, 67, 69, 78; on values and
science, 310–11, 381; and Vienna
Circle, 1
Nathanson, Jerome, 109
Needham, Joseph, 60
neo-conservatism, 166
neo-Thomism, 11
Neurath, Marie, 32, 66, 267, 294, 309,
347, 366, 367
Neurath, Otto, xv, 4, 11, 58, 63, 80,
123, 347, 358; against absolutism,